D0866994

The origins and relationships
of lower invertebrates

Proceedings of an International Symposium
held in London, September 1983

The Systematics Association

Special Volume No 28

The origins
and relationships
of lower invertebrates

Edited by

S. Conway Morris
Department of Earth Sciences, University of Cambridge

J. D. George
Department of Zoology, British Museum (Natural History)

R. Gibson
Department of Biology, Liverpool Polytechnic

H. M. Platt
Department of Zoology, British Museum (Natural History)

Published for the SYSTEMATICS ASSOCIATION by

CLARENDON PRESS · OXFORD

1985

Oxford University Press, Walton Street, Oxford OX2 6DP

Oxford New York Toronto
Delhi Bombay Calcutta Madras Karachi
Kuala Lumpur Singapore Hong Kong Tokyo
Nairobi Dar es Salaam Cape Town
Melbourne Auckland

and associated companies in
Beirut Berlin Ibadan Mexico City Nicosia

Oxford is a trade mark of Oxford University Press

Published in the United States
by Oxford University Press, New York

British Library Cataloguing in Publication Data
The origins and relationships of lower
invertebrates—(The Systematics Association
special volume, ISSN 0309-2593; no. 28)
1. Invertebrates—Evolution   2. Phylogeny
I. Morris, S. Conway   II. Series
592'038   QL362
ISBN 0-19-857181-X

Typeset by DMB (Typesetting), Oxford
Printed in Great Britain by
St Edmundsbury Press,
Bury St Edmunds, Suffolk.

Preface

'Anything said on these questions lies in the realm of fantasy' was Libbie Hyman's pithy conclusion, having provided a trenchant review of certain speculations concerning metazoan phylogeny in the penultimate volume of her monumental series *The invertebrates*. More than 20 years have elapsed since her appraisal was published, and in the intervening period a series of meetings such as those in Pacific Grove, Erlangen, and Hull has reviewed progress. This latter commodity has not always been over-abundant in discussions on the origins and relationships of lower invertebrates, and it was with trepidation that we attempted to organize a review of even a moiety of this enormous subject area. Inevitably coverage is incomplete, but our intention was in part to complement the earlier Systematics Association meeting held in Hull on *The origin of major invertebrate groups* (1979: Systematics Association Special Volume 12), by concentrating on the lower (not minor!) phyla and balancing its palaeontological emphasis with a more neontological slant. Whilst the papers, not unexpectedly, provided no clear overall conclusion to the main theme, the meeting did provide a timely forum, and the emphasis given to ultrastructural studies and cladistic analysis provides pointers to areas that are certain to show rapid development and progress.

The meeting was held in the British Museum (Natural History) on 7-9 September 1983, with over 150 people attending from 18 countries. The formal sessions included 28 lectures, in addition to which more than 20 poster sessions provided a further focus for discussion. Of the original invited contributions 23 are published here. Talks were also given by T. Cavalier-Smith, K. Fauchald, M. E. Fransen, R. M. Kristensen, and S. A. Wainwright, but for a variety of reasons (e.g. publication elsewhere), written versions of their contributions are not included here.

With four editors the potential for confusion is at a premium, but we have endeavoured to impose some uniformity of style and terminology. One particular point deserves mention: the spelling of Platyhelminthes. Despite its well-established usage, an appeal has been made recently to refer to the group as Plathelminthes on the grounds of a historical priority. The rules devised by the ICZN, perhaps wisely, are not applicable to higher taxonomic categories, and we see little advantage in changing the existing spelling.

It is appropriate to record our thanks for help received from many quarters, especially to the Trustees of the BM(NH) for the use of the Lecture Hall and Conversazione Room. The British Council and the National Science Foundation (USA) generously provided support for a number of the overseas speakers. The Council of the Systematics Association provided the initial impetus for the meeting and provide continued support and encouragement. The skills of John Taylor and Helen Boxhall in the cartographic office of the Department of Earth Sciences, Open University were invaluable, while Adele Prouse and Phyl Fisher (Department of Earth Sciences, Cambridge University) have skilfully redrafted a number of diagrams. Secretarial help from Beryl West (Open University) and Mary Askham (Cambridge University) has been invaluable. We also acknowledge with gratitude the comments by a number of reviewers.

January 1984   S. C. M.
                                                            J. D. G.
                                                              R. G.
                                                            H. M. P.

Contents

Contributors

PETER AX

II Zoologisches Institut und Museum, Universität Göttingen, Berliner Strasse 28, D-3400 Göttingen, Federal Republic of Germany.

ROBERT D. BARNES

Department of Biology, Gettysburg College, Gettysburg, Pennsylvania 17325, USA.

GUNNAR BERG

Department of Zoology, University of Göteborg, Box 250 59, S-400 31 Göteborg, Sweden.

P. R. BERGQUIST

Department of Zoology, University of Auckland, Private Bag, Auckland, New Zealand.

P. J. S. BOADEN

The Queen's University of Belfast, Marine Biology Station, The Strand, Portaferry, County Down BT22 1PF, Northern Ireland, UK.

PIERRE CLÉMENT

Laboratoire d'Histologie et Biologie Tissulaire, Université Claude Bernard-Lyon 1, 43 Boulevard du 11 Novembre, F69622 Villeurbanne Cedex, France.

S. CONWAY MORRIS

Department of Earth Sciences, University of Cambridge, Downing Street, Cambridge CB2 3EQ, UK.

ULRICH EHLERS

II Zoologisches Institut und Museum, Universität Göttingen, Berliner Strasse 28, D-3400 Göttingen, Federal Republic of Germany.

G. R. HARBISON

Woods Hole Oceanographic Institution, Woods Hole, Massachusetts 02543, USA.

MEREDITH L. JONES

Department of Invertebrate Zoology, National Museum of Natural History, Smithsonian Institution, Washington, D.C. 20560, USA.

JACOB VAN DER LAND

Rijksmuseum van Natuurlijke Historie, Postbus 9517, 2300 RA Leiden, Netherlands.

S. LORENZEN

Zoologisches Institut, Christian-Albrechts-Universität, Olshausenstrasse 40-60, D-2300 Kiel 1, Federal Republic of Germany.

MARLENE MAINITZ

Universitäts-Hautklinik, Alserstrasse 4, A-1090 Wien, Austria.

C. METTAM

Department of Zoology, University College, PO Box 78, Cardiff CF1 1XL, UK.

CLAUS NIELSEN

Zoological Museum, University of Copenhagen, Universitetsparken 15, DK-2100, Copenhagen Ø, Denmark.

ARNE NØRREVANG

Institute of Cell Biology and Anatomy, University of Copenhagen, Universitetsparken 15, DK-2100, Copenhagen Ø, Denmark.

P. J. W. OLIVE

Dove Marine Laboratory and Department of Zoology, University of Newcastle-upon-Tyne, Newcastle-upon-Tyne NE1 7RU, UK.

MARY E. RICE

Department of Invertebrate Zoology, National Museum of Natural History, Smithsonian Institution, Washington, D.C. 20560, USA.

REINHARD M. RIEGER

Department of Biology and Curriculum of Marine Sciences, University of North Carolina, Wilson Hall 046A, Chapel Hill, North Carolina 27514, USA.

ELAINE A. ROBSON

Department of Pure and Applied Zoology, University of Reading, Whiteknights, Reading RG6 2AJ, UK.

JULIAN SMITH III

Department of Zoology, University of Maine, Orono, Maine 04469, USA.

BEATE SOPOTT-EHLERS

II Zoologisches Institut und Museum, Universität Göttingen, Berliner Strasse 28, D-3400 Göttingen, Federal Republic of Germany.

WOLFGANG STERRER

Bermuda Biological Station, Ferry Reach 1-15, Bermuda.

SETH TYLER

Department of Zoology, University of Maine, Orono, Maine 04469, USA.

JEAN VACELET

Station Marine d'Endoume, LA 41 CNRS, Rue de la Batterie-des-Lions, 13007 Marseille, France.

WILFRIED WESTHEIDE

FB5, Spezielle Zoologie, Universität Osnabrück, D-4500 Osnabrück, Federal Republic of Germany.

1. Coralline sponges and the evolution of Porifera

JEAN VACELET

Station Marine d'Endoume, Marseille, France

Abstract

It has long been believed that much of the fossil record of the phylum Porifera is restricted to those forms with a skeleton of fused spicules. While such skeletons occur in the three classes of recent sponges, they are not common and most Recent sponges will not fossilize readily. About 15 Recent species have an unusual solid calcareous skeleton, bearing a striking resemblance to that of various corals. These 'coralline sponges' are found presently in submarine caves and cryptic habitats of warm-temperate and tropical zones, mainly in coral reefs. They are the survivors of ancient reef builders such as stromatoporoids, tabulate 'corals' and sphinctozoans, which were believed to be extinct and the nature of which was disputed. As most of these enigmatic organisms were previously allocated to the phylum Cnidaria, these findings greatly enlarged the fossil record of the Porifera, and may change our views on evolution in this phylum.

It was proposed in 1970 and 1973 that these survivors should be regarded as a fourth class of Porifera, the sclerosponges or the ischyrosponges. In the author's opinion, this view cannot be maintained. These sponges display both a high diversity and strong similarities with some 'normal' Recent sponges devoid of a calcareous skeleton; their incorporation into the pre-existing orders of the classes Calcarea and Demospongea seems to be more appropriate. Although a calcareous, massive skeleton is of prime importance in their organization and in their role as reef builders, it appears to have little phylogenetic significance in the phylum Porifera. This view is supported by various lines of evidence, including the existence in *Merlia* of two species which differ only in a calcareous skeleton being present in one and absent from the other. The calcareous skeleton is considered to be an archaic feature, present in most Palaeozoic sponges and probably appearing independently in different lineages. A major proportion of the Recent sponge fauna could have descended from these diverse calcified forms, losing their rigid skeleton in the course of evolution. Such a disappearance of the calcareous skeleton possibly occurred when competition between calcified sponges and reef builders, such as scleractinian corals, turned to the advantage of the latter in the Jurassic and Cretaceous.

Introduction

Fossil sponges first appeared in the Cambrian, there being no fossil record from the Precambrian, although the Ediacaran fauna has a number of invertebrates of more advanced grades of organization without hard skeletons. Fossil sponges are known only from their hard parts, and until recently it was believed that these were mostly skeletons of fused calcareous or siliceous spicules. Such a fossil record gives a strongly biased view of the ancient fauna. Sponges with skeletons of fused spicules exist today in the three classes of Recent sponges (Hexactinellida, Demospongea and Calcarea), but they are only a minor component of the sponge fauna, and the vast majority of Recent sponges will not fossilize.

I would like to focus attention on the history and meaning of calcification in sponges, rather than on speculations concerning the early history of this presumably most primitive of phyla. Discoveries of living coralline sponges have shown that calcification is not only represented by the well known calcareous spicules of the Calcarea, and that a skeleton of fused spicules is not the only structure that can fossilize. Fifteen species of Recent sponges are unusual in having a solid calcareous skeleton not composed of spicules (Vacelet 1964, 1983; Hartman and Goreau 1970, 1975). Such a skeleton bears striking resemblances to that of various cnidarians, and could fossilize. These coralline sponges are 'living fossils', survivors of enigmatic Palaeozoic and Mesozoic reef-builders such as the stromatoporoids, chaetetids and sphinctozoans. Most of these abundant reef-builders were previously classified in the Cnidaria. It appears that workers did not take into account most of the poriferan fossil record in their phylogenetic constructions, and that living remnants of an ancient sponge fauna exist in Recent seas. It is thus possible to compare the tissue and spicule complement of the 'living fossils' with that of other modern forms. This is a fascinating exercise, as the classification and phylogeny of the Porifera are based on soft-tissue and spicule characteristics, most of which are lost in the fossil record.

The coralline sponges

Although these sponges have only been noticed recently they are not uncommon, and their characters can be easily studied (Table 1.1). Their habitat is, however, most unusual: they all live in submarine caves or in cryptic habitats of coral reefs. Most are from tropical areas, with the exception of two species living in caves in the Mediterranean and around Madeira. It seems that this cryptic environment, in which few species can thrive, has been a refuge where relic species have been protected from competition with more modern forms.

Two main facts emerge:
(i) These unusual sponges are diverse;
(ii) They display strong similarities to some Recent non-calcified sponges. This is surprising as their fossil relatives date back to the Palaeozoic or early Mesozoic periods, and as their calcareous skeleton is now so unusual in sponges.

Two of these relic species have living tissue and spicules similar to those found in the class Calcarea. In *Petrobiona massiliana* Vacelet and Lévi, a Mediterranean cave dweller, the skeleton is a very hard and solid calcareous mass, on which the living tissue and calcareous spicules are located. The living tissue, spicules and reproductive pattern are typical of one of the two subclasses known in the Calcarea, the Calcaronea. The second species, *Murrayona phanolepis* Kirkpatrick, is a tropical sponge, and its characters are those of the other subclass of the Calcarea, the Calcinea. Thus two sponges exist which, without their solid calcareous skeleton, would be regarded as normal Calcarea, but which would not be close relatives and would be classified in two different subclasses. However, they share a special type of spicule, the 'tuning-fork' triactine, which is also found in the pharetronids, relic species of the Calcarea with a fused spicule skeleton which are known from the early Mesozoic. This spicule is probably a character of a common ancestor of the two subclasses, lost in modern species, but remaining in the relic forms of both subclasses.

An interesting point is that these two sponges, although relic species and so presumably more primitive than recent Calcarea, are actually less simple. Their aquiferous system belongs to the complex leuconoid type, while many Recent Calcarea display a simple syconoid or asconoid type. In *Petrobiona massiliana*, the fertilization and nutrition of the ovocyte takes place with the help of a carrying cell, according to the conventional pattern described in the Calcaronea (Duboscq and Tuzet 1937), but the process here is clearly more elaborate, with a whole choanocyte chamber absorbed by the ovocyte through both the carrying cell and the envelope of the spermatocyst (Vacelet 1964). The

Table 1.1. Characters and affinities of living sponges with a solid calcareous skeleton

	Murrayona : one sp.	*Petrobiona* : one sp.	Cerato-porellidae : six spp.	*Astrosclera* : one sp.
Spicules	Calcareous, free. Calcarea, Calcinea	Calcareous, free. Calcarea, Calcaronea	Siliceous, in part enclosed in skeleton. Demospongea, affinities Agelasidae	As Cerato-porellidae. May be absent in the Pacific forms
Calcareous skeleton and mode of formation	Calcite, reticu-late. Spherulitic and trabecular. Formation ?	Calcite, massive. Spherulitic and trabecular. Formation ?	Aragonite, Penicillate spherolithes. Formation ?	Aragonite. Spherulitic, regular. Formation : intracellular
Trace of aquiferous system on skeleton ?	Yes	No	Yes, astrorhizae	Yes, astrorhizae
Affinities of soft parts	Calcarea, subclass Calcinea	Calcarea, subclass Calcaronea	Demospongea, affinities Agelasidae	Demospongea, affinities Agelasidae
Repro-ductive affinities	Calcarea, subclass Calcinea	Calcarea, subclass Calcaronea	? Perhaps Demospongea Ceractino-morpha for one sp.	Demospongea, Ceractino-morpha
Rich symbiotic microflora	No	No	Yes	Yes
Resting bodies	No	Yes	No	No
Geographic distribution	Indopacific	Mediterranean	Five Caribbean spp., one Pacific sp.	Indopacific
Corres-ponding fossil groups	? Probably some Inozoan pharetronids	? Probably some Inozoan pharetronids	Triassic Cerato-porellidae and some Inozoan pharetronids	Some stromato-poroids and Inozoans

Vaceletia : one sp.	*Calcifibrospongia :* one sp.	*Merlia :* one sp.	*Acanthochaetetes :* one sp.
Absent	Siliceous, in part enclosed in skeleton. Demospongea, Ceractinomorpha	Siliceous, free. Demospongea with uncertain affinities	Demospongea, Tetractinomorpha (Spirastrellidae)
Aragonite, external. Stacking of microcrystals. Mineralization of a non collagenous matrix	Fibro-radiated aragonite. Mineralization of a collagenous matrix	Calcite and aragonite. Vertical trabeculae. May be absent. Formation?	Calcite. Stacking of microlamellae. Formation?
Yes, pores and siphon	Yes, astrorhizae	No, Tabulae	Yes, astrorhizae. Tabulae
Demospongea, Ceractinomorpha	Demospongea,?	Demospongea, Tetractinomorpha	Demospongea, Tetractinomorpha
Demospongea, Ceractinomorpha	?	?	?
Yes	?	No	No
No	No	Yes	Yes
Indopacific	Caribbean	Circumtropical and Mediterranean	Pacific Ocean
Some sphinctozoans	Mesozoic stromatoporoids	Stromatoporoids	Chaetetidae and some Tabulomorpha

simplicity of modern forms compared with the living fossils will be discussed below.

All the other coralline sponges have spicules and living tissues similar to those of the class Demospongea. Their spicules, when present, are siliceous, and indicate affinities with diverse subclasses or orders of demosponges.

The sphinctozoans are chambered sponges, believed to have been extinct since the Cretaceous period, and included in the class Calcarea. I discovered one surviving species, *Vaceletia crypta* (Vacelet), which is about 10 mm high, and has an external skeleton (Vacelet 1979, Fig. 1.1(a)). Thus the sponge has to build successive chambers in order to grow. The living tissue is inside the chambers, and the filtered water is expelled through a central siphon (Fig. 1.1(b)). This sponge is not a calcarean: it is devoid of spicules, as are its Triassic fossil relatives, and the larva and living tissue are both similar to those of the subclass Ceractinomorpha of the demosponges. The microstructure and mode of formation of the skeleton are most unusual among calcified sponges, as a new chamber is formed by mineralization of an organic template.

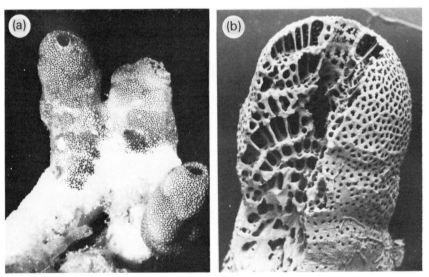

Fig. 1.1. A colonial sphinctozoan, *Vaceletia crypta*. (a) An overall view of a living specimen (× 3.4). (b) Section through the calcareous skeleton showing the perforate central siphon and dome-shaped chambers in which the living tissue is located (× 15).

The resulting feltwork of aragonitic needles is identical to the skeleton of some well preserved Triassic examples. Other fossil sphinctozoans, however, have a very different microstructure, and some even have calcareous spicules (Reid 1968). This indicates that the sphinctozoans

are not a natural group, but are merely a grade of organization characterized by an external skeleton which appeared more than once in the
history of the Porifera.

Although sphinctozoans were recently described from the upper
Cambrian, they are mainly found after the Carboniferous period. An
apparent external skeleton is also known in another group of enigmatic
organisms, the Archaeocyatha, which although important reef-builders
in the lower Cambrian, had disappeared by the upper Cambrian.
Their organization resembles that of sphinctozoans, but the main partitions are vertical and the organization is usually more elaborate.
They are usually regarded as a distinct phylum of primitive animals,
different from but closely related to the Porifera. Owing to their similarities with the sphinctozoans and to their probable filter-feeding
habits, it has recently been suggested (Debrenne and Vacelet in press)
that they are sponges and constitute an extinct class of the Porifera.
This would place the appearance of a calcareous skeleton in sponges
back to the early Cambrian, and sponges would represent one of the
earliest radiations of invertebrate hard parts.

In another species of coralline sponge, *Astrosclera willeyana* Lister
(Fig. 1.2), which has been considered a survivor of the stromatoporoids, important Palaeozoic reef-builders, the living tissue shows

Fig. 1.2. *Astrosclera willeyana*. The apical living part of this calcified sponge shows
astrorhizae imprinted into the skeleton by excurrent canals (× 3).

relationships with the subclass Ceractinomorpha, as well as with the living Sphinctozoa. The microstructure of its skeleton is, however, different, because the skeleton is made of aragonitic spherules which are also found in a number of fossils, including some chaetetids, stromatoporoids and some sphinctozoans. These spherules are secreted intracellularly.

The Ceratoporellidae, which are related to *Astrosclera*, constitute the only family of calcified sponges which is not monotypic. Six species are known, mostly from Caribbean caves, although one species is from the Pacific. Their siliceous spicules are entrapped in the aragonitic skeleton, and their remains are found in some well-preserved fossils. They have spicule and soft-tissue affinities with the family Agelasidae, a family of non-calcified sponges which are most abundant in the Caribbean region and the systematic position of which is not clear among the demosponges.

In another species from the Caribbean, *Calcifibrospongia actinostromarioides* Hartman, the skeleton is calcitic and displays strong morphological similarities to that of Mesozoic stromatoporoids. The spicules are similar to those found in the order Haplosclerida, subclass Ceractinomorpha.

Acanthochaetetes wellsi Hartman and Goreau belongs to the chaetetids, an important group of the so-called tabulate corals, which were reefbuilders in the Palaeozoic and Mesozoic eras. The skeleton is calcitic and microlamellar. The soft parts and the siliceous spicules are similar to those found in the order Hadromerida, which belongs to the subclass Tetractinomorpha, whereas the other demosponge-like calcified sponges belong to the subclass Ceractinomorpha. In fact, this sponge would fit perfectly well in the family Spirastrellidae were it not for its calcareous skeleton. So, this calcified sponge and at least some fossil chaetetids appear to be closely related to a common family of Recent sponges.

The genus *Merlia* is characterized by a unique spiculation. There are two types of specimen, which may represent different species (Vacelet 1980), *M. normani* Kirkpatrick, with a calcareous skeleton, and *M. deficiens* Vacelet, without a calcareous skeleton. All the other characters are similar, and it is not certain that they are different species. Both have a wide circum-tropical distribution. One must conclude that the remarkable calcareous skeletons of these sponges, which are the only remains available to palaeontologists, are of little phylogenetic importance.

Discussion

Interpretations of these organisms have been diverse. It is wise to disregard the opinion that they are chimeras, dead corals perforated by boring sponges, as was suggested at the beginning of the century when few species and specimens were known. To-day there are two main interpretations.

In the first (Hartman and Goreau 1970; Termier and Termier 1973), emphasis is put on the calcareous skeleton. The sponges which share this remarkable feature are separated in a new class of the Porifera, the sclerosponges or ischyrosponges. This scheme has won wide acceptance, because it can be used easily by palaeontologists, who only have the skeleton available. However, the sclerosponges are certainly polyphyletic, and an unnatural group.

In the second interpretation (Vacelet 1979, 1981), each of these calcified sponges, and when possible its fossil relatives, is classified in the class Calcarea or Demospongea, based on a range of characters. Table 1.2 shows that a calcareous skeleton has commonly appeared in most of the lineages of the Porifera since the early Palaeozoic period. The only class in which such a skeleton apparently never occurred is in the Hexactinellida. Recent work (Mackie and Singla 1983) has shown that Hexactinellida are very different from other sponges due to their syncytial organization. They may constitute a separate sub-phylum or even, according to Bergquist (this volume), a separate phylum. Their siliceous skeleton is very complex and elaborate, and moreover there is an extinct group of presumed calcified sponges, the Archaeocyatha, in which the calcareous skeleton is also unusually complex. Is there any relationship between these two groups? It may be tentatively suggested that hexactinellids are archaeocyathids which have lost their calcareous skeleton, a process of evolution which seems to have occurred commonly in the other classes of Porifera.

In the subclass Homosclerophorida of the Demospongea there is no known calcareous skeleton, but in all other subclasses of both Calcarea and Demospongea, at least one living species possesses a solid calcareous skeleton. As these living calcified sponges are similar or sometimes apparently identical to some fossils, we can place these fossil groups in the classification of Recent Porifera. So it seems likely that the chaetetids belong to the tetractinomorph Demospongea (and most probably to the order Hadromerida), while the stromatoporoids, the ceratoporellids and a part of the sphinctozoans belong to the ceractinomorph Demospongea. However, it must be noted that these fossil groups, which are known mainly from their skeletal remains, may be polyphyletic, and that a number of lineages must be extinct. For instance, some fossil sphinctozoans are true Calcarea, probably belonging to the subclass Calcaronea, while most of the fossil forms and their living representatives are demosponges. It must not be forgotten that only 15 living species are known, whereas the fossil forms number in the hundreds and range over five hundred million years.

Table 1.2. A classification of modern sponges listing only those with solid calcareous skeletons, including some indications of their affinities to fossil groups

Class Hexactinellida
Siliceous spicules with three axes

{ No known calcified form

Class Calcarea
Calcareous spicules

Subclass Calcinea : Murrayonidae (1 sp.) : ? pharetronids (in part)
Subclass Calcaronea : Petrobionidae (1 sp.) : ? pharetronids (in part)

Class Demospongea
Siliceous spicules with one or four axes

Subclass Homoscleromorpha : no known calcified form

Subclass Tetractinomorpha { Acanthochaetetidae (1 sp.) : Chaetetidae and some Tabulomorpha
Merliidae (1 sp.) : uncertain affinities with tabulates and stromatoporoids

Subclass Ceractinomorpha { Ceratoporellidae (six spp.) : pharetronids (in part)
Astroscleridae (one sp.) : stromatoporoids (in part)
Cryptocoelidae (one sp.) : sphinctozoans (in part)
Calcifibrospongiidae (one sp.) : stromatoporoids (in part)

It seems likely that this type of calcareous skeleton was common from the early Palaeozoic. Owing to its diversity in composition and microstructure, both in the relic species and in fossils, it is necessary to suppose that it appeared more than once and in different lineages. This type of skeleton allowed sponges to be the main carbonate reef-builders in the Palaeozoic and early Mesozoic. These calcified sponges, however, were accompanied by close relatives devoid of a calcareous skeleton, which are either known in the fossil record by a skeleton of fused spicules or have left no remains. In the Mesozoic, mainly in the Jurassic, the scleractinian corals proved to be more successful as reef-builders (Newell 1972), and the importance of sponges as reef-builders declined. The calcareous skeleton then disappeared in most sponges, a process which resulted in the Recent fauna, in which most species rely mainly on a spicular and collagenous skeleton. However, a few remnants of the calcified fauna managed to survive in the dark cavities of coral reefs, where they are protected from competition.

It should be emphasized that calcified fossil Porifera and Cnidaria have been confused and are still sometimes difficult to separate. The gross morphology of the solid, calcareous skeleton can be very similar in the two phyla, although the microstructure, which is more diverse in the Porifera, seems always to be different.

The coral-like skeleton appeared and then disappeared independently in most lineages of the Porifera. Calcified sponges are, and apparently were, not very different from non-calcified ones, and it appears that in a number of cases, two closely related lines, one calcified and the other non-calcified, evolved side by side. It would be interesting to explore whether such an interpretation could apply to other phyla, in which more or less closely related groups possess or do not possess a calcareous skeleton. An example may be the scleractinian corals and the actinians. Some temporary gaps in the fossil record of certain groups could possibly be explained by a disappearance of the calcareous skeleton. An example could be the stromatoporoids, which disappeared as fossils from the Carboniferous to the Jurassic. They possibly survived during this gap of more than one hundred million years as non-calcified animals.

In Porifera simple forms have apparently followed more elaborate ones. It would be expected that the calcified sponges, which are remnants of a·very old fauna and which are now restricted to monotypic orders or families living in a restricted habitat, would display more simple biological and morphological features than the modern forms, which are presumably more highly evolved. Surprisingly the situation is reversed. This is not particularly evident for the species which are related to the demosponges, but is very evident in the Calcarea. This is interesting, more especially as this class contains forms which may be considered as the simplest sponges. The asconoid grade of organization is found in both lineages of the Calcarea (genus *Clathrina* in the Calcinea, and *Leucosolenia* in the Calcaronea). Such sponges are merely tubes lined by choanocytes, and a consistent interpretation would be that they are survivors of the first stages of organization of sponges established in the Precambrian, from which other grades originated, but which have not been fossilized because of their fragility. In such an interpretation, these genera also would be panchronistic forms, but with a far longer history than calcified forms, being relics of the Precambrian. This is unlikely, for the asconoid grade of organization is connected in all evolutionary stages to the other grades found in the well-diversified modern Calcarea. The abundance and diversity of these grades of organization in variable shallow water habitats where competition is severe, indicate a group in expansion which underwent a recent burst in evolution, and not a group of panchronistic forms (Borojevic 1979). Whatever lines of evolution can be imagined, the result is a simplification of the organization in the Calcarea, with complex primitive forms which are nearly extinct and in evolutionary stasis, and simple, modern forms which recently successfully diversified. Thus, in the Porifera, the simplest phylum of modern invertebrates, it is likely that the simplest forms are not primitive, but highly evolved species.

References

Borojevic, R. (1979). Evolution des spongiaires Calcarea. *Colloq. Int. C.N.R.S.* 291, Biologie des spongiaires, 527-30.

Debrenne, F. and Vacelet, J. (in press). Archaeocyatha: is the sponge model consistent with their structural organization? Fourth International Symposium on Fossil Cnidaria, Washington, D.C.

Duboscq, O. and Tuzet, O. (1937). L'ovogenèse, la fécondation et les premiers stades du développement des éponges Calcaires. *Archs Zool. exp. gén.* 79, 157-316.

Hartman, W. D. and Goreau, T. F. (1970). Jamaican coralline sponges: their morphology, ecology and fossil relatives. *Symp. zool. Soc., Lond.* 25, 205-43.

—— —— (1975). A Pacific tabulate sponge, living representative of a new order of sclerosponges. *Postilla* 167, 1-21.

Mackie, G. O. and Singla, C. L. (1983). Studies on hexactinellid sponges, I. Histology of *Rhabdocalyptus dawsoni* (Lambe, 1873). *Phil. Trans. R. Soc.* B301, 365-400.

Newell, N. D. (1972). The evolution of reefs. *Scient. Am.* 226, 54-65.

Reid, R. E. (1968). *Tremacystia, Barroisia* and the status of Sphinctozoida (Thalamida) as Porifera. *Paleont. Contr. Univ. Kansas* 34, 1-10.

Termier, H. and Termier, G. (1973). Stromatopores, Sclérosponges et Pharétrones: les Ischyrospongia. *Annls Mines géol., Tunis* 26, 285-97.

Vacelet, J. (1964). Étude monographique de l'éponge calcaire Pharétronide de Méditerranée, *Petrobiona massiliana* Vacelet & Lévi. Les Pharétronides actuelles et fossiles. *Recl Trav. Stn mar. Endoume* 50, 1-125.

—— (1979). Description et affinités d'une éponge Sphinctozoaire actuelle. *Colloq. Int. C.N.R.S.* 291, Biologie des spongiaires, 483-93.

—— (1980). Squelette calcaire facultatif et corps de régénération dans le genre *Merlia*, éponges apparentées au Chaetétidés fossiles. *C. r. hebd. Séanc. Acad. Sci., Paris* 290, 227-30.

—— (1981). Éponges hypercalcifiées ('Pharétronides', 'Sclérosponges') des cavités des récifs coralliens de Nouvelle-Calédonie. *Bull. Mus. natn. Hist. nat., Paris* 3A, 313-51.

—— (1983). Les éponges calcifiées et les récifs anciens. *Pour la Science* 68, 14-22.

2. Poriferan relationships

P. R. BERGQUIST

*Department of Zoology, University of Auckland,
New Zealand*

Abstract

A diagnosis of the Porifera based upon the possession of flagellated choanocytes, organized into a single layer which serves the purpose of driving a water current, recognises a uniquely poriferan characteristic. This definition of the phylum can no longer embrace the Hexactinellida which are distinctive in major aspects of histology and physiology, most significantly in having in part a syncytial rather than cellular organization. The Hexactinellida are regarded as a distinct phylum, the Symplasma.

Within the Porifera discovery of the Sclerospongiae and Sphinctozoa has posed problems of relationship at the class level. Within the largest class, the Demospongiae, relationships between orders have been recently clarified. Emphasis upon reproductive characteristics, ultra-structural histology, and comparative biochemistry has permitted a coherent classification to be developed. Examples of the systematic reorganizations suggested are the recognition of the Verongida as a separate order, the separation of the Nepheliospongida from the Haplosclerida, and the transfer of *Agelas* from the Ceractinomorpha to the Tetractinomorpha.

Reproductive patterns which have contributed to the justification for all rearrangements now pose problems of definition at sub-class level. The Ceractinomorpha can no longer be viewed as having uniformly larviparous reproduction; rather it is argued that primitive oviparity is retained in two orders. The Tetractinomorpha always has included forms with oviparous and larviparous reproduction but further sub-division of this group is likely.

Considering the relationship of sponges to other multicellular organisms, it is necessary to decide whether sub-kingdom status, either as

Parazoa or Enantiozoa, is warranted and, if not, to justify inclusion of Porifera within the Metazoa. Since the presence of choanocytes alone does not diagnose sponges and 'reversal of layers' does not occur at metamorphosis, arguments for sub-kingdom status are not persuasive. Metazoan attributes are apparent in many details of structure and physiology; these include membrane junctions, cellular recognition and basic immune systems, and connective tissue matrix structure.

What is a poriferan?

To discuss relationshps within and between phyla, it is necessary to have a clear diagnosis of the groups under consideration. In writing *Sponges* (1978) I was unwise enough to begin with a clear statement: 'It is easy to define a sponge', proceeding to comment that 'a sponge is a sedentary filter-feeding metazoan which utilizes a single layer of flagellated cells, the choanocytes, to pump a unidirectional water current through its body.' This was not a great number of specifics on which to diagnose a phylum, but with the state of knowledge of sponge biology as it then stood it was not possible without considerable qualification to make further generalizations which embraced all groups customarily included in the Porifera. The definition which I gave stressed not just the possession of flagellated choanocytes, but also their organization into a single layer which had the primary function of creating a water current. It is this feature which is a unique poriferan venture in evolution.

To emphasize other criteria, such as presence of a cellular, perforated pinacoderm, the lack of a basement membrane to stabilize the internal (flagellated), and both internal and external (pinacodermal) epithelia, the presence of a connective tissue matrix (mesohyl) supporting skeletal and mobile cellular elements, leads always to a dilemma over the Hexactinellida. All of these additional features are similar in three classes, Demospongiae, Calcarea and Sclerospongiae, which are distinct from other Metazoa. However, in the case of the Hexactinellida earlier reports raised the possibility that the hexacts did not have either a cellular pinacoderm or mesohyl matrix, and most significantly that they were in major part syncytial, being made up of a trabecular network of which the flagellated surface was an extension. It was pointless to speculate on this question without the evidence of modern ultrastructural observations. I commented, however, that should these differences be verified the Hexactinellida could with justification be regarded, like the fossil Archaeocyatha, as a separate evolutionary development, a view held earlier by Bidder (1929, 1930).

Implicit in my earlier diagnosis of Porifera is the assumption that choanocytes are nucleate, discrete cells and not anucleate flagellated extensions of an underlying syncytium. The recent discovery (Reiswig 1979; Mackie and Singla 1983) that the Hexactinellida are indeed syncytial requires either that this definition be modified or that the separation of the hexacts into a distinct higher category be formalized. Following detailed ultrastructural and histological investigation with accompanying physiological work, particularly on the characteristics of impulse conduction in the syncytial network (Mackie, Lawn, and Pavans de Ceccatty 1983), an impressive case for high level systematic separation of the Hexactinellida from the Porifera has been compiled. There are many points of distinction but Reiswig and Mackie (1983), who make the case for redefinition of the phylum Porifera, place greatest emphasis upon syncytial as opposed to cellular construction both in the trabecular network and in the flagellated layer of Hexactinellida. These authors have redefined the Porifera to accommodate these peculiarities and have proposed two sub-phyla, Symplasma with a single class Hexactinellida, and Cellularia for the remaining three classes. On the basis of their work I prefer to make a case for removal of the Hexactinellida from the Porifera. This is not the place to make an extended formal diagnosis of the phylum, indeed Reiswig and Mackie (1983) have done this, but with the Hexactinellida removed the concept of Porifera as indicated originally stands.

The major points distinguishing hexacts must first be the syncytial construction of the general body trabeculum and dermal membrane. Associated with the latter is the absence of a continuous cellular pinacoderm either as a dermal structure or lining internal canals. This, plus the absence of myocytes, means that there are no contractile structures. Second, the syncytial nature of the choanoderm (choanosyncytium), in conjunction with the compound nature of this region where there are anucleate differentiated flagellate effectors, basal collar bodies lacking nuclei in the mature state, and inner and outer trabecular reticulation, all contribute to the 'flagellated layer' showing a marked divergence from the true poriferan condition (Fig. 2.1).

The connective tissue matrix is also distinctive; it is a structure termed the mesolamella, composed of thin sheets of collagen fibrils and in being continuous through entire sponges providing a suspensory network for attachment of trabecular tissues. Cellular elements, such as archaeocytes and thesocytes, appear to be relatively immobile in contrast to the equivalent cells in the Porifera proper.

An important discovery relates to the structures involved in impulse conduction in Hexactinellida (Pavans de Ceccatty and Mackie 1982; Mackie, Lawn, and Pavans de Ceccatty 1983). These take the form of 'plug' connections, perforate septa which arise from Golgi structures, attach to cell membranes and become inserted into syncytial bridges. These 'plugs' form a partial barrier to translocation of materials be-

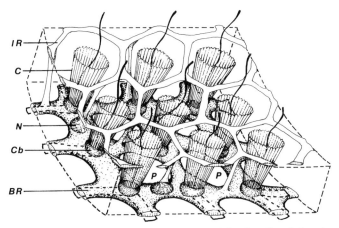

Fig. 2.1. A diagrammatic representation of a section of the flagellated chamber wall of a hexactinellid sponge. Cytoplasmic bridges or stolons join the collar bodies which make up the choanosyncytium and these structures underlie the basal reticulum. Abbreviations: BR, basal reticulum of the trabecular syncytium; C, collar; Cb, collar body; IR, inner or secondary reticulum of the trabecular syncytium; N, nucleus of the trabecular syncytium; P, prosopyle. (After Reiswig and Mackie 1983).

tween distinct specialized areas of the cytoplasm and resemble the 'pit connections' of red algae more closely than any type of junction found in the animal kingdom (Fig. 2.2). Septate junctions as found in Calcarea (Green and Bergquist 1979, 1982) possibly occur in hexacts but have not been conclusively demonstrated. It is clear from physiological work on impulse conduction and ciliary arrest behaviour that in these respects the Hexactinellida are very much more sophisticated than other sponges and that the trabecular system, a syncytium with connecting bridges and porous plugs, can function as a through conducting system. The Porifera, which lack gap junctions or any demonstrably equivalent structures, cannot obtain a sufficient level of electrical coupling for impulse conduction.

The thrust of my argument has been to exclude the Hexactinellida from the Porifera. Dealing with this in some detail is necessary for three reasons: first, because those workers who have clarified hexactinellid structure and function have made a significant contribution to invertebrate biology; second, it is appropriate to report the creation of a new phylum 'Symplasma' in this volume; and lastly, the separation of the two phyla makes the task of discussing poriferan relationships easier.

Relationships within the Porifera

The central theme of my own research is to document better and to understand the relationships within the largest class of Porifera, the

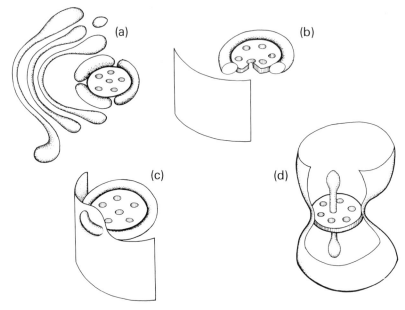

Fig. 2.2. Production of a hexactinellid 'plug' from Golgi structures (a) and its insertion into a cell membrane of intracellular bridge (b, c) and final positioning with membrane and plug fusion complete (d). Membraneous saccules lodged in the plug pores extend on both sides of the disc. (After Mackie and Singla 1983).

Demospongiae. The Calcarea is a small group, clearly defined on histological, skeletal, and developmental grounds. The recently established Sclerospongiae, while having attributes of both Demospongiae and Calcarea in their possession of calcareous and siliceous skeletal components, are closest to the Demospongiae, but have yet to be definitely related to either group (but see Vacelet, this volume). I will concentrate on the Demospongiae, poriferans with a siliceous one- to four-rayed spicule skeleton in conjunction with, or supplementary to, an organic skeleton of collagen dispersed in the mesohyl or secreted as filaments or fibres.

This large group, accounting for around 95 per cent of recent Porifera, has always proved difficult to classify and the emphasis on spicule attributes which characterized most classifications prior to 1950, though having a certain utilitarian value, served to obscure reproductive, biochemical or histological relationships. It remained for Levi (1956) to propose a rearrangement of the classification of the Demospongiae which recognized the significance of differing reproductive patterns as well as skeletal morphology. His separation into two subclasses (Ceractinomorpha, characterized by a larviparous mode of reproduction and production of solid parenchymella larvae, and Tetractinomorpha, characterized by both oviparous and larviparous reproduction and production of either blastula or parenchymella larvae) required the orders of Demospongiae to be regrouped. It was soon apparent that this was a more coherent and natural arrangement. In establishing this classification Levi noted that the Porifera were the last major group of organisms in which the orders were not clearly defined; stimulated by Levi's work we have subsequently moved to a position where ordinal groups are clearer and some relationships between orders can be recognized. This has resulted, however, in problems at sub-class and higher levels, and also with the Calcarea. Exciting new discoveries, such as the recent Sphinctozoa and the enigmatic Sclerospongiae, pose similar problems in interpretations of relationships.

In Levi's original proposition the subclass Ceractinomorpha was a well-characterized group of five related orders which, so far as was known, were all larviparous with similar larvae. The Tetractinomorpha was a group of five orders which never clearly formed a single group; further, because many species were oviparous, little information was available on reproductive patterns.

While the greatest need was to gain more information on reproductive patterns, particularly in the tetractinomorph orders, two parallel lines of investigation were suggested by advances in electron microscopy (EM) and biochemistry. First, histological attributes had been used very little in phyletic comparisons because sponge cells and matrix were very difficult to characterize by optical microscopy. They could, however, be characterized by EM. Also, details of larval structure could profitably be investigated using scanning electron microscopy (SEM). Second, it had long been recognized that Demospongiae expressed great diversity in many aspects of their biochemistry, with novel compounds of many categories being recorded, most notably by Bergmann (1949, 1962). The development of routine, high resolution separation techniques meant that patterns of occurrence of various categories of metabolites could be investigated across large species samples.

Contributions along all of these lines (Levi and Boury-Esnault 1979) have permitted Levi's original arrangement to be refined. At present three sub-classes are recognized, the Ceractinomorpha constituted largely as before but with seven rather than five orders, the Verongiinae and Petrosiidae having been given ordinal status.

The Tetractinomorpha contains five orders; one of the original orders (Epipolasida) is no longer recognized, three of the original groups remain (Axinellida, Hadromerida, Choristida); one family, the Tetillidae, has been elevated to ordinal status as the Spirophorida, and the Agelasida, previously placed within the Ceractinomorpha, has been added.

The Homoscleromorpha, with a single order, the Homosclerophorida, has been segregated from the Tetractinomorpha, these sponges having blastula larvae and a distinctive skeleton and histology. The detailed arguments for the recognition of these categories are provided by Bergquist and Hartman (1969), Levi (1973), Bergquist (1978, 1980), Bergquist, Hofheinz and Oesterhelt (1980) and Bergquist and Wells (1983).

I cannot here detail the basis of this classification, which now achieves some phylogenetic coherence, but I can emphasize how EM and biochemistry have contributed significantly to the arrangement. Three groups of sponges provide examples which can be briefly dealt with.

The present status of the Verongida as an order within the Ceractinomorpha acknowledges that the group is distinct from the Dictyoceratida, within which it previously had sub-family status. The reason for the earlier placement was that both groups had fibre rather than spicule skeletons. The distinctions sustaining ordinal status have been drawn on the basis of their unique and complex histology as evidenced by EM, their oviparous as opposed to larviparous reproduction, and their divergent biochemistry. All Verongida synthesize strongly antibacterial brominated phenolic compounds. They also contain high percentages of their lipids in the form of sterols, some of which have a novel structure. These characters stand in strong contrast with the Dictyoceratida which incubate parenchymella larvae, have a simple basic histology, and utilize alternative pathways of isoprenoid metabolism to produce a range of antibacterial terpenoid compounds. As a consequence of the latter attribute their sterol percentage is low, around 0.1-2.0 per cent of the lipid compared with 18-25 per cent in Verongida.

These differences and others separate the Verongida as a homogeneous group unrelated to other Ceractinomorpha. The occurrence of oviparity in the Verongida upsets the coherence of the subclass as originally defined, and it has been suggested (Bergquist 1978) that subclass status for the Verongida would ultimately be justified. I stress here that the first strong indication of distinctness for the verongiid sponges came as a result of biochemical studies on free amino acids (Bergquist and Hartman 1969).

Also now recognized within the Ceractinomorpha is the order Nepheliospongida, the genera of which were previously dispersed between several haplosclerid families. These sponges emphasize solid spicule skeletons rather than the light spicule/fibre combinations typical of the Haplosclerida. When material of the Devonian *Nepheliospongia* was redescribed and shown to be related to Recent sponges, such as *Petrosia*, an ancient separation between the Recent solid skeleton and fibrous skeleton groups was established. At the same time studies of larval biology in Haplosclerida established that the group generally incubated larvae with either long posterior or total uniform ciliation. However, no larvae were ever found in *Petrosia* (Bergquist, Sinclair, Green and Silyn-Roberts 1979) but Jamaican specimens of *Xestospongia*, a related genus, were found to be oviparous. There are further indications of distinctness; three genera, *Petrosia, Strongylophora*, and *Calyx*, yielded as major metabolites novel C29 sterols with extended side chains which often included a cyclopropene ring. No such sterols have been found in any haplosclerid sponges. On the basis of this palaeontological, structural, biochemical, and reproductive divergence, the order Nepheliospongida was established to receive this group of genera (Bergquist 1980).

The recognition of two new orders from within the assemblage of species previously placed within the Ceractinomorpha, both characterized by oviparity, poses diagnostic problems at the sub-class level. The two oviparous groups show no relationship to each other, nor do they have any affinity with oviparous groups within the Tetractinomorpha. The Nepheliospongida share some affinities with certain Haplosclerida, but the Verongida have no obvious relationships except to the Dendroceratida with respect to fibre structure. If oviparity is the primitive condition in Demospongiae, and its generality suggests this, then the Ceractinomorpha must be viewed as a sub-class where the ancient condition is retained in two quite separate lines. The dominant and more recent evolutionary trend has, however, been towards larviparity. An argument for retaining a single subclass is found in sterol chemistry. Most Verongida contain a range of novel C29 sterols termed aplystanes. These, in common with the cyclopropane and cyclopropene structured sterols found in some Nepheliospongida, have an unusual 26-methyl substitution and an identical configuration at the 24 (R) position. The only other known occurrence of aplystane sterols in this subclass is in one species of the Haplosclerida. The indication is that a similar biosynthetic mechanism exists in the three orders. The verongiid line has emphasized C26-methylation among sterol metabolites and produces many variations. Some Nepheliospongida are able to incorporate cyclopropane (-ene) rings in sterol side chains, others contain C30 sterols which could be elaborations of the typical verongiid didehydroaplysterol. Certainly one haplosclerid retains the ability to perform 26-methylation, but in all other respects sterol patterns in the order are very similar to those of other larviparous Ceractinomorpha. Further searches should be made for 26-methyl sterols in Ceractinomorpha, for the retention of this biosynthetic alternative and oviparity may both be primitive attributes.

It is reasonable to question the decision to leave the Verongida within the Ceractinomorpha and a case can be made for their separation. I remain convinced, however, that the Nepheliospongida and Haplosclerida are related. It cannot be argued that the Haplosclerida should be separated from the Halichondrida and Poecilosclerida, therefore the subclass can no longer be represented as possessing a homogeneous reproductive mode.

A final example concerns the genus *Agelas*, which is perhaps the most enigmatic of Recent Porifera. The Agelasida is a monogeneric order placed within the Tetractinomorpha where it must stand as closest to the Axinellida. *Agelas* was placed within the Ceractinomorpha (Poecilosclerida) until discrepancies in free amino acid pattern indicated a closer alliance with some Axinellida. Further biochemical work on the distribution of bromopyrrole compounds (oroidin) and sterols indicated a strong alliance between *Agelas* and some Axinellida; oroidins occur only in these two groups and the co-occurrence of a Δ7 ring structure and stanols is unusual. Reiswig (1976) then discovered that *Agelas* is oviparous; the combination of characters thus led to *Agelas* being referred, as a separate family, to the Axinellida (Bergquist 1978). The problem remains that the skeleton of *Agelas* is an amazing mosaic. Superficially the fibrous dominance produces a dictyoceratid effect, but the fibre organization is not dictyoceratid. Echinating spicules, with verticillate spining, are also present. Similar but not identical spicules occur in one or two Poecilosclerida and in one sclerosponge. There are arguments both ways but Hartman (1980) judged the distinction to be sufficient to sustain ordinal status. Biochemistry in this case provided an accurate guide to the correct taxonomic location of *Agelas*; the question of the level of separation from other Axinellida awaits resolution by ultrastructural study. It is interesting to note that the Clathriidae within the Poecilosclerida are the group where spicules similar to those of *Agelas* occur. Some genera also share a novel category of carotenoid with *Agelas*. This relationship is still being investigated.

These examples suffice to show that in the Porifera, where systematists are handicapped by paucity of definitive characters on which to evaluate relationships, biochemistry and ultrastructure can and will continue to contribute significantly.

The metazoan nature of sponges

No discussion of poriferan relationships would be complete without reference to the status of the phylum relative to others in the lower levels of the animal kingdom. I will refer briefly to the arguments for separation of Porifera from Metazoa as a separate sub-kingdom, either Parazoa or Enantiozoa, and will then make the case for inclusion of the Porifera within the Metazoa. In this context I prefer to leave the status of the Symplasma for future discussion.

The argument by Sollas (1884) for sub-kingdom status for Porifera as the Parazoa rested primarily upon the presence of choanocytes in sponges and the similarity of these cells to the choanoflagellates. However, choanocytes occur in several metazoan phyla and their presence in sponges is on its own no basis for a wide separation of sponges from other multicellular organisms. The concept of a sub-kingdom Enantiozoa arose from Delage's conviction that the solid sponge parenchymella larva, at metamorphosis, underwent a curious inversion of larval layers to produce the adult cellular arrangement (Delage 1892). This inversion involved the inward migration of flagellated cells to later form the feeding layer or choanoderm. Internal larval cells were thus left to form the outer lining of the adult body. Delage placed an emphatic embryological interpretation on this sequence, stating that in sponge development the two body layers were reversed, ectoderm invaginating into endoderm to form the digestive cavity. This is quite the opposite of the metazoan case. Many workers have used this as a basis for maintaining a separate status for the Porifera, but few have sought factual information to substantiate first, whether this inward migration does occur and, second, whether sponge larval cell 'layers' can in any sense be equated with, for example, coelenterate larval regions, or germ layers as understood in other groups. Bergquist (1978) dealt with this subject but at that time there was no proof that flagellated cells did not migrate inwards despite strong indications that the concept was false (Meewis 1938; Tuzet 1963, 1973; Brien 1967). Recent work (Bergquist and Glasgow 1984) combining EM and cell aggregation experiments on settling and metamorphosing larvae of *Halichondria* has finally provided evidence that larval ciliated cells are both shed and phagocytosed at metamorphosis, that choanocytes differentiate internally from archaeocytes as Meewis (1938) had suggested, and that larval flagellated cells, when separated from internal cells, have no capacity to aggregate and differentiate. We contend that 'inversion of layers' does not occur and this peculiar process thus cannot be used to sustain sub-kingdom status for the Porifera.

Recent work on sponge physiology, development, histology, and biochemistry argues for metazoan status for the Porifera. Indeed, in many attributes sponges show the basic expression of processes refined and developed by other metazoan groups.

Pavans de Ceccatty (1974), writing on the contractile system of sponges and comparing this in terms of structure and function with coelenterates, remarked that sponges provide 'a witness to the prehistory of the nervous system'. I would go further and note that the same comment applies with respect to other features such as cellular recognition systems, immune systems, development of occluding membrane junctions, and connective tissue development.

Earlier reference was made to the presence of septate junctions in Porifera; since neither freeze-fracture nor lanthanum staining has been carried out on these structures they cannot unequivocally be equated with any particular junction type, of which there are many in lower invertebrates. A point of great interest, however, concerns parallel membrane junctions (Green and Bergquist 1979, 1982). These are long distances of evenly spaced membranes with separation distances between 10 and 20 nm depending on the species. No septa occur to preserve the regular membrane space. This raises the possibility that, contrary to accepted belief, septa do not dictate membrane spacing; rather membrane spacing depends on some as yet unknown 'membrane spacing factor'. Septa may have evolved in response to a need, not constantly present in sponges, to partition and seal intracellular space. The wide range of septal patterns in invertebrate junctions, in conjunction with the constancy of intercellular spacing, supports this argument. Under such conditions a simple change in the 'spacing factor' to permit membranes to come together at intervals would achieve the transition from a septate invertebrate to a chordate tight junction pattern. No alternative theory explains this transition. The structural simplicity preserved in sponges permits this viewpoint to be taken on the evolution of more complex metazoan junctions.

The sponge story with respect to cellular recognition biochemistry is long, complex, and still being written. Cellular recognition with subsequent adhesion and/or separation is basic to three essential metazoan attributes, the establishment of multicellularity, the ability to recognize and distinguish self and non-self and thus preserve the integrity of the individual, and the ability to undergo multicellular morphogenesis. In sponges we have on the one hand a whole range of proteoglycan and glycoprotein compounds which are membrane-located and which act to stabilize adhesion between cells of the same strain or species (Burger, Burkart, Weinbaum and Jumblott 1978). These are the cell-cell recognition molecules and elucidation of the components of the system and the biochemical controls on its operation is still continuing. The question of the precise role of such a system *in vivo* remains of great interest to sponge biologists. On the other hand there is evidence for the existence of diffusible protein molecules which act to control very specifically the level of 'stickiness' two cells will express towards each other. These molecules, which promote adhesiveness between 'like' cells and depress adhesiveness between 'unlike' cells, have been termed interaction modulation factors (IMFs; Curtis 1978). The coincidence of adhesion promotion by IMFs and acceptance of grafts between sponges of the same species, in conjunction with the converse, adhesion depression and graft rejection, has led to the interesting proposal that IMFs are at some level involved with histocompatibility (Curtis 1980; Evans, Kerr and Curtis 1980).

Along quite another line of investigation, in what has been described as the 'most exciting new area of transplantation studies' (Hildemann, Bigger and Johnston 1979), several workers have demonstrated that sponges express highly discriminating transplantation immunity, possibly with a specific memory component, and distinct cytotoxic incompatibility interactions.

The experimental work in this entire area is still incomplete and many questions remain. However, the basic facts are established; self–non-self discriminatory systems with many attributes of the immune systems of higher organisms are expressed in sponges. The idea that IMFs are terminal fragments of the histocompatibility antigens and that the role of this entire system is morphogenetic control, is considered by some to be one of the most stimulating of recent time (Curtis 1980). The basic, simple expression of a cell/tissue system seen in the Porifera provides here a preview of evolutionary developments in higher Metazoa.

Space does not permit me to deal with the sponge matrix and to establish its status as a true connective tissue. Garrone (1978) has already done this well. My argument is that with a true connective tissue, basic self discrimination, reproductive patterns which in elementary processes are not unique, membrane characteristics and intracellular actin systems which foreshadow those of higher organisms, and contractile systems responsive to endogenous and exogenous stimuli, the Porifera are properly placed at the base of the Metazoa.

References

Bergmann, W. (1949). Comparative biochemical studies on the lipids of marine invertebrates, with special reference to the sterols. *J. mar. Res.* 8, 137-76.

—— (1962). Sterols: their structure and distribution. In *Comprehensive biochemistry* (eds M. Florkin and H. S. Mason) Vol. 3, pp. 103-62. Academic Press, New York.

Bergquist, P. R. (1978). *Sponges.* Hutchinson, London.

—— (1980). The ordinal and subclass classification of the Demospongiae (Porifera); appraisal of the present arrangement and proposal of a new order. *N.Z. Jl Zool.* 7, 1-6.

—— and Glasgow, K. J. (1984). Developmental potential of sponge larval cells. *N.Z. Jl Zool.* (in press).

—— and Hartman, W. D. (1969). Free amino acid patterns and the classification of the Demospongiae. *Mar. Biol.* 3, 247-68.

—— and Wells, R. W. (1983). Sponge chemotaxonomy: the development and current status of the field. In *Marine natural products*, (ed. P. Scheuer) Pt 5, pp. 1-50. Academic Press, New York.

—— and Hofheinz, W., and Oesterhelt, G. (1980). Sterol composition and the classification of the Demospongiae. *Biochem. syst. Ecol.* 8, 422-35.

—— Sinclair, M. E., Green, C. R., and Silyn-Roberts, H. (1979). Comparative morphology and behaviour of sponge larvae. *Colloq. int. C.N.R.S.* 291, Biologie des spongiaires, pp. 103-11.

Bidder, G. P. (1929). Sponges. In *Encyclopedia Britannica* (14th edn) Vol. 21, pp. 254-61.

—— (1930). On the classification of sponges. *Proc. Linn. Soc., Lond.* 141, 44-7.

Brien, P. (1967). Les éponges—leur nature metazoaire—leur gastrulation—leur état colonial. *Ann. Soc. R. zool. Belg.* 97, 197-235.

Burger, M. M., Burkart, W., Weinbaum, G., and Jumblott, J. (1978). Cell-cell recognition: molecular aspects. Recognition and its relation to morphogenetic processes in general. In *Cell-cell recognition* (ed. A. S. G. Curtis) pp. 1-24. Society for Experimental Biology, Symposium 32. Cambridge University Press, Cambridge.

Curtis, A. S. G. (1978). Cell-cell recognition: positioning and patterning systems. In *Cell-cell recognition* (ed. A. S. G. Curtis) pp. 51-82. Society for Experimental Biology, Symposium 32. Cambridge University Press, Cambridge.

—— (1980). Histocompatibility systems: cell recognition and cell adhesion. In *Cell adhesion and mobility* (eds A. S. G. Curtis and J. D. Pitts) pp. 273-92. Cambridge University Press, Cambridge.

Delage, Y. (1892). Embryogenèse des éponges siliceuses. *Archs Zool. exp. gén.* 10, 345-498.

Evans, C. W., Kerr, J., and Curtis, A. S. G. (1980). Graft rejection and immune memory in marine sponges. In *Phylogeny of immunological memory* (ed. M. J. Manning) pp. 27-34. Elsevier, New York.

Garrone, R. (1978). Phylogenesis of connective tissue. Morphological aspects and biosynthesis of sponge intercellular matrix. In *Frontiers of matrix biology* (ed. L. Robert) Vol. 5, pp. 1-250. Karger, Basle.

Green, C. R. and Bergquist, P. R. (1979). Cell membrane specializations in the Porifera. *Colloq. int. C.N.R.S.* 291, Biologie des spongiaires, 153-7.

—— (1982). Phylogenetic relationships within the Invertebrata in relation to the structure of septate junctions and the development of occluding junctional types. *J. Cell Sci.* 53, 279-305.

Hartman, W. D. (1980). Systematics of the Porifera. In *Living and fossil sponges* (eds J. W. Wendt, W. D. Hartman, and F. Wiedenmayer) pp. 24-51. University of Miami, Florida.

Hildemann, W. H. Bigger, C. H., and Johnston, I. S. (1979). Histocompatibility reactions and allogeneic polymorphism among invertebrates. *Transplant. Proc.* 11, 1136-42.

Levi, C. (1956). Étude des *Halisarca* de Roscoff. Embryologie et systématiques des Demosponges. *Archs Zool. exp. gén.* 93, 1-181.

—— (1973). Systématique de la classe des Demospongiaria (Demosponges). In *Traité de Zoologie*, Vol. 3, *Spongiaires* (ed. P.-P. Grasse) pp. 37-631. Masson et Cie, Paris.

—— and Boury-Esnault, N. (eds) (1979). Biologie des spongiaires. *Colloq. int. C.N.R.S.* 291.

Mackie, G. O. and Singla, C. L. (1983). Studies on hexactinellid sponges. I. Histology of *Rhabdocalyptus dawsoni* (Lambe, 1873). *Phil. Trans. R. Soc.* B301, 365-400.

—— Lawn, I. D., and Pavans de Ceccatty, M. (1983). Studies on hexactin-ellid sponges. II. Excitability conduction and coordination of responses in *Rhabdocalyptus dawsoni* (Lambe, 1873). *Phil. Trans. R. Soc.* B301, 401-8.

Meewis, H. (1938). Contribution à l'étude de l'embryogenèse des Myxo-spongidae *Halisarca lobularis* (Schmidt). *Archs Biol.* 50, 3-66.

Pavans de Ceccatty, M. (1974). The origin of the integrative systems: a change in view derived from research on coelenterates and sponges. *Perspect. Biol. Med.* 17, 379-90.

—— and Mackie, G. O. (1982). Genèse et evolution des interconnexions syncytiales et cellulaires chez une éponge hexactinellide en cours de reaggre-gation aprés dissociation *in vitro*. *C. r. hebd. Séanc. Acad. Sci., Paris* 294, 939-44.

Reiswig, H. M. (1976). Natural gamete release and oviparity in Caribbean Demospongiae. In *Aspects of sponge biology* (eds F. W. Harrison and R. R. Cowden) pp. 99-112. Academic Press, New York.

—— (1979). Histology of Hexactinellida (Porifera). *Colloq. int. C.N.R.S.* 291, Biologie des spongiaires, 173-80.

—— and Mackie, G. O. (1983). Studies on hexactinellid sponges. III. The taxonomic status of Hexactinellida within the Porifera. *Phil. Trans. R. Soc.* B301, 419-28.

Sollas, W. J. (1884). On the development of *Halisarca lobularis* (O. Schmidt). *Q. Jl microsc. Sci.* 24, 603-21.

Tuzet, O. (1963). The phylogeny of sponges according to embryological, histological and serological data, and their affinities with the Protozoa and Cnidaria. In *The lower Metazoa. Comparative biology and phylogeny* (eds. E. C. Dougherty, Z. N. Brown, E. D. Hanson and W. D. Hartman) pp. 129-48. University of California Press, San Francisco.

—— (1973). Introduction et place des spongiaires dans la classification. In *Traité de Zoologie*, Vol. 3, *Spongiaires* (ed. P.-P. Grasse) pp. 1-26. Masson et Cie, Paris.

3. The trochaea theory: an example of life cycle phylogeny

CLAUS NIELSEN

Zoological Museum, Universitetsparken 15, Copenhagen, Denmark

and

ARNE NØRREVANG

Institute of Cell Biology and Anatomy, Universitetsparken 15, Copenhagen, Denmark

Abstract

The metazoan ancestor was a holopelagic blastaea with monociliate cells which gave rise to a holopelagic gastraea. From the gastraea, one line led to the Cnidaria, which retained the monociliate cells but became mostly pelago-benthic. Another line led to the trochaea, which remained holopelagic but had a ring of multiciliate cells with compound cilia, the archaeotroch, forming a downstream-collecting system around the blastopore. From the trochaea, one line led to the Gastro-neuralia, in which the adults became benthic, lost the archaeotroch and crept on the single cilia along the fused lateral lips of the blastopore; the larvae elaborated the archaeotroch into a prototroch, metatroch, and telotroch. Another line led from the trochaea to the holopelagic tornaea, which acquired a new mouth surrounded by a ring of single cilia on monociliate cells, the neotroch, forming an upstream-collecting system, while the archaeotroch was retained as a locomotory organ around the anus (blastopore); the pelago-benthic descendants of the tornaea are the Notoneuralia.

Introduction

Interest in phylogeny has in recent years been stimulated both by a wealth of new information, e.g. on embryology, ultrastructure and biochemistry, and by a series of papers on theoretical systematics. In this paper we shall try to combine a re-evaluation of the more classical information with new ideas and information.

Two principles have been especially important in our reasoning: (i) whole life cycles must be considered; and (ii) changes in structure must be placed in a functional context. We also emphasize the concept, most consistently advocated by Jägersten (1972), that changes from pelagic to benthic life (and vice versa) must be correlated with fundamental changes in function and hence in structure. Inevitably, a number of our conclusions coincide with those of various earlier authors but to avoid obscuring our theory we have almost completely omitted any discussion of the historical background.

Nielsen's (1979) paper on larval ciliary bands and metazoan phylogeny represents an earlier stage of the present theory. It emphasized the presence of downstream-collecting ciliary systems in gastroneuralian larvae. It also focussed on the origin of the paired ventral nerves from a circumblastoporal nerve in the gastroneuralians as opposed to the dorsal central nervous system of the notoneuralians. The main stumbling block of the early theory is that it failed to explain the origin of the ciliary system of the notoneuralian larvae.

The main pathways of metazoan evolution

This paper provides an outline of the series of changes in structure and function which we believe occurred during the evolution from early metazoans to the three main extant groups: Cnidaria, Gastroneuralia (Protostomia) and Notoneuralia (Deuterostomia). This theory is illustrated in Fig. 3.1. The allocation of certain phyla to these groups will become clear from the text, but a discussion of all the animal phyla will be the topic of a separate paper (Nielsen, in press).

Our theory depends on the following series of basic assumptions:

1. *Ontogeny to some extent reflects phylogeny.* This is now accepted by almost all authors and a relevant discussion may be found in Jägersten (1972).

2. *The metazoans constitute a monophyletic group.* The basic cytological features of all metazoan cells are very uniform and also a number of more specialized cells such as spermatozoa and cyrtocytes show so many detailed similarities that polyphyletic evolution seems very improbable.

3. *When holopelagic ancestors became benthic a pelagic larval stage was retained.* This gave rise to the pelago-benthic life cycles so characteristic of many metazoan phyla (Jägersten 1972). Originally these larvae were planktotrophic, but in many lines the eggs acquired so much yolk that

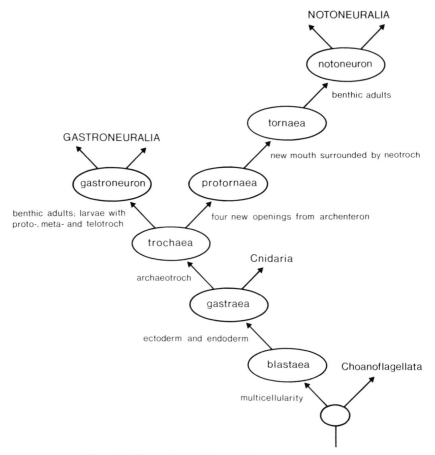

Fig. 3.1. The early evolution of the animal kingdom.

the larvae became lecithotrophic and lost the organs used in feeding. This loss appears to be irreversible (Strathmann 1978) but structures used in feeding by the adult stages may develop early and so become functional within the larva which then becomes secondarily plankto-trophic (adultation of Jägersten 1972). Consequently, those feeding structures occurring (almost) exclusively in the pelagic larvae of several phyla must be interpreted as structures inherited from a common an-cestor, and they are therefore of considerable phylogenetic importance.

4. *Adultation is an important evolutionary mechanism.* This is especially so for the development in the larvae of structures which originally evolved as adaptations for the benthic phase (Jägersten 1972).

5. *Highly specialized adults may have very 'primitive' larvae and vice versa.* In species with pelago-benthic life cycles the larvae and the adults adapt to quite different environments, and there is nothing to indicate for

example that a strongly modified larva (e.g. lecithotrophic or parasitic) or even direct development cannot lead to a primitive adult.

Blastaea: the metazoan ancestor

There are so many similarities between certain metazoan cells and choanoflagellates that we believe that the metazoan ancestor was a colony of choanoflagellate-like cells. The first metazoan evolved when the connections between cells in the colony became so intimate that a division of labour occurred. The presence of a coeloblastula stage in the ontogeny of at least some species in almost all phyla indicates that this ancestral metazoan was equivalent to the 'blastaea' of Haeckel (1874). The monociliate (or flagellate) cells of the blastaea fed by endocytosis and most of them had collars of shorter or longer microvilli comparable with those of present-day choanoflagellates and choanocytes. The original blastaea had no axis of orientation but further coordination between the cells led to the fixation of a preferred direction of swimming.

Evolution in the plankton: gastraea–trochaea–protornaea–tornaea

The following sequence of postulated evolutionary events is illustrated in Fig. 3.2. With a preferred direction of swimming an antero-posterior axis became established. Conditions for endocytosis must have been more favourable at the posterior leeward side of the organism and this shelter became enhanced by a flattening of the posterior side. An invagination of this side would have given further advantages for food uptake and made a differentiation of two types of somatic cells possible, i.e. locomotory cells (ectoderm) at the anterior and lateral parts of the organism and digestive cells (endoderm) at the invaginated archenteron. All cells were monociliate but certain specializations may soon have arisen. This organism represents the well-known gastraea (Haeckel 1874). The cells at the anterior pole of the gastraea were the first to come into contact with new chemical and mechanical stimuli during swimming and became the apical sense organ characteristic of most invertebrate larvae. The other cells of the ectoderm were not specialized at first, but with increasing body size the cilia of a ring of cells around the blastopore specialized to form a downstream-collecting system. The cells of this ring were probably monociliate at first, but subsequently became multiciliate and coordinated as compound cilia (syncilia) which enhanced their efficiency both in feeding and locomotion. This organism, the trochaea, thus had an apical sense organ, a

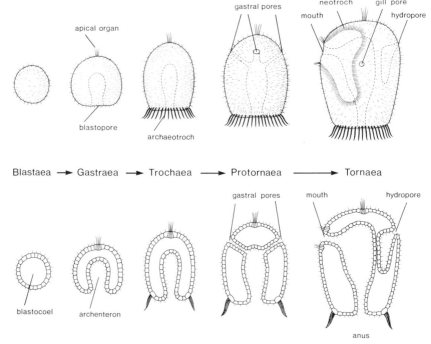

Fig. 3.2. Evolution in the plankton.

monociliate ectoderm, a ring of cells with compound cilia (the arch-aeotroch) around the blastopore, and monociliate cells forming a cir-cumblastoporal ciliary zone: the cilia of this zone transported particles caught by the archaeotroch to the ciliated archenteron.

As the next step in the evolution of the holopelagic organisms we believe that four new openings formed between the upper part of the archenteron and the exterior, a structure resembling the four apical canals from the distal part of the gut in many ctenophores. The result-ing organism, which we call the protornaea, at first had the blastopore functioning both as mouth and anus but the new openings made it possible to establish a unidirectional flow, with particles entering through one of the new openings and leaving through the blastopore. A new particle-collecting organ, the neotroch, differentiated around the new mouth; it consisted of a band of monociliate cells constituting an upstream-collecting system. The area between the neotroch and the mouth, the oral field, was covered with small cilia which transported the particles to the mouth. The archaeotroch lost its feeding function but retained a locomotory function. The two lateral openings from the archenteron probably functioned like the primitive 'gill pores' of *Cephalodiscus*, allowing excess water brought in with food particles to pass out from the oesophagus (Gilmour 1982). The canal leading to the fourth (aboral) opening lost the connection with the gut, increased in size and gave rise to an anterior coelomic pouch connected with the exterior via a hydropore. We call this organism the tornaea.

It should be stressed that it fits our theory equally well if the four new openings (mouth, two gill pores, and hydropore) came into existence sequentially. However, the development of the hydropore and mouth in certain holothurians and enteropneusts makes it probable that at least these two openings became established simultaneously.

The nervous system of all these pelagic organisms had as their main components an apical 'brain' and a ring of nerves along the cells of the archaeotroch coordinating its ciliary movements. A number of nerves (? 4 or 8) connected the apical organ with this circumblastoporal ring.

Pelagic–sessile life cycle: Cnidaria

The cnidarians evolved directly from the gastraea and retained the monociliate cells and the two-layered bodyplan, only modified slightly by the presence of the supporting mesogloea between the two layers. Embolic gastrula stages are found in many types of both anthozoans and medusozoans, and the solid planula larva which is often regarded as characteristic of cnidarians has probably evolved in several lines. It is possible that the early cnidarians were holopelagic, with a gastraea larva and an adult medusa. However, a benthic polyp stage, and thus a pelago-benthic life cycle, must have evolved very soon.

Pelagic-creeping life cycles

1. Gastroneuralia

The ancestral gastroneuralian, which we call the gastroneuron, was a trochaea in which the larva remained pelagic whereas the adult became benthic and began to creep with the blastopore in contact with the sub-stratum (Fig. 3.3). The adult both crept and collected food particles with the single cilia surrounding the blastopore (the circumblastoporal zone), while the archaeotroch, which was used for swimming and particle collection in the larva, was lost. Movement on the substratum was initially at random, but a defined direction of creeping and thus a new anterior pole subsequently became established. The blastopore elongated along the new antero-posterior axis and unidirectional flow became established for food particles which then entered the archenteron at the anterior end of the blastopore and were ejected at the

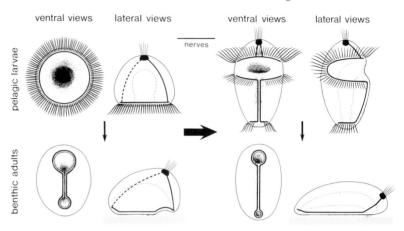

Fig. 3.3. The evolution from an early gastroneuron (left) to a typical gastroneuralian (right). Modified from Nielsen (1979).

posterior end. The lateral lips of the blastopore fused, forming a tube-shaped gut and leaving the mouth and the anus as the only remaining parts of the blastopore. The single cilia along the fused blastopore lips formed a creeping organ, the gastrotroch. The typical nervous system of gastroneuralians resulted from the lateral closure of the blastopore: the ring nerve following the archaeotroch cells formed a loop around the new mouth, a pair of midventral longitudinal nerves, and a loop around the anus. The apical organ, which moved towards the new anterior pole, was still connected to this ventral system via at least one pair of anterior nerves.

The larva of the early gastroneuralian was a pelagic, planktotrophic trochaea-like organism but the specializations mentioned above soon became manifest in the larval stage through adultation. The blastopore closed partially as in the adult and the part of the archaeotroch surrounding the mouth extended into a pair of lateral loops, almost encircling the apical end. The anterior part of this loop became the large prototroch, which is used both for swimming and particle collecting. The posterior part became the smaller metatroch used only in feeding. In some polychaete larvae the metatroch is connected to the prototroch mid-dorsally and is incomplete mid-ventrally, leaving space for the gastrotroch, the ventral part of the circumblastoporal ciliary zone. The prototroch and metatroch are downstream-collecting systems orientated according to their derivation from the archaeotroch. The lateral parts of the archaeotroch along the fused blastopore lips were lost. The gastrotroch (neurotroch) creates a rejection current, which carries away unwanted particles, in some cases through a midventral interruption of the telotroch. The compound cilia around the anus became the telotroch, which of course had lost its particle-collecting function, but whose movements are in accordance with its derivation from the archaeotroch. The original gastroneuralian larva is thus the trochophore.

2. Notoneuralia

The ancestral notoneuralian, which we call the notoneuron, was a tornaea in which the larva remained pelagic whereas the adult became benthic, creeping on the area between the mouth and the anus (the 'ventral' side, not the same as 'ventral' in the gastroneuralians, see Fig. 3.4). In the adult the neotroch was used for feeding and was probably drawn out on tentacles to enlarge its length and thereby its particle-collecting capacity. The 'gill pores' were not specialized for filter feeding. The archaeotroch and its accompanying nerve lost their function in the benthic adult and were only retained in the larva. Initially, the apical organ formed the brain but a nervous concentration subsequently extended along the dorsal midline.

The larva of the notoneuron was a pelagic, planktotrophic tornaea-like organism resembling the tornaria, with an upstream-collecting system consisting of single cilia around the mouth and a ring of compound, locomotory cilia, the archaeotroch, around the anus.

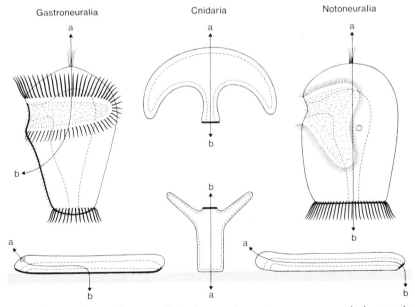

Fig. 3.4. Comparison of the axes of the three major metazoan groups. a-b denotes the apical-blastoporal axis. The blastoporal area is indicated by a heavy line. It is obvious that the 'ventral' side of the gastroneuralians is not homologous with the 'ventral' side of the notoneuralians.

Consequences for metazoan classification

Our theory directly implies the following dichotomous arrangement of the major metazoan groups:

Gastraeozoa ⎰ Cnidaria
⎱ Trochaeozoa ⎰ Gastroneuralia
⎱ Notoneuralia

The theory also implies that bilaterality evolved twice among the trochaeozoans and that the term Bilateria therefore must be abandoned. A discussion of the allocation of all animal phyla to their proper places in the phylogenetic tree will be the topic of a forthcoming paper.

Discussion

In all theories on metazoan phylogeny, two basic questions are intimately connected: 'Which forms were the ancestors of the metazoans?' and 'Are the metazoans a monophyletic group?'.

The search for ancestral forms has centered around two protozoan groups, ciliates and flagellates. The scheme of protistan radiation proposed by Sleigh (1979) shows the Acoela, and from there the higher metazoans, as descendants of ciliate-like ancestors with closed mitosis and the Cnidaria and Porifera as descendants of flagellates with open mitosis. The review by Taylor (1976) shows, however, that this characteristic varies in several groups, in slime moulds even within the life cycle of one species, and we therefore reject this idea.

Theories focusing on ciliates or ciliate-like organisms as ancestors envisage the origin of the first metazoan through a cellularization of a large ciliate or ciliate-like organism. The theories considering a flagellate as the more likely ancestor envisage the evolution of a division of labour among the cells of a flagellate colony. Cellularization theories mostly emphasize similarities between ciliates and acoel turbellarians, which were believed to be at least partially plasmodial. It has, however, been shown that all cell layers of the acoels are cellular and all the species studied so far have holoblastic cleavage, so that if cell membranes were absent in certain tissues this would probably be the result of fusion of cells (i.e. a syncytium). We can see nothing to support these theories and they are now rejected by almost all authors.

Flagellate theories regard variously shaped colonies of benthic or pelagic flagellates as the metazoan ancestor. The ancestor is generally believed to have been monociliate and many authors see a choanoflagellate colony as a plausible ancestral form, because choanocyte-like cells have been observed in many phyla (for references see Salvini-Plawen 1978). The presence of diplosomal bodies in choanoflagellates and many monociliate metazoan cells also supports this idea (Hibberd 1975; Rieger 1976).

Two main variants of the flagellate theory prevail: the plakula theory where the ancestral metazoan is considered to have been a flat, two-layered organism, and the blastaea theories which take a spherical colony as their starting point. According to the plakula theory (e.g. Grell 1974) the ancestral metazoan was derived from a discoid, creeping, two-layered flagellate colony in which a division of labour evolved between a layer of digestive cells facing the substratum and an upper layer of non-digestive cells; reproductive cells were probably located between the two layers. This organization is very close to the existing placozoan *Trichoplax*. During evolution, a preferred creeping direction became established and the lateral areas of the organism fused to produce an organism with a tube-shaped gut. The theory does not account for the evolution of the main metazoan types: Cnidaria, Gastroneuralia, and Notoneuralia. It also disregards the fact that a plakula stage has not been observed in any metazoan ontogeny but instead blastula and gastrula stages are observed in almost all the phyla at the developmental stage where, according to the theory, one should expect a plakula stage. Another factor which makes this theory rather improbable is that it implies that the pelago-benthic life cycle which is so widespread among the metazoan phyla, is entirely a result of secondary (caenogenetic) specializations. The plakula theory does not give any hint of how the highly characteristic planktotrophic larval types evolved.

According to the blastaea theories, which go back to Haeckel (1874, 1875), a hollow spherical colony of flagellates was the metazoan ancestor. The organization of the green alga *Volvox* is often used to visualize the morphology and development of the ancestral form but this does not mean that *Volvox* or similar autotrophic organisms should be regarded as the immediate ancestors of the metazoans. Most of the recent supporters of these theories believe that the ancestor was a choanoflagellate which formed colonies similar to those of *Sphaeroeca* and *Proterospongia*.

Blastaea theories fall into two main groups: planuloid and gastraea. According to the former (Ivanov 1973; Salvini-Plawen 1978) a compact, planula-like organism evolved from the blastaea either through delamination or ingression of cells from the surface layer. Later, a mouth opening formed and the organism thus represented an ancestral acoelous turbellarian from which all bilateral animals could have evolved. In the planuloid, nutrients were taken up by the cells of the outer layer and transported to the inner cells; at a later stage the inner cells became rearranged in the shape of an archenteron. This theory again fails to account for the evolution of Gastroneuralia and Noto-neuralia and of the pelago-benthic life cycle. Furthermore, we find it difficult to envisage the gradual evolutionary change from an organism in which particles are taken up by the outer cells and nutrients trans-ported to the inner cells to an organism in which inner cells take up particles which have entered through a new opening in the outer cell layer. The adaptive steps leading to the evolution of a mouth opening in the outer cell layer are not accounted for.

According to the gastraea theory (Haeckel 1875; Remane 1967; Jägersten 1972) embolic gastrulation recapitulates evolution from a spherical blastaea to a cup-shaped gastraea with locomotory ectoderm and nutritive endoderm. The continuous changes in morphology and function leading to the gastraea are easy to follow and the widespread occurrence of embolic gastrulation in the invertebrates (Jägersten 1972) lends further support. With Haeckel (1875) and Remane (1967) we believe that the gastraea was pelagic and radially symmetrical; Jägersten (1972) proposed that the evolution proceeded via bilateral, benthic organisms, i.e. bilateroblastaea-bilaterogastraea, and although this theory must be accepted as possible on morphological and func-tional grounds, we find it less probable because such stages are either unknown or very rare ontogenetically.

In the gastraea theory the evolution of mesoderm and coelom oc-curred in subsequent stages, and these structures are regarded as homologous throughout the so-called Bilateria. We stress that although mesoderm is always situated between ectoderm and endoderm its origin is highly variable, e.g. from the 4d-cell in the spiralians and from various parts of the archenteron in echinoderms and chordates. We therefore believe that mesoderm is *not* homologous throughout the metazoans. Therefore, the coeloms of spiralians and notoneuralians cannot be homologous. We also find it unlikely that all the bilateral phyla which show no trace of a coelom at any stage should have de-scended from coelomate ancestors. We therefore reject all theories which include a creeping, bilateral gastraea with gastric pouches evolv-ing into coelom as the ancestor of bilateral metazoans, e.g. the entero-coel and the archicoelomate theories.

The present version of the gastraea theory, which we call the trochaea theory, coincides with the ideas of Haeckel (1874, 1875) and Remane (1967) concerning the evolution of the gastraea and the Cnidaria. The gastraea and Cnidaria are characterized by having only monociliate cells, whereas the trochaea was characterized by the presence of a ring of multiciliate cells with compound cilia forming a downstream-collecting system around the mouth (blastopore). A single cilium has limited length and power, but the creation of compound cilia permits the units to be longer and more powerful (Blake and Sleigh 1975).

Although the present theory (Fig. 3.1) differs from earlier ones in the derivation of the bilateral phyla we nevertheless recognize the same two main groups, Gastroneuralia (Protostomia) and Notoneuralia (Deuterostomia). Evolution of the gastroneuralian ancestor, gastro-neuron, with a pelago-benthic life cycle and a trochopore larva, from the trochaea was outlined in an earlier paper (Nielsen 1979), and in our opinion it is easy to trace the continuous changes in structure and function leading from the holopelagic gastraea-trochaea to the pelago-benthic gastroneuron. The early gastroneuralians probably had very little mesoderm (as do most trochophores) and it appears easy to visual-ize the evolution of different types of mesodermal organization, e.g. the compact mesenchyme of the platyhelminthes and the metameric coelomic pouches of the annelids (Clark 1964).

The derivation of the notoneuralian ancestor, the notoneuron, ap-pears less obvious, and we have therefore proposed the existence of an evolutionary stage, the protornaea, between the trochaea and the noto-neuron. This hypothethical form retained the blastopore and had four openings from the apical part of the archenteron; one opening later became the mouth, the two neighbouring became a pair of 'gill pores' and the fourth became the hydropore. There is embryological evidence to support the formation of mouth and hydropore from the archenteron in the late gastrula stage, whereas the gill pores develop later and do not function in the larva.

We feel uncertain about the functional changes leading to the estab-lishment of these four new openings. It is, of course, possible that the different types of openings evolved in sequence instead of simul-taneously, but this would not necessitate major changes in our theory and the problems of explaining how the new openings which were formed remain the same. However, secondary connections between the gut and the exterior, i.e. other than the blastopore, are observed in a number of phyla. In ctenophores, the apical end of the archenteron has four pockets, two of which serve as anal pores; this is rather close to the situation in the protornaea. In some hydro- and scyphomedusae, each radial canal opens in a pore near the tentacle and Hyman (1940) observed that carmine particles mixed in the food of the hydromedusa *Aequorea* were expelled from the gastral system through these pores. In the polyclad turbellarians *Yungia* and *Cycloporus*, Lang (1884) described numerous anal pores. In the nemertean *Malacobdella* a mouth opening forms at the site of the blastopore but this primary mouth later closes again and a new mouth opening breaks through at the anterior (Ham-marsten 1918).

It is thus clear that secondary openings functioning as mouth or anus have become established in several phyla and the consistency of the development of mouth, hydropore, and gill openings as secondary connections between the apical end of the archenteron and the exterior in notoneuralian embryology also lends credibility to the existence of the protornaea/tornaea. The evolution from the holopelagic tornaea to the pelago-benthic notoneuron, with a pelagic tornaria larva and a creeping, benthic adult involves only minor changes such as the reduction of the archaeotroch in the adult.

The essence of the trochaea theory can be expressed in the following diagram:

Holopelagic: blastaea → gastraea → trochaea → protornaea → tornaea
                          ↓   ↓   ↓
Pelago-benthic: Cnidaria Gastroneuralia   Notoneuralia

References

Blake, J. R. and Sleigh, M. A. (1975). Hydromechanical aspects of ciliary propulsions. In *Swimming and flying in nature* (eds T. V. T. Wu, C. J. Brokaw, and C. Brenner) Vol. 2, pp. 185-209. Plenum Press, New York.

Clark, R. B. (1964). *Dynamics in metazoan evolution: the origin of the coelom and segments.* Clarendon Press, Oxford.

Gilmour, T. J. M. (1982). Feeding in tornaria larvae and the development of gill slits in enteropneust hemichordates. *Can. J. Zool.* 60, 3010-20.

Grell, K. G. (1974). Vom Einzeller zum Vielzeller, Hundert Jahre Gastraea-Theorie. *Biol. uns. Zeit* 4, 65-71.

Haeckel, E. (1874). Die Gastraea-Theorie, die phylogenetische Classification des Tierreichs und die Homologie der Keimblätter. *Jena. Z. Naturwiss.* 8, 1-55.

—— (1875). Die Gastrula und die Eifurchung der Thiere. *Jena. Z. Naturwiss.* 9, 402-508.

Hammarsten, O. D. (1918). Beitrag zur Embryonalentwicklung der *Malacobdella grossa* (Müll.). *Arb. zootom. Inst. Univ. Stockh.* 1, 1-96.

Hibberd, D. J. (1975). Observations on the ultrastructure of the choano-flagellate *Codosiga botrytis* (Ehr.) Saville-Kent with special reference to the flagellar apparatus. *J. Cell Sci.* 17, 191-219.

Hyman, L. H. (1940). *The invertebrates: Protozoa through Ctenophora.* McGraw-Hill, New York.

Ivanov. A. (1973). *Trichoplax adhaerens,* a phagocytella-like animal. *Zool. Zh.* 52, 1117-31 [in Russian].

Jägersten, G. (1972). *Evolution of the metazoan life cycle.* Academic Press, London.

Lang, A. (1884). Die Polycladen (Seeplanarien) des Golfes von Neapel und der angrenzenden Meeresabschnitte. *Fauna Flora Golf. Neapel* 11, 1-688.

Nielsen, C. (1979). Larval ciliary bands and metazoan phylogeny. *Fortschr. zool. Syst. Evolutionsforsch.* 1, 178-84.

—— (In press). Animal phylogeny in the light of the trochea theory. *Biol. J. Linn. Soc.*

Remane, A. (1967). Die Geschichte der Tiere. In *Die Evolution der Organismen* (ed. G. Herberer) Vol. 1, pp. 589-677. Gustav Fischer, Stuttgart.

Rieger, R. M. (1976). Monociliated epidermal cells in Gastrotricha: Significance for concepts of early metazoan evolution. *Z. zool. Syst. Evolutionsforsch.* 14, 198-226.

Salvini-Plawen, L. v. (1978). On the origin and evolution of the lower Metazoa. *Z. zool. Syst. Evolutionsforsch.* 16, 40-88.

Sleigh, M. A. (1979). Radiation of the eukaryote Protista. In *The origin of major invertebrate groups* (ed. M. R. House) pp. 23-53. Academic Press, London.

Strathmann, R. R. (1978). The evolution and loss of feeding larval stages of marine invertebrates. *Evolution* 32, 894-906.

Taylor, F. J. R. (1976). Flagellate phylogeny: a study in conflicts. *J. Protozool.* 23, 28-40.

4. Covariability of reproductive traits in marine invertebrates: implications for the phylogeny of the lower invertebrates

P. J. W. OLIVE

Dove Marine Laboratory and Department of Zoology,
The University, Newcastle upon Tyne, UK

Abstract

Jägersten (1972) and Franzén (1956, 1970) have postulated that plank-totrophic larval development and the possession of simple spermatozoa associated with the discharge of small oocytes freely into water are 'primitive' conditions for the Metazoa. A review of the covariation of these reproductive traits with others show that these so-called primitive traits are not independent variables. They are associated in particular with large-bodied organisms which accumulate gametocytes in the body cavities or intercellular spaces, discharging them to the exterior in annual synchronized spawning crises. Diminutive organisms rarely, if ever, show the supposedly primitive traits. It is concluded that plankto-trophy arose in relatively large-bodied organisms. It is difficult to sus-tain the view that the pelago-benthic life cycle was established once by a common metazoan ancestor of meiobenthic proportions. It seems likely that the capacity to sustain a planktotrophic mode of develop-ment has been exploited by the descendants of a number of organisms which reached a grade of organization which permitted the accumu-lation and storage of developing gametocytes.

Introduction

Certain reproductive conditions of the marine invertebrates have come to be regarded as 'primitive'. The most notable among these are development via a ciliated, planktotrophic, and pelagic larval phase (Jäger-sten 1972) and external fertilization with a simple round-headed spermatozoan of a characteristic type which occurs in most lower invertebrate phyla (Franzén 1956, 1970). These two theses were drawn together by Jägersten (1972) to identify what he interpreted as 'original' and 'altered' conditions in the marine Invertebrata (Table 4.2). In this paper I shall try to determine in what sorts of organisms we find a full set of 'primitive' features and in what organisms we find 'altered' conditions. Do these traits occur independently of all others or do they form part of a recognizable reproductive strategy?

Defining sets of reproductive traits: identifying the covariables

The overall reproductive biology of an organism is the product of a large number of different reproductive traits, each of which appears as a variable (Stearns 1980). Several of the more important ones are listed in Table 4.1 and the question arises: are all of these reproductive traits independent variables, or is there a finite number of sets of these reproductive traits which act as covariables? If all the traits listed in Table 4.1 were independent variables, an almost infinite variety of combinations could occur which would be impossible to classify. Theoretical and empirical studies suggest, however, that this is not so. The most widely accepted theoretical framework is that developed by MacArthur and Wilson (1967) from Dobzhansky (1950) which supposes the existence of two fundamentally different modes of natural selection which either maximize the intrinsic rate of population increase (*r*-selection) or the competitiveness of individuals and thus the carrying capacity in a saturated environment (K-selection). The theory has been used to predict functional correlates of *r*- and K-selection (Pianka 1970), which are sets of supposedly covariable reproductive traits. There are a number of objections to the *r*/K-selection theory, and in particular to the way it has been used (Parry 1981). Stearns (1976) pointed out that different models can predict the evolution of the same sets of reproductive traits but under different circumstances and it is by no means certain that the 'functional correlates' identified by Pianka are covariables. It is in fact increasingly evident that for marine invertebrates they are not. The problem of defining a mathematical theory to predict optimal reproductive strategies is a formidable one and a satisfactory general theory has not yet been formulated (see Stearns, in press, for further discussion).

In an earlier review of the reproductive biology of the Polychaeta (Olive and Clark 1978) an intuitive classification of their reproduction

Table 4.1. Reproductive traits of marine invertebrates

Trait	Pelagic		Non pelagic
1. Development		planktotrophic / lecithotrophic / mixed	
2. Egg size	Small c. 50 μm	→	Large >1000 μm
3. Fecundity	High 10⁶	→	Low 1
4. Brood frequency	Low 1 p.a.	→	High, Many p.a. almost continuous
5. Broods per lifetime	One		Many
6. Longevity (Generation time)	Perennial, Many years	Annual	Subannual, A few days or weeks
7. Body size	Large, Length >1000 mm	↕	Small <1 mm
8. Spermatozoa	Simple		Advanced
9. Fertilization	External without sperm storage		Internal or with sperm transport or storage
10. Reproductive effort†	Large		Small
	Made later		Made early

†Reproductive effort can be defined as $\dfrac{E_G}{E_S+E_G}$ or $\dfrac{\Delta E_G}{\Delta E_S+\Delta E_G}$

where E_G is the energy allocated to germinal tissues, E_S the energy allocated to somatic tissues, ΔE_G the annual increment in germinal tissues, and ΔE_S the annual increment in somatic tissues.

into a number of distinct categories was attempted, and in a more recent review (Olive 1984) I have extended that classification. Fauchald (1983) adopted a different system of classification based on a smaller suite of species, but incorporating rigorously determined values for a number of important reproductive traits. Systems of classification such as these are only possible if the following conditions are satisfied:

(i)   the existence of covariable reproductive traits; and

(ii) discontinuous rather than continuous distributions of organism frequency with respect to the major axes of classification.

Covariable reproductive traits

1. Egg size and developmental mode

The relationship between egg size and developmental mode in the invertebrate phyla is widely recognized (Thorson 1946; Chia 1974; Hermans 1979). Smaller eggs tend to give rise to ciliated planktotrophic pelagic larvae, whereas larvae developing from larger eggs either have an abbreviated non-feeding pelagic phase or have direct non-pelagic development. The relationship is not a simple one—an important transition occurs at a spherical egg diameter between 100 and 180 μm, equivalent to a volume of 500-3 000 μm$^3 \times 10^3$. The position of the transition varies between groups: it is relatively low in bivalve molluscs, but for many groups oocytes with a diameter less than 140 μm almost always develop into a ciliated planktotrophic larva.

It is possible for such larvae to be brooded or encapsulated during the earliest stages of embryonic and larval development and a distinction must be made between 'mixed' developmental strategies in which brooding or encapsulation precede a pelagic phase and those in which there is no pelagic phase (Pechenik 1979).

Larvae developing from oocytes greater than 180 μm in diameter rarely feed in the plankton. The range of oocyte diameter for the Sipuncula, Priapulida, Echiura and Polychaeta falls into a range between 50 μm spherical diameter to 300 μm. A very small number of Polychaeta may have oocyte diameters in excess of this range (Schroeder and Hermans 1975; Fauchald 1983), but in some other groups, notably the Echinodermata, much larger oocytes up to 2 mm in diameter are possible. Larger eggs of this type are, however, unusual and are invariably associated with direct development and usually with some type of brooding behaviour (Chia 1968, 1970).

2. Developmental mode and adult size

There is an association between egg size, developmental mode and adult size which finds expression in two ways.

1. Large organisms may have either small eggs and planktotrophic development or large eggs and direct development, but extremely small organisms rarely, if ever, exhibit planktotrophic development.

2. Where closely related brooding and non-brooding forms co-occur, the smaller form is usually the brooder (Strathmann and Strathmann 1982).

The association between brooding and small body size has been particularly well documented for the Echinodermata (Chia 1968, 1970; Dartnall 1969; Schoener 1972; Menge 1975; Emson and Crump 1979; Hendler 1979; Emson and Wilkie 1984). The production of very large oocytes with a diameter greater than 1 mm can occur however in relatively large echinoids, as in the deep-sea echinothurid *Araeosoma owstoni* (Amemiya, Shinkai, and Uemura, 1984). A similar relationship has been observed in several groups of Mollusca (Fretter 1948; Sellmer 1967; Gallardo 1977; Hoagland 1977; Pearse 1979; Sastry 1979; Beauchamp, 1984). The relationship between extreme size and non-pelagic development is obscured however in the Opisthobranchia (Clark 1975; Todd 1979; De Freese and Clark 1983). There appears, however, to be a general trend which links suppression of the pelagic larval stage to small body size which recurs throughout the invertebrate phyla. It can be recognized, for instance, in the Cnidaria (Fadlallah and Pearse 1982a, b), Sipuncula (Åkesson 1958, 1961; Gibbs 1975; Rice, M. E. 1978, 1983), Priapulida (Nørrevang and Van der Land 1983) and Polychaeta (Olive and Clark 1978; Olive 1984).

The association between small body size and non-pelagic development is particularly apparent among the very smallest representatives of the invertebrate phyla in which pelagic development seems to be precluded. The interstitial fauna of marine sand share a characteristic suite of reproductive specializations including hermaphroditism, copulation, cuticular insemination, and other advanced forms of insemination, low fecundity, direct development, and frequently various types of encapsulation or brood care (Swedmark 1954; Westheide 1971). These characteristics apply particularly to the interstitial Polychaeta and to the Archiannelida. Among the latter, pelagic trochophore larvae are found only in the Polygordiidae and some of the Protodrilidae, and these organisms are significantly larger than the other 'Archiannelida' (Jägersten 1952; Jouin 1971).

Contra-indications to the general relationship between body size and developmental mode among closely related organisms are not common, but Christie (1982) has recently shown that the smaller of the diminutive scale worms, *Pholoe minuta* and *P. anoculata*, have small eggs and pelagic development whereas the slightly larger *P. anoculata* has larger eggs and direct development. Sexual reproduction with the production of planktotrophic larvae occurs in some very small ophiuroids described by Emson and Wilkie (1984) but in these the dominant mode of reproduction is asexually by fission.

3. Developmental mode, fecundity, brood frequency and body size

There is a clear trade-off between egg size/developmental mode and fecundity, since for organisms of comparable size relatively fewer of the larger eggs can be produced. This relationship is compounded by the association of direct development with small body size. Large organisms have the capacity to spawn single very large broods whereas diminutive organisms are not able to do so. Reproductive output in them can only be increased by increasing brood frequency.

Among the free-living marine invertebrate fauna the transition from low brood frequency and high fecundity to high brood frequency and low fecundity is a covariable with the transition from larger body size to small body size. This may be due to a design constraint operating on small organisms since they are unable to store gametes in large numbers prior to their discharge. It is not a design constraint, however, which precludes larger organisms from the production of small broods at frequent intervals. This is a normal feature of the reproductive biology of macro-invertebrates in freshwater and terrestrial environments which does not seem to occur among the invertebrates living in marine environments. The combination of large body size, high fecundity and high brood frequency is a characteristic of endoparasitic invertebrates.

The planktotrophic larvae of marine invertebrates in the temperate zone are able to exploit a seasonally recurring food resource in the phytoplankton succession and this is probably one of the selective advantages of this mode of reproduction and development. The discharge of a single cohort of gametocytes in an annual spawning crisis is therefore particularly appropriate for organisms with a pelagic mode of development and it is a dominant mode of reproduction of all groups of macrobenthic invertebrates in the temperate zone. It is usually associated with a high degree of between-individual synchrony of reproduction and in the cases where development is not pelagic, between-individual synchrony may be lost, as has occurred in the case of the polychaete *Cirratulus cirratus* (Olive 1970). The simultaneous discharge of a stock of gametocytes developed and stored during a period of up to 12 months or in some cases an entire lifetime, is a characteristic of most large polychaetes, sipunculans, nemerteans, priapulids, molluscs, and solitary corals. The transition to a more continuous mode of gamete discharge is a gradual one. Many Sipuncula discharge the gametes in a series of spawnings during a period of several weeks (Rice, M. E.

1975), and progressive discharge of gametes occurs in the polychaete *Nicolea zostericola* (Eckelbarger 1975). In these cases the coelomoducts may have the capacity to select mature gametes from a maturing pool. In *Harmothoe imbricata*, two batches of oocytes are spawned in succession but the periods of vitellogenesis are quite separate (Daly 1972). In the Mollusca there is a marked tendency for the progressive discharge of gametes during a spawning period which overlaps extensively with periods of gamete production (e.g. Seed 1976). In the Muricidae there is a clear relationship between adult size and clutch frequency (Spight, Birkeland, and Lyons 1974). Large adults spawn large clutches annually for several years whereas smaller species spawn small clutches repeatedly during a single breeding season.

4. Repeated breeding and the biology of fertilization

Marine invertebrates which breed once per year usually adopt a spawning mechanism in which gametocytes are discharged into the surrounding sea water, fertilization being external. In these circumstances a high degree of fertilization can be achieved through a combination of the following features:

1. Extremely high numbers of male and female gametes.

2. Well-synchronized gametogenic and spawning activity leading to near simultaneous discharge by all members of a population.

3. Pheromonal stimulation of male or female spawning.

4. Behavioural changes leading to the closer association of gravid or spawning individuals. The epitokal transformations and swarming behaviour of the Polychaeta are examples of this behaviour.

5. Deposition of spawn at surfaces where mixing of sperm and eggs will be brought about by water movements.

In smaller organisms which spawn repeatedly over an extended period, reproductive efficiency demands storage of sperm by females so that each batch of eggs may be fertilized without the simultaneous discharge of gametes by males. This consequence of small size and repeated breeding is particularly clear in the Polychaeta where seminal receptacles are uncommon and copulation is rare. The Spirorbinae and Fabricinae, which are repeated breeders, have well-developed seminal receptacles (Daly and Golding 1977; Daly 1978) and these structures also occur frequently in the smaller-bodied Spionidae (Rice, S. A. 1978, 1980). Copulation is inferred in the *Capitella* species complex and spermatozoa have been observed lodged in the female genital tracts of *Capitella capitata* (e.g. Eisig 1887). Specialized copulatory mechanisms are also highly characteristic of the interstitial Polychaeta (Stecher 1968; Westheide 1971, 1978, 1979).

The adaptive landscape—a way of viewing the reproductive traits of invertebrates

The identification of these sets of covariables makes it possible to represent the combinations of reproductive traits exhibited by the Invertebrata as the position occupied by a species within a volume defined by three mutually perpendicular axes:

(i)   body size (covariable with longevity or generation time);
(ii) egg size (covariable with developmental mode);
(iii) fecundity (inversely covariable with brood frequency).

The three axes (Fig. 4.1) define a volume which is not equally occupied by living invertebrates. A large number of reproductive characteristics or traits seem to be determined by the position within this framework which a species occupies. The 'reproductive strategies' or 'life diagram patterns' described by Olive and Clark (1978) and Fauchald (1983) are equivalent to groups of species occupying different positions within this space.

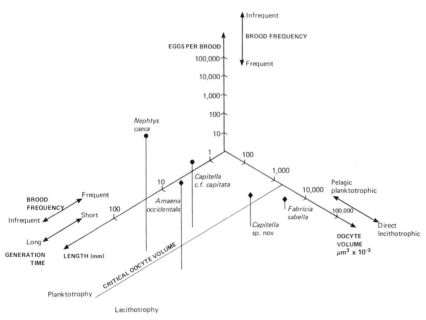

Fig. 4.1. Principal axes which describe most of the variable traits which constitute the 'reproductive strategy' of an organism. Note the logarithmic scales. The volumes described by these axes are not equally occupied by all organisms. The few species included in this figure are represented as points (circles for planktotrophic development, diamonds for lecithotrophic development). If data were available, each species should be represented as that part of a plane at right angles to the oocyte volume axis defined by the lifetime size/fecundity relationships.

Figure 4.2 is an attempt to represent the occupied parts of this volume. The shape drawn was derived empirically and refers particularly to the Polychaeta. The exact shape for this and other phyla must be derived from a broad survey of the covariability of fecundity, egg size and body size. In most species, egg size is fixed and each species in effect occupies that part of a plane perpendicular to the oocyte volume axis defined by the lifetime body size/fecundity relationships. The general features indicated in Fig. 4.2, however, are applicable to all the lower invertebrate phyla.

The first zone (zone 1) is that occupied by larger-bodied organisms having high fecundity and discrete (i.e. once per year) reproduction. Two clear trade-offs can be recognized within the zone; that between semelparity and iteroparity, and that between pelagic development and direct or lecithotrophic development.

Organisms in zone 1 may breed either once per lifetime or several times per lifetime. Those which breed once will have correspondingly higher fecundity than those breeding at annual intervals. Polychaeta belonging to these two categories have been described as monotelic and polytelic (Clark and Olive 1973; Olive and Clark 1978); polytely being a particular form of iteroparity. The trade-off between iteroparity and semelparity operates independently of that between egg size/developmental mode and fecundity. Within zone 1 there occurs a transition between organisms with pelagic larvae and those with non-pelagic larvae. The latter will tend to have lower fecundity and relatively larger oocyte volume and they will therefore occupy a lower, inner and more right-handed position in Fig. 4.2.

The perennial species included by Fauchald (1983) in his classification of Polychaeta reproduction all had moderately large oocyte diameters and direct development; this was because his sample included mainly representatives of the Onuphiidae as perennial species. The majority of perennial polychaetes, in fact, have smaller eggs and pelagic planktotrophic development. No Polychaeta have been described which, like the large terrestrial Oligochaeta, produce albumen-rich cocoons, each containing only one or two embryos at frequent intervals. This pattern of reproduction is uncommon among larger marine invertebrates except perhaps in the deep sea.

A second zone (zone 2) is that occupied by smaller-bodied representatives of the invertebrate phyla. Either direct or pelagic development is possible with associated trade-offs between oocyte diameter and fecundity. There is, however, a distinct trend towards increased brood frequency and partial abbreviation of the pelagic phase. The generation time of organisms in this zone will be less than that in zone 1 and is often about a year.

The complex of species associated with *Capitella capitata* is located in

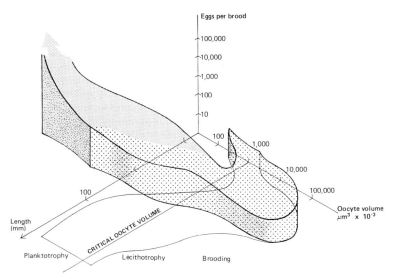

Fig. 4.2. An empirical suggestion of the likely shape of that part of the volume defined in Fig. 4.1 which is occupied by living Polychaeta. A similar shape may apply to other marine invertebrate groups. Note that planktotrophy (faint stipple) occurs where oocyte volume is less than c. $1000 \ \mu m^3 \times 10^3$.

zone 2 and these species illustrate particularly well the trends associated with decreasing body size.

The smallest invertebrate organisms occupy zone 3 (Fig. 4.2). The lifespan of such organisms is usually less than a year and they rarely have pelagic larvae; fecundity per brood is low, often reaching the theoretical minimum of one (as in the Nerillidae; Jouin 1968). Egg size is nevertheless relatively small—a constraint imposed by their small size. Organisms described as multi-annual by Fauchald (1983) are included in this zone. Representatives of most phyla occupying this zone have been regarded by many authors as exhibiting highly specialized reproductive biology.

Implications for the phylogeny of the lower invertebrates

The analysis of the reproductive traits of invertebrates developed above reveals a strong association between pelagic planktotrophic development and large-bodied organisms having a capacious coelom suitable for the development and storing of large numbers of gametocytes to be discharged during a single annual spawning crisis. Even in the acoelomate phyla such as the Nemertea, the Cnidaria and in effect the Mollusca, a large body size seems to be one of the characteristics required for the production of a sufficiently large number of offspring for pelagic planktotrophic development to be viable. The critical size at which planktotrophic development is no longer possible must be determined for each taxonomic group; it may be relatively small but will exclude the smallest members of each phylum.

The primitive conditions identified by Jägersten (1972) occur predominantly in zone 1 and these traits must have evolved in relatively large-bodied organisms. In view of this the hypothesis embodied by Jägersten's concept of primitive and altered conditions can be extended (Table 4.2).

Small-bodied organisms have restricted body spaces, which in truly interstitial forms usually disappear altogether (see Westheide, this volume), consequently large numbers of gametes cannot be matured simultaneously and a discrete mode of reproduction is precluded. In such conditions pelagic development, with its associated high mortality rates, is unlikely to be successful. Further consequences of this link between small size, lecithotrophic development and repeated breeding

Table 4.2. Implications of the hypothesis that planktotrophic pelagic development and external fertilization are 'primitive conditions'

Original conditions	Altered conditions
1. Eggs freely discharged into water, free pelagic development.	Eggs not freely discharged.
2. External fertilization (sperm type primitive according to Franzén 1956, 1970).	Internal fertilization (often with sperm storage) (sperm type modified).
3. Small quantity of yolk in the egg.	Large quantities of yolk in the egg.
4. Total equal cleavage.	Unequal cleavage.
5. Blastula with blastocoel.	Blastocoel more or less obliterated.
6. Gastrulation by invagination.	Gastrulation by other types.
7. Planktotrophy in the larva.	Lecithotrophy in the larva.
8. Autotrophic egg production.	Heterotrophic egg production.
9. Discrete, once per year reproduction with strong seasonality, between and within individual synchrony of gametogenesis.	Repeated often more or less continuous reproduction with reduced seasonality, loss of between and within individual synchrony of gametogenesis.
10. Long term storage of accumulated germ cells	Repeated spawning of small batches of germ cells without accumulation.
11. Large body size, often with capacious coelom or intercellular spaces.	Small body size, partially obliterated coelom. Accumulation of gametes in intercellular spaces not possible.

1-7 Abbreviated from Jägersten 1972.

are: adoption of specialized modes of rapid oocyte development (see Eckelbarger 1983 for a review of this in Polychaeta); the adoption of specialized methods of sperm transfer, and the consequent development of spermatozoa of an advanced type. The situation in moderately sized organisms is much more variable, but nevertheless some of the supposedly 'altered' conditions are usually found and the pelago-benthic life cycle is unlikely to have arisen in diminutive organisms. This interpretation represents a major divergence from the bilaterogastrea theory of Jägersten (1972) which supposes that the pelago-benthic life cycle was established by a common ancestor of the Metazoa which was a diminutive pelagic organism.

According to my view, the pelagic larval phase has arisen in large-bodied organisms which were able to retain the developing gametocytes in sufficient numbers to permit a discrete and epidemic spawning mechanism. Accordingly, the different types of ciliated larvae represent parallel evolution (cf Nielsen and Nørrevang, this volume).

There are groups of organisms, however, which seem to be united by the possession of a similar type of ciliated larva. Two alternative conclusions can be drawn. Either:

(i)   phyla which seem to be united by possession of a characteristic and uniform larval type (e.g. Sipuncula, Echiura, Annelida, and Mollusca) have evolved from a single large-bodied ancestral type of organism; or

(ii)   morphologically and functionally similar ciliated larvae can arise independently in response to similar design requirements (parallel evolution) and similarity of larval types is not necessarily an indication of close phylogenetic relationship.

Strathmann (1978a,b) confirms that the ciliated planktotrophic larvae of the Oligomera (brachiopods, ectoprocts, phoronids, hemichordates, echinoderms) and those of the Spiralia (annelids, echiurans, sipunculans, molluscs, entoprocts) are extremely ancient and very conservative structures. There has been a progressive loss of the ciliated larval stage in many of these groups and, once lost in any one line of evolution, it is not easily regained, although, as in the Annelida, a number of different types of feeding free-swimming larvae may have evolved. Accordingly, the second of the two hypotheses outlined here seems to be improbable and it may be suggested that groups of phyla are united in the possession of a characteristic larval type. It may also be suggested that the common ancestral organism in which the characteristic larvae was developed was not likely to have been an organism of meiobenthic proportions.

Not all of the groups having pelagic larvae are large-bodied organisms, the principal exceptions being the Entoprocta and the Bryozoa, yet in both of these pelagic primary larvae are said to occur. Why should this be so? One answer might be that those organisms which exhibit a ciliated planktotrophic larva and which are not themselves large-bodied organisms, are derived from organisms which had large-bodied representatives and have retained the pelagic larva despite the exigencies of small body size.

The Bryozoa would seem to fit this model. Farmer (1977) has suggested that the characteristic cyphonautes larva may have evolved as a modification of an actinotroch-like larva for planktotrophic development during the evolution of colonial ectoprocts from a solitary phoronid-like ancestor. Planktonic cyphonautes larvae now occur only in very few Bryozoa (Ström 1977; Zimmer and Woollacott 1977). They occur only in two groups of the marine Bryozoa—the Anasca and the Ctenostomata. In the Anasca the cyphonautes larvae are restricted to the families Electridae and Membraniporidae. Among the Ctenostomata the occurrence of the cyphonautes larvae is very erratic.

Two hypotheses can be suggested to account for the 'retention' of the cyphonautes larvae by Bryozoa. The first may be called the 'small individual zooid-large collective colony hypothesis'; the second the 'restricted habitat location hypothesis'. According to the first, a primitive mode of reproduction can be retained due to the collective size of the colony being such that the colony is in effect a large-bodied organism. This hypothesis may apply to the Membraniporidae; *Membranipora membranacea*, for instance, forming extensive colonies of many thousands, perhaps millions, of zooids.

This hypothesis does not apply to cases of planktotrophy found in the Electridae and the Ctenostomata, and, particularly in the latter, planktotrophy seems to be associated with the colonization of specialized habitats. Thus, planktotrophy is found in *Hypophorella expansa* which inhabits the linings of polychaete tubes. This hypothesis also seems to apply to the Entoprocta, all of which are diminutive sessile organisms which do not form colonies but which do have 'ciliated larvae'. According to Jägersten (1972) and others, the larvae can be regarded as 'trochopore-like' but Mariscal (1975) writes 'in the case of some of the pedicillinid larvae at least the larva is best described as a miniature swimming adult calyx'. Reproductive conditions in the Entoprocta are highly specialized; very few gametes develop at any one time and there is evidence that the developing embryos receive nutrients directly from the parents (Mariscal 1975). Fertilization is internal and according to Franzén (1983) the structure of the eggs and spermatozoa have changed considerably from the primitive conditions in marine invertebrates.

The loxosomatid entoprocts are found in extremely specialized habitats; over half of them are associated with the tubes of polychaetes, and others have been found living symbiotically with ectoprocts, sipunculans, echinoderms and ascidians (Mariscal 1975).

The ciliated larva should be regarded as a special attribute of the Entoprocta developed and retained to permit the colonization of highly dispersed localized habitats.

Conclusions

Many of the numerous reproductive traits of marine invertebrates are fixed attributes not subject to short term variation. Others, notably fecundity and reproductive effort, are more variable. Most of the 12-15 reproductive traits which together constitute the reproductive strategy of an organism are not independent variables. In particular the sets of reproductive traits identified as 'advanced' by Jägersten and Franzén are dominant among the smallest representatives of all the lower invertebrate phyla, whereas conditions identified as 'primitive' are only found in large-bodied organisms or in association with the adoption of other specialized conditions (such as fission) in diminutive organisms.

The phylogenetic implication of this thesis seems to be that either:

(i) all contemporary invertebrate phyla are descended from a single relatively large-bodied ancestral group of organisms which had adopted a reproductive mode including discrete reproduction and planktotrophy; or

(ii) the so-called 'primitive' reproductive traits have evolved independently in several lines of evolution as organisms achieved a grade of organization which permitted the retention of gametocytes in large numbers and discrete spawning which are prerequisites for successful planktotrophic development.

It is difficult to reconcile the ideas propounded here with phylogenetic schemes which represent pelagic-ciliated larva-like organisms as stem groups for the Metazoa (e.g. see Nielsen and Nørrevang, this volume).

These inferences rest ultimately on an assumption that the factors responsible for the evolution of life history traits are the same now as they were when the invertebrate phyla were established. The validity of this assumption will only be known when a rigorous theoretical framework for the understanding of the evolution of life histories has been established.

References

Åkesson, B. (1958). A study of the nervous system of the Sipunculideae with some remarks on the development of two species, *Phascolion strombi* Montagu and *Golfingia minuta* Keferstein. *Unders. Öresund* 38, 1-249.

—— (1961). The development of *Golfingia elongata* Keferstein (Sipunculida). *Ark. Zool.* 13, 511-31.

Amemiya, S., Shinkai, T., and Uemura, I. (1984). Giant chromosome in the sperm nucleus of echinothurid sea urchins. In *Advances in invertebrate reproduction* (eds W. Engels and A. Fischer) Vol. 3. Elsevier-North Holland Press, Amsterdam. p. 556.

Beauchamp, K. (1984). The reproductive ecology of a hermaphrodite clam, *Lasaea subviridis*. In *Advances in invertebrate reproduction* (eds W. Engels and A. Fischer) Vol. 3. Elsevier-North Holland Press, Amsterdam. p. 557.

Chia, F. S. (1968). Some observations on the development and cyclic changes of the oocytes in a brooding starfish, *Leptasterias hexactis. J. Zool., Lond.* 154, 453-61.

—— (1970). Some observations on the histology of the ovary and RNA synthesis in the ovarian tissues of the starfish *Henricia sanguinolenta. J. Zool., Lond.* 162, 287-91.

—— (1974). Classification and adaptive significance of developmental patterns in marine invertebrates. *Thallassia jugosl.* 10, 121-30.

Christie, G. (1982). The reproductive cycles of two species of *Pholoe* (Polychaeta: Sigalionidae) off the Northumberland coast. *Sarsia* 67, 283-92.

Clark, R. B. (1975). Nudibranch life cycles in the north west Atlantic and their relationship to the ecology of fouling communities. *Helgoländer wiss. Meeresunters.* 27, 28-69.

—— and Olive, P. J. W. (1973). Recent advances in polychaete endocrinology and reproductive biology. *Oceanogr. mar. Biol. Ann. Rev.* 11, 176-223.

Daly, J. M. (1972). The maturation and breeding biology of *Harmothoe imbricata* (Polychaeta: Polynoidae). *Mar. Biol.* 12, 53-66.

—— (1978). Growth and fecundity in a Northumberland population of *Spirorbis spirorbis* (Polychaeta: Serpulidae). *J. mar. biol. Ass. U.K.* 58, 177-90.

—— and Golding, D. W. (1977). A description of the spermatheca of *Spirorbis spirorbis* (L.) (Polychaeta: Serpulidae) and evidence for a novel mode of sperm transmission. *J. mar. biol. Ass. U.K.* 57, 219-27.

Dartnall, A. J. (1969). A viviparous species of *Patiriella* (Asteroidea, Asterinidae) from Tasmania. *Proc. Linn. Soc. N.S.W.* 96, 39-49.

De Freese, D. E. and Clark, R. B. (1983). Analysis of reproductive energetics of Florida Opisthobranchia (Mollusca: Gastropoda). *Int. J. Inv. Reprod.* 6, 1-10.

Dobzhansky, T. (1950). Evolution in the tropics. *Am. Scient.* 38, 208-21.

Eckelbarger, K. J. (1975). A light and electron microscopical investigation of gametogenesis in *Nicolea zostericola* (Polychaeta: Terebellidae). *Mar. Biol.* 30, 353-70.

—— (1983). Evolutionary radiation in polychaete ovaries and vitellogenic mechanisms: their possible role in life history patterns. *Can. J. Zool.* 61, 487-504.

Eisig, H. (1887). Die Capitelliden des Golfes von Neapel. *Fauna Flora Golf. Neapel* 16, 1-906.

Emson, R. and Crump, R. G. (1979). Description of a new species of *Asterina* (Asteroidea), with an account of its ecology. *J. mar. biol. Ass. U.K.* 59, 77-94.

Emson, R. and Wilkie, I. (1984). Reproductive strategies of small brittle-stars. In *Advances in invertebrate reproduction* (eds W. Engels and A. Fischer) Vol. 3. Elsevier-North Holland Press, Amsterdam, p. 576.

Fadlallah, Y. H. and Pearse, J. S. (1982a). Sexual reproduction in solitary corals: Overlapping oogenic and brooding cycles and benthic planulas in *Balanophyllia elegans. Mar. Biol.* 71, 223-32.

—— —— (1982b). Sexual reproduction in solitary corals: Synchronous gametogenesis and broadcast spawning in *Paracyathus stearnsii. Mar. Biol.* 71, 233-9.

Farmer, J. D. (1977). An adaptive model for the evolution of the ectoproct life cycle. In *Biology of bryozoans* (eds R. M. Woollacott and R. L. Zimmer) pp. 487-517. Academic Press, New York.

Fauchald, K. (1983). Life diagram patterns in benthic polychaetes. *Proc. biol. Soc. Wash.* 96, 160-77.

Franzén, Å. (1956). On spermiogenesis, morphology of the spermatozoan and biology of fertilisation among invertebrates. *Zool. Bidr. Upps.* 31, 355-482.

—— (1970). Phylogenetic aspects of the morphology of the spermatozoa and spermiogenesis. In *Comparative spermatology* (ed. B. Baccetti) pp. 29-46. Academic Press, New York.

—— (1983). Bryozoa Entoprocta. In *Reproductive biology of invertebrates.* Vol. 1. *Oogenesis* (eds K. G. and R. G. Adiyodi) pp. 561-9. John Wiley, New York.

Fretter, V. (1948). The structure and life history of some minute prosobranchs of rock pools, *Skeneopsis planorbis* (Fabricius), *Omalgyra atomus* (Philippi), *Rissoella diaphana* (Alder) and *Rissoella opulina* (Jeffreys). *J. mar. biol. Ass. U.K.* 27, 597-632.

Gallardo, C. S. (1977). Two modes of development in the morphospecies *Crepidula dilata* (Gastropoda: Calyptraeidae) from southern Chile. *Mar. Biol.* 39, 241-51.

Gibbs, P. E. (1975). Gametogenesis and spawning in a hermaphroditic population of *Golfingia minuta* (Sipuncula). *J. mar. biol. Ass. U.K.* 55, 69-82.

Hendler, G. (1979). Sex-reversal and viviparity in *Ophiolepis kieri*, n. sp., with notes on viviparous brittlestars from the Caribbean (Echinodermata: Ophiuroidea). *Proc. biol. Soc. Wash.* 92, 783-95.

Hermans, C. O. (1979). Polychaete egg sizes, life histories and phylogeny. In *Reproductive ecology of marine invertebrates* (ed. S. E. Stancyk) pp. 1-9. University of South Carolina Press, Columbia, SC.

Hoagland, K. E. (1977). Systematic review of fossil and recent *Crepidula* and discussion of evolution of the Calyptraeidae. *Malacologia* 16, 353-420.

Jägersten, G. (1952). Studies on the morphology, larval development and biology of *Protodrilus. Zool. Bidr. Upps.* 29, 425-511.

—— (1972). *Evolution of the metazoan life cycle.* Academic Press, New York.

Jouin, C. (1968). Sexualité et biologie de la reproduction chez *Mesonerilla Remane et Meganerilla* Boaden (Archiannelida, Nerillidae). *Cah. Biol. mar.* 9, 31-52.

—— (1971). Status of the knowledge of the systematics and ecology of the Archiannelida. *Smithson. Contr. Zool.* 76, 47-56.

MacArthur, R. H. and Wilson. E .O. (1967). *Theory of island biogeography.* Princeton University Press, Princeton, NJ.

Mariscal, R. N. (1975). Entoprocta. In *Reproduction of marine invertebrates* (eds A. C. Giese and J. S. Pearse) Vol. 2, pp. 1-41. Academic Press, New York.

Menge, B. L. (1975). Brood or broadcast? The adaptive significance of different reproductive strategies in the two intertidal sea-stars *Leptasterias hexactis* and *Pisaster ochraceus*. *Mar. Biol.* 31, 87-100.

Nørrevang, A. and Land, J. van der (1983). Priapulida. In *Reproductive biology of invertebrates* (eds K. G. and R. G. Adiyodi) Vol. 1, pp. 269-82. John Wiley, New York.

Olive, P. J. W. (1970). Reproduction of a Northumberland population of the polychaete *Cirratulus cirratus*. *Mar. Biol.* 5, 259-73.

—— (1984). Environmental control of reproduction in Polychaeta. *Fortschr. Zool.* Vol. 29, pp. 17-38 (eds A. Fischer and H. D. Pffanenstiel). Fischer Verlag, Stuttgart.

—— and Clark, R. B. (1978). Physiology of reproduction. In *Physiology of annelids* (ed. P. J. Mills) pp. 271-368. Academic Press, London.

Parry, G. D. (1981). The meanings of *r*- and K-selection. *Oecologia* 48, 260-4.

Pearse, J. S. (1979). Polyplacophora. In *Reproduction of marine invertebrates* (eds A. C. Giese and J. S. Pearse) Vol. 5, pp. 27-35. Academic Press, New York.

Pechenik, J. A. (1979). The role of encapsulation in invertebrate life histories. *Am. Nat.* 114, 859-70.

Pianka, E. R. (1970). On *r*- and K-selection. *Am. Nat.* 104, 592-7.

Rice, M. E. (1975). Sipuncula. In *Reproduction of marine invertebrates* (eds A. C. Giese and J. S. Pearse) Vol. 2, pp. 67-127. Academic Press, New York.

—— (1978). Morphological and behavioural changes at metamorphosis in the Sipuncula. In *Settlement and metamorphosis of marine invertebrate larvae* (eds F. S. Chia and M. E. Rice) pp. 83-102. Elsevier, New York.

—— (1983). Sipuncula. In *Reproductive biology of invertebrates* (eds K. G. and R. G. Adiyodi) Vol. 1, pp. 283-96. John Wiley, New York.

Rice, S. A. (1978). Spermatophores and sperm transfer in spionid polychaetes. *Trans. Am. microsc. Soc.* 97, 160-70.

—— (1980). Ultrastructure of the male nephridium and its role in spermatophore formation in spionid polychaetes (Annelida). *Zoomorphologie* 95, 181-94.

Sastry, A. N. (1979). Pelecypoda (excluding Ostreidae). In *Reproduction of marine invertebrates* (eds A. C. Giese and J. S. Pearse) Vol. 4, pp. 113-292. Academic Press, New York.

Schoener, A. (1972). Fecundity and possible mode of development of some deep-sea ophiuroids. *Limnol. Oceanogr.* 17, 193-9.

Schroeder, P. C. and Hermans, C. O. (1975). Annelida: Polychaeta. In *Reproduction of marine invertebrates* (eds A. C. Giese and J. S. Pearse) Vol. 3, pp. 1-213. Academic Press, New York.

Seed, R. (1976). Ecology. In *Marine mussels, their ecology and physiology* (ed. B. L. Bayne). I.B.P. Handbook No. 10, pp. 13-66, Cambridge University Press, Cambridge.

Sellmer, G. (1967). Functional morphology and ecological life history of the gem clam, *Gemma gemma* (Eulamellibranchia; Veneridae). *Malacologia* 5, 137-223.

Spight, T. M., Birkeland, C., and Lyons, A. (1974). Life histories of large and small murexes (Prosobranchia: Muricidae). *Mar. Biol.* 24, 229-42.

Stearns, S. C. (1976). Life history tactics: a review of the ideas. *Q. Rev. Biol.* 51, 3-47.

—— (1980). A new view of life history evolution. *Oikos* 35, 266-81.

—— (in press). The tension between adaptation and constraint in the evolution of life histories. In *Proceedings of the third international symposium of invertebrate reproduction* (eds W. Engels and A. Fischer). Elsevier-North Holland Press, Amsterdam.

Stecher, H. J. (1968). Zur Organisation und Fortpflanzung von *Pisione remota* Southern (Polychaeta: Pisionidae). *Z. Morph. Ökol. Tiere* 61, 347-410.

Strathmann, R. R. (1978a). The evolution and loss of feeding larval stages of marine invertebrates. *Evolution* 32, 894-906.

—— (1978b). Progressive vacating of adaptive types during the Phanerozoic. *Evolution* 32, 907-14.

—— and Strathmann, M. F. (1982). The relationship between adult size and brooding in marine invertebrates. *Am. Nat.* 119, 91-101.

Ström, R. (1977). Brooding patterns of bryozoans. In *Biology of bryozoans* (eds R. M. Wollacott and R. L. Zimmer) pp. 23-56. Academic Press, New York.

Swedmark, B. (1954). Étude du developpement larvaire et remarques sur la morphologie de *Protodrilus symbioticus* Giard (Archiannelida). *Ark. Zool.* 6, 511-22.

Thorson, G. (1946). Reproduction and larval development of Danish bottom invertebrates with special references to the planktonic larvae in the sound (Øresund). *Meddr. Kommn Danm. Fisk.-og Havunders, Ser. Plankton* 4, 1-523.

Todd, C. D. (1979). Reproductive energetics of two species of clavid nudibranchs with planktotrophic and lecithotrophic larval strategies. *Mar. Biol.* 53, 57-68.

Westheide, W. (1971). Interstitial Polychaeta (excluding Archiannelida). In *Proceedings of the first international conference on meiofauna* (ed. N. C. Hulings) *Smithson. Contr. Zool.* 76, 57-70.

—— (1978). Ultrastructure of the genital organs in interstitial Polychaeta: 1. Structure, development and function of the copulatory stylets in *Microphthalmus* cf *listensis. Zoomorphologie* 91, 101-18.

—— (1979). Ultrastructur der Genital organe interstitieller Polychaeten II. Mänliche kopulatiasorgane mit intrazellularen stilettstaben in einer *Microphthalmus. Zool. Scr.* 8, 111-18.

Zimmer, R. L. and Woollacott, R. M. (1977). Structure and classification of gymnolaemate larvae. In *Biology of bryozoans* (eds R. M. Woollacott and R. L. Zimmer) pp. 91-142. Academic Press, New York.

5. Speculations on coelenterates

ELAINE A. ROBSON

Department of Pure and Applied Zoology, University of Reading, Reading, UK

Abstract

Fossil Cnidaria existed in the Cambrian and ancestral forms are attributed to the Precambrian. From the work of Werner and others Medusozoa are thought to have diverged from Anthozoa at an early stage, ancestral polyps then showing tetraradial symmetry. Scyphozoa and Conulata represent the oldest Medusozoa, and probably Octocorallia the earliest Anthozoa. The classes of Cnidaria have been distinct for so long that only characters common to the phylum may offer clues about its historical relationships. In certain features, however, functional convergence can be discerned. Cnidae, as an example, are an intrinsic element evolved possibly from gland cells but not from any of the analogues found among Protista. In *Hydra* cnidae develop from stem cells which give rise also to nerve cells and (probably) to germ cells. The Cnidaria are distinct from more complex animals in possessing fewer types of differentiated cells. They are limited also by their characteristic epithelial organization and associated cardinal features such as radial symmetry and axial polarity. If more taxa of extinct 'diploblastic' and 'triploblastic' animals existed in the Cambrian than those of which we are now aware, the cnidarian pattern may represent one of a number of avenues by which increased histological complexity evolved.

Introduction

The study of evolution and of phylogenetic relationships can be seen as a progressive 'research programme' in the dynamic sense defined by Lakatos's (1978) view of science. Within such a field of knowledge, however, what Ziman (1968) terms the consensible view is at the same time important: 'The recognition that scientific knowledge must be public and consensible (to coin a necessary word) allows one to trace out the complex inner relationships between its various facets.' The present symposium illustrates this well.

To appraise current research into phyla and their possible relationships with other major groups calls for selective presentation. This chapter begins with a summary of expert views of the origins and affinities of Cnidaria, and current understanding of a few of their characters is then discussed.

In order to perceive the profile of existing knowledge and its lacunae more clearly, and to obtain better answers to phylogenetic questions, it is no idle task to consider, were it feasible, how the spectrum of extinct cnidarian populations might have been sampled at intervals during and since the Precambrian. If extinct species could be obtained live and studied both *in situ* and in the laboratory, a panoply of new evidence would become available to systematists, physiologists, and cell biologists: the relationships of early Cnidaria to other Precambrian taxa might be clearer if their molecular genetics could be studied as well as their morphology. To devise a plausible and efficient sampling programme which would span cnidarian history from the Precambrian to the present is a demanding but worth-while exercise: for if taxonomic lineages were sought, it might be necessary to ascertain not only sampling intervals but also ecological locations (e.g. plankton, benthos or reef) together with geographical distribution of taxa in the light of evolving oceans and continents. Further, the numbers of specimens required to describe new species and to plan laboratory work on life-histories, behaviour, or biochemical attributes would need careful thought. So too would the formulation of crucial phylogenetic hypotheses whose predictions might be tested in such unusual circumstances.

Origins

The Cnidaria featured in the 1978 symposium on 'Origins of major invertebrate groups': it is clear from reviews of palaeontological evidence by Scrutton (1979) and Brasier (1979) that fossil forms now attributed to the Cnidaria were important in the Lower Cambrian and before (see also Conway Morris, this volume). The earliest are medusoids possibly akin to Scyphozoa, while other radial organisms suggest that chondrophores (Hydrozoa) and soft corals may also have existed. By the Cambrian there are traces attributed to burrowing anemones, ?tabulate corals, possibly early Rugosa, and Conulata. Some important groups, such as hydroids and most of the Rugosa, are not known before the Ordovician. The scientific traveller to the Precambrian, however, might hope for some chance of finding their antecedents.

The phylum Cnidaria (Hatschek 1888: in Hyman 1940) includes four or five classes which remain distinct at least as far back as the

Ordovician. The Anthozoa, including today all the sea anemones and
diverse hard and soft corals, also encompass extinct Rugosa and Tabu-
lata (Scrutton 1979; *Zoological Record* of 1984). They are sessile polyps,
frequently colonial, which are characterized by a regular arrangement
of supporting muscular septa or mesenteries. These confer biradial
symmetry on the polyps, a feature lacking in other Cnidaria. Bilateral
symmetry becomes established in the late planula larva when the first
mesenteries appear (e.g. Doumenc 1976). Anthozoa flourish in almost
every marine environment and, most important of all, they have produced coral reefs of both ancient and modern types, an achievement
attributed in part to symbiotic association with dinoflagellates (Hill and
Wells in Moore 1956; Rosen 1981). Calcified skeletons or spicules are
characteristic of many Anthozoa but are absent in anemones. Hand
(1966) proposed that actiniarian and corallimorpharian anemones were
descended from calcified corals. Ancient lineages within the Anthozoa
would include octocorals and Ceriantharia (Carlgren 1944; Schmidt
1974; Schmidt and Zissler 1979).

As the other modern Cnidaria include a medusoid phase in the life-
history, Petersen (1979) has suggested that they form a subphylum
Medusozoa. Werner's (1973) far-reaching analysis shows how the
classes Scyphozoa, Cubozoa, and extinct Conulata may be related to
the Hydrozoa (Fig. 5.1). The Hydrozoa, perhaps the most diverse of
these, encompass the thecate and athecate hydroids, many of whose
sessile polypoid colonies produce free medusae, the freshwater hydras,
siphonophores, which are highly differentiated pelagic colonies budded
from an original polyp (Totton 1965; Pugh 1983), as well as Narcomedusae, Trachymedusae, and Limnomedusae, and the calcified

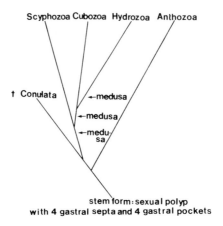

Fig. 5.1. Diagrammatical survey of the relationships of the classes of Cnidaria (after
Werner 1973)

Milleporida and Stylasterina, the former with transient medusae. A review is given by Bouillon (in press).

The Scyphozoa include the larger medusae, nearly all of which arise from a sessile scyphistoma polyp with characteristic tetraradial symmetry, usually by strobilation. Werner (1973) has illuminated the life-history of Coronatae in showing how medusae are produced by different species of *Stephanoscyphus* polyps. The extinct Conulata are related to them (see Chapman 1966; Werner 1979). Werner (1975, 1983) separated the Cubozoa (formerly Cubomedusae) as a new class according to the complex features of the medusae and their distinctive life-history, in which the larval polyp changes directly into a medusa. The Scyphozoa are considered to represent the most basic class of Cnidaria, with the Cubozoa representing an evolutionary link between the Scyphozoa and Hydrozoa. The Anthozoa would also be derived from a tetramerous sessile polyp, which diverged from the common ancestors at an early stage (Werner 1973). Such ancestral forms are to be sought in the Precambrian.

Relationships

Many Cnidaria, thanks to features such as radial symmetry, complexity of growth patterns, transparency, or brilliant colouration, have given scientists and artists alike the impression of beautiful but often delicate animals (e.g. Allman 1871; Saville-Kent 1893; Haeckel 1899). The phylum is so widely distributed and successful that its assets must outweigh apparent fragility. What characters do all its members have in common? If we are to consider also how Cnidaria might be related to other phyla we should try to decide which of their characters have been maintained because they are particularly important, and which have been retained because little selective pressure has fallen on them. The features in question are all those which define the phylum (see Moore 1956; Hyman 1940 is more succinct): biradial or radial symmetry; body wall consisting of ectoderm, endoderm, and mesogloea; enteron with one opening, the mouth, which is surrounded by tentacles; well-developed muscular action; nerve net; stinging capsules (cnidae or nematocysts); a planula larva is characteristic; and multiplication is often effected by budding.

In discussing the relationships of the Cnidaria one cannot forget that they are relatively simple animals. The terms simple and complex are useful in drawing attention to features which are best perceived by comparison, but some yardstick for estimating complexity of organization is needed: e.g. it is possible to extrapolate conclusions from the relative complexity of nervous systems in different Metazoa because their parameters can to some extent be specified (Robson 1975). Bonner (1965) quantifies a broad view of phyla by equating complexity of organization with differentiation or division of labour. As shown in Fig. 5.2 his insight brings out the fact that coelenterates possess fewer types of cells than most other animal phyla, and sponges fewer still. This important quality also emerges from Chapman's (1974) review of cnidarian histology. How it is governed by the genome is a matter for further research. It could possibly be relevant that Cnidaria, like sponges, show a low range of nuclear DNA content (Ayala 1976).

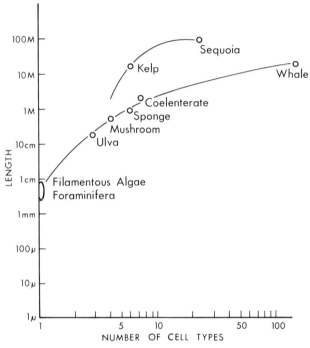

Fig. 5.2. Graph showing a very rough estimate of the maximum size known for a given number of cell types, which is used as a measure of complexity (after Bonner 1965).

In a possible phylogeny of the animal kingdom Barnes (1980, Fig. 3.1) presents the generally agreed view of how Cnidaria feature among other invertebrate phyla. They arise in parallel with Ctenophora from a planuloid ancester but like present-day sponges are, as it were, out on a limb. Werner (1973) is 'of the opinion that they must be considered as a blind-ending side branch from which no other group of invertebrates can be derived'. Even Haeckel (e.g. 1874), an early pioneer in studies of the Cnidaria and sponges, offered a not dissimilar view. His scheme was monophyletic: his Gastraea theory, now discarded, invoked similarities between developing stages and ancestral adult forms, so that the gastrula of a worm or of a fish would represent the ancestral coelenterate stage (de Beer 1958; Clark 1964). At the same time Cnidaria (Acalephae and Zoophyta), linked with sponges, formed one of six lines of extant Metazoa.

The antecedents of the Cnidaria, indeed of the Metazoa, are still thought of as Protista. Sleigh (1979) would derive Cnidaria and sponges from zooflagellates different from those leading to Ciliophora and to acoel turbellarians (see also Hand 1963). As is well known, Hadzi (1963) suggested that ciliates gave rise to Acoela (implying a process of cellularization), and from these he derived the Anthozoa. Pantin (1960) also looked upon Anthozoa as the most primitive class of Cnidaria, but he demolished the idea that diploblastic animals, with their characteristically epithelial organization of tissues, were evolutionarily simplified.

It is difficult to extrapolate an ancestral form in the Cnidaria if only because the extant classes have been evolving independently for so long. The early fossil record is remarkable, but incomplete in at least two respects. Not only may the allocation of certain fossils even to the right class be fraught with difficulty, but it seems hard to rule out the possibility that the early Cnidaria included unknown extinct taxa which did not survive beyond the Cambrian. There are precedents for this view from well-preserved early invertebrates among the arthropods (Whittington 1979) and molluscs (Yochelson 1979). The suggestion that more kinds of ancient 'coelenterates' may have existed than those of which we are now aware is of no value without supporting fossil evidence, and so it may never be substantiated. The premises, however, seem not unreasonable compared, for example, to those needed to estimate the probability that life may have originated several times (Raup and Valentine 1983).

Characteristics

The extent to which characters of Cnidaria may indicate common ground with other phyla is considered in the case of cnidae, epithelial organization, and the determination of polarity and symmetry.

1. Cnidae

One of the hallmarks of Cnidaria are the stinging capsules or nematocysts (Mariscal 1974). They are used in feeding, defence, and adhesion to substrates; some of the former inject powerful venom. The seminal work of Weill (1934) classifies nematocysts into morphological categories (between 20 and 30 are now recognized) whose diversity is partly functional. The nematocyst profile (cnidom) may be characteristic at

generic or species level. Schmidt (1974), from his survey of cnidae
within the Anthozoa, has drawn phylogenetic conclusions which are
supported also by a study of spermatogenesis (Schmidt and Zissler
1979).

How did nematocysts arise? Hadzi (1963) offered an ingenious explanation which has not been fully evaluated: an ancestral gland cell
became transformed by a series of steps into a cnidoblast producing a
nematocyst.

I next summarize briefly what is known about nematocysts and
consider analogues of cnidae in Protozoa (reviews in Hovasse 1965;
Mariscal 1974; Uspenskaya 1977).

Nematocysts are collagenous structures. The capsule and hollow
thread are secreted by a cnidoblast, via the Golgi and endoplasmic
reticulum. The capsule rudiment is continuous with the developing external tube, which is surrounded by microtubules. The tube invaginates
into the capsule, showing three helical pleats, and later acquires barbs
(Fig. 5.3). The capsule is closed by an operculum or flaps (see Holstein
1981 for *Hydra*; Schmidt 1981 for octocorals). Particularly in the
Hydrozoa (e.g. Westfall 1970) a complex arrangement of rodlets and
fibrils surrounds the capsule, and the centriole is at the base of a pronounced cnidocil, presumed to be sensory. When the nematocyst dis-
charges the thread everts, capsular fluid emerging at the tip (usually),
and the capsule is later extruded from the cell (Fig. 5.5).

The cnidocyte is a receptor and it responds directly to stimuli, but
the threshold may be modulated via synaptic neurons or by epithelial
potentials (Lubbock and Shelton 1981). Recent work (Lubbock and
Amos 1981; Lubbock, Gupta, and Hall 1981; Gupta and Hall 1984)
reveals that the capsule moves to the cell surface (see also Yanagita
1973) and as it comes into contact with the external medium Ca^{2+}
is lost. This somehow increases the osmotic pressure of capsular fluid,

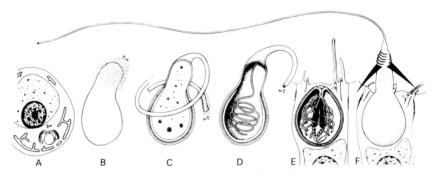

Fig. 5.3. Scheme of nematocyst morphogenesis in *Hydra*. A-E, stages in development
of a stenotele; F, an exploded capsule. From Holstein (1981), with permission of
Academic Press.

causing a rapid influx of water, the hydrostatic pressure of the capsule increases and the thread everts. It is suggested that the discharge of nematocysts may be a modified version of other exocytic processes, for example, those of glandular secretion, which is an interesting corroboration of Hadzi's idea (see Holstein and Tardent 1984).

The tentacles of ctenophores adhere by means of unique colloblasts, which are specialized glandular cells; but nothing yet known about them is reminiscent of cnidae (Franc 1978).

Certain predators of Cnidaria are able to sequester second-hand nematocytes for their own defence (e.g. aeolid nudibranchs, see Harris 1973; some Turbellaria, see Karling 1966).

Analogues of cnidae

Much can be learned about nematocysts by examining analogous devices in Protozoa. There is no evidence for homology (Pantin 1951) in the following examples.

(i) *Remanella multinucleatus.* In this psammophilous ciliate Raikov (1978) has described 2 μm pyriform capsules containing a short eversible thread, near the ends of the body. The thread is coiled vertically and its concentric structure and that of the bilaterally compressed capsule are fairly distinctive.

(ii) *Polykrikos schwartzi.* Complex nematocyst-like structures are known in this dinoflagellate (Westfall, Bradbury, and Townsend 1983), but there is no recent account of their discharge in the living state or of their function. *P. schwartzi* is a heterotrophic species without chloroplasts.

(iii) *Microsporidia.* These are a widespread order of parasitic Protozoa, in which a spore with a long eversible thread is the infective stage (e.g. Vávra 1976; Larsson 1981). In *Nosema apis* a spore (3 to 5 μm long) everts the long polar filament in the gut of the host, and the amoeboid sporaplasm containing the nucleus passes down it. Formation of the concentrically layered fibrous tube and of associated structures (i.e. membranous polaroplast and posterior vacuole), although due to Golgi and endoplasmic reticulum, recalls little that is seen in cnidoblasts. The spore capsule contains chitin (Vávra 1976). Activation and the mechanism of discharge are not fully understood: swelling of the polaroplast is thought to cause an increase in pressure sufficient to evert the tube.

(iv) *Myxosporidia.* The best analogues of nematocysts are the polar capsules of Myxosporidia, an order of protozoan parasites of fish, amphibians and certain invertebrates. The infective stage is a resistant spore, formed by two valves. Inside are two polar capsules, each with an eversible adhesive thread which on discharge serves to anchor the spore to the gills or gut of a new host. The valves, once unsealed, later allow the amoeboid sporoplasm to pass into the host (although this has not yet been observed). The polar capsules and valves are of unknown proteinaceous composition.

Polar capsules are small (2 to 10 μm) but their size range overlaps that of small nematocysts (e.g. in Hydrozoa) and their round or oval shape is somewhat comparable (Figs 5.5-5.7). Their structure and development were illustrated very clearly by Bütschli (1881) and electron microscopy has greatly extended current knowledge (e.g. Lom and de Puytorac 1965; Lom 1969; Desser and Paterson 1978; Uspenskaya 1982; Pulsford and Matthews 1982; Desser, Molnar, and Weller 1983). Within the capsule the tube is coiled transversely and cross-sections in the electron microscope reveal a characteristic figure-of-eight profile, due to two helical pleats or grooves. A well-defined opercular 'stopper' is present. Each polar capsule develops within a special cell, indeed much as a nematocyst since the capsule and an outgrowing external thread are laid down by the endoplasmic reticulum and Golgi apparatus. Similarly, the tube at this stage is surrounded by microtubules, and later it is withdrawn into the capsule (Fig. 5.4). But centrioles are absent, and there is nothing corresponding to a cnidocil. The two spores may discharge independently (Fig. 5.6). The control of discharge and its mechanism are still being investigated (Lom, 1964), and Uspenskaya (1982) has suggested that activation depends on the withdrawal of Ca^{2+}.

These protozoans are considered to be multicellular (Grassé and Lavette 1978). The cell lineage of pansporoblast development is a regular sequence by which two spores are formed within an envelope cell, each with six nuclei (Current and Janovy 1977). There are two capsulogenic cells, two valvogenic cells and a binucleate sporoplasm. The two polar capsules may develop out of phase and are not linked by cytoplasmic connexions (thus unlike e.g. cnidoblasts in *Hydra*). The valves of myxosporidian spores do not resemble any structures found in Cnidaria or other Metazoa.

Polar capsules are a stable and useful attribute of the Myxosporidia, and the same holds for other analogues of cnidae. It may be concluded that in the Cnidaria, nematocysts are a comparably independent or intrinsic feature.

2. Epithelial organization

The most far-reaching and limiting characteristic of Cnidaria is the epithelial organization of tissues and their relation to mesogloea, for

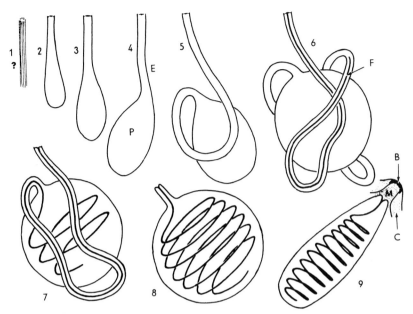

Fig. 5.4. Interpretation of polar capsule development in Myxosporidia, stages 1-9, showing growth and invagination of the tube, and the mature capsule. From Lom and de Puytorac (1965), with permission of Editions du Centre National de la Recherche Scientifique.

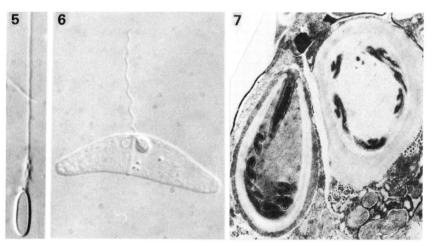

Fig. 5.5. Discharged nematocyst of *Hydractinia echinata*, × 1400.
Fig. 5.6. Spore of *Ceratomyxa arcuata*, showing one of the two polar capsules discharged, × 1400.
Fig. 5.7. Undischarged polar capsules of *Myxobolus exiguus* in longitudinal section, × 1400. From Pulsford and Matthews (1982), with permission of Blackwell Scientific Publications Ltd.

herein lies the basis of their simplicity. A good example of how physiological and behavioural attributes reflect the special arrangements of
ectoderm, endoderm, and mesogloea is the neuromuscular system. The
musculo-epithelial basis for organization of muscle is unique (Pantin
1960) and so, in many respects, are nerve nets. Coelenterates are often
referred to as the simplest animals with a nervous system (Robson
1975). The development of epithelial conduction pathways and nerve
nets together has allowed remarkably complex behaviour to evolve
(Mackie 1970). It is these highly selected characters which capture
attention and current work on the physiology of the cnidarian nervous
system reveals an increasing number of features also present in
vertebrates (see Shelton 1982; and also Anderson 1980; Anderson and
Schwab 1982; Martin and Spencer 1983; Grimmelikhuijzen 1983).

The dynamic aspects of tissue organization in Cnidaria are best
understood from work on *Hydra* (Tardent and Tardent 1980; Lenhoff
1983). *Hydra* can be described in terms of three main cell lineages, each
of which contains a population of stem cells from which differentiated
cells are derived (Campbell and Bode 1983). These are ectodermal
cells, endodermal cells (though lineage relationships of the gland cells
are not yet known), and interstitial cells and their derivatives. The pool
of interstitial cells in the ectoderm gives rise to cnidoblasts and to nerve
cells as they are needed (Fig. 5.8) and to gametes. The control of
mitosis and differentiation depends on position in the animal, on the
balance of cell numbers, and on factors which regulate growth and cell
migration (David 1983).

It is possible to eliminate interstitial cells by treatment with colchicine
and other agents, and such *Hydra* therefore lack nerve cells and nematocysts as well (Marcum and Campbell 1978). Hand-reared specimens
survive for months and show a normal budding rate. The morphogens
controlling head and foot differentiation are still present (Schaller,
Rau, and Bode 1980; Schaller, Grimmelikhuijzen, and Schmidt 1983).
In Weissman's terms, these epithelial specimens are entirely somatic,
since the germ cells seem to have been lost (see Tardent 1978).

In order to relate the organization of cnidarian tissues even to that of
ctenophores and sponges further developmental and genetic evidence
seems needed.

3. Polarity and symmetry

Axial polarity in *Hydra* is usually explained in terms of positional information (Wolpert, Hornbruch, and Clarke 1974; Gierer 1977). It appears to be determined by the two epithelia (Smid and Tardent 1982).
It has been shown in hydroids that polarity of the planula larva is
related to the site of polar body formation and of fertilization (Freeman

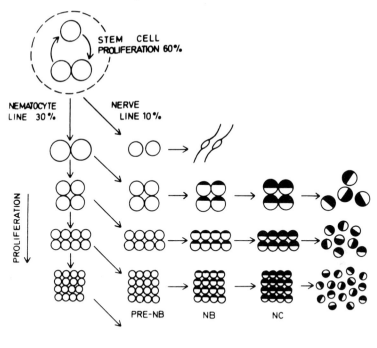

Fig. 5.8. Schematic representation of stem cell proliferation and differentiation in *Hydra*. Cells shown in clusters are held together by cytoplasmic bridges. Pre-NB, NB and NC are stages in differentiation of nematocytes. Stem cells and early committed cells in the nematocyte and nerve pathways are morphologically indistinguishable and constitute a class of large undifferentiated interstitial cells. From David (1983), with permission of Churchill Livingstone.

1981; Freeman and Miller 1982). Recent work on Hydrozoa with direct development, however, shows that in Trachylina and Siphonophora embryonic regions are determined cytoplasmically at a much earlier stage, as in ctenophores; in *Nanomia* bilateral symmetry and the oral-aboral axis are specified at the two-cell stage (Freeman 1983). The potential significance of these findings should not be overlooked.

Radial or biradial symmetry is not an easy subject to investigate. The selective advantages it confers on anemones and medusae are readily appreciated, but the 'design potential' of radiate organization is evidently limited, even where it has permitted the development of many forms of colonial growth. The few explanations put forward include radially organized domains (*Tubularia*: Campbell and Campbell 1968), primary meristematic zones (Anthozoa: Grebel'nyi 1981), and a polar coordinate model (*Bunodactis*: Shostak 1981). One can see that radial symmetry based on an epithelial pattern of tissue organization allows physiologically viable animals to exist within a considerable size range, and the combination works sufficiently well, as it were, not to have been selected against in the course of time.

Speculation and proof

The extent to which models for *Hydra* can be applied to other Cnidaria (e.g. Scyphozoa, Anthozoa) is a matter for further research but they provide valuable insight into some of the unique features of this phylum. If complexity, estimated in terms of cell differentiation, is reflected in the presence of only three cell lineages (at least in *Hydra*), one might expect homeostasis to be due to physiological interaction of different types of cell throughout the animal. In this phylum specialized organs are notably few. The increased physiological specialization seen in more complex animals should be reflected in a larger number of cell lineages: for want of a better term, I suggest that coevolution of different groups of differentiated cells, in physiologically viable and interacting clusters or lineages, is an important aspect of diversification in the lower Metazoa. If the early Cnidaria were diploblastic in a sense that we would recognize, with their limited range of differentiated cells representing only two or three cell lineages, it would not ensue that a more complex, triploblastic arrangement of tissues could have arisen only once. If many lines of early Metazoa became extinct these could perhaps have included not only unknown 'coelenterate' forms but a number of unfossilized and relatively unsuccessful triploblasts as well.

If proof were in the Precambrian I would like to ascertain whether apparently cnidarian species had nematocysts, nerve nets, and interstitial cells. Were any colonial? Did they already form calcareous spicules? I would like to examine plankton and benthos for invertebrates we shall never know about to see in how many ways they invented triploblastic tissues, and to discover ancient pluricellular Protista with incipient mesogloea although I suspect that it might be far from easy to recognize a phylum in its infancy even with hindsight.

Acknowledgements

I wish to thank Dr A. Pulsford and Dr R. A. Matthews for introducing me to the study of Myxosporidia, and the Director and staff of the Marine Biological Association, Plymouth, for laboratory facilities.

References

Allman, G. J. (1871). *Monograph of the gymnoblastic or tubularian hydroids.* Ray Society, London.

Anderson, P. A. V. (1980). Epithelial conduction: its properties and functions. *Prog. Neurobiol.* 15, 163-203.

—— and Schwab, W. E. (1982). Recent advances and model systems in coelenterate neurobiology. *Prog. Neurobiol.* 19, 213-36.

Ayala, F. J., ed. (1976). *Molecular evolution.* Sinauer Associates, Massachusetts.

Barnes, R. D. (1980). *Invertebrate zoology* (4th edn). Saunders College, Philadelphia.

Beer, G. de (1958). *Embryos and ancestors* (3rd edn). Clarendon Press, Oxford.

Bonner, J. T. (1965). *Size and cycle.* Princeton University Press, Princeton, NJ.

Bouillon, J. (in press). Les hydrozoaires. In *Traité de zoologie, Cnidaires, Cténaires* (ed. P.-P. Grassé) Vol. 3. Masson et Cie, Paris.

Brasier, M. D. (1979). The Cambrian radiation event. In *The origin of major invertebrate groups* (ed. M. R. House). Systematics Association Special Vol. 12, pp. 103-59. Academic Press, London.

Bütschli, O. (1881). Beiträge zur Kenntnis der Fischpsorospermien. *Z. wiss. Zool.* 35, 629-51.

Campbell, R. D. and Bode, H. R. (1983). Terminology for morphology and cell types. In *Hydra: research methods* (ed. H. M. Lenhoff) pp. 5-14. Plenum Press, New York.

—— and Campbell, F. (1968). *Tubularia* regeneration: radial organization of tentacles, gonophores, and endoderm. *Biol. Bull. mar. biol. Lab., Woods Hole* 134, 245-51.

Carlgren, O. (1944). Das System und die Entwicklungslinien der Anthozoen zugleich einige Bemerkungen über Pax' Bearbeitung dieser Tiergruppe in der "Tierwelt der Nord- und Ostsee". *K. fysiogr. Sällsk. Lund Förh.* 14, 40-55.

Chapman, D. M. (1966). Evolution of the scyphistoma. In *The Cnidaria and their evolution* (ed. W. J. Rees). *Symp. zool. Soc. Lond.* 16, 51-75.

—— (1974). Cnidarian histology. In *Coelenterate biology* (eds L. Muscatine and H. M. Lenhoff) pp. 1-92. Academic Press, New York.

Clark, R. B. (1964). *Dynamics in metazoan evolution: the origin of the coelom and segments.* Clarendon Press, Oxford.

Current, W. L. and Janovy, J. (1977). Sporogenesis in *Henneguya exilis* infecting the channel catfish: an ultrastructural study. *Protistologica* 13, 157-67.

David, C. N. (1983). Stem cell proliferation and differentiation in *Hydra.* In *Stem-cells. Their identification and characterisation* (ed. C. S. Potten), pp. 12-27. Churchill Livingstone, Edinburgh.

Desser, S. S., Molnar, K., and Weller, I. (1983). Ultrastructure of sporogenesis of *Thelohanellus nikolskii* Akhmerov, 1955 (Myxozoa; Myxosporea) from the common carp, *Cyprinus carpio. J. Parasitol.* 69, 504-18.

—— and Paterson, W. B. (1978). Ultrastructural and cytochemical observations on sporogenesis of *Myxobolus* sp. (Myxosporidia: Myxobolidae) from the common shiner *Notropis cornutus. J. Protozool.* 25, 314-26.

Doumenc, D. A. (1976). Etude ultrastructurale des stades parenchymula et actinella de l'actinie *Cereus pedunculatus. Archs Zool. exp. gén.* 117, 295-324.

Franc, J.-M. (1978). Organization and function of ctenophore colloblasts: an ultrastructual study. *Biol. Bull. mar. biol. Lab., Woods Hole* 155, 527-41.

Freeman, G. (1981). The role of polarity in the development of the hydrozoan planula larva. *Wilhelm Roux Arch. EntwMech. Org.* 190, 168-84.

—— (1983). Experimental studies on embryogenesis in hydrozoans (Trachylina and Siphonophora) with direct development. *Biol. Bull. mar. biol. Lab., Woods Hole* 165, 591-618.

—— and Miller, R. L. (1982). Hydrozoan eggs can only be fertilized at the site of polar body formation. *Devl Biol.* 94, 142-52.

Gierer, A. (1977). Biological features and physical concepts of pattern formation exemplified by *Hydra*. *Curr. Top. devl Biol.* 11, 17-59.

Grassé, P.-P. and Lavette, A. (1978). La Myxosporidie *Sphaeromyxa sabrazesi* et nouvel embranchement des Myxozoaires (Myxozoa). Recherches sur l'état pluricellulaire primitif et considerations phylogénétiques. *Annls Sci. nat. Zool.* 20, 193-285.

Grebel'nyi, S. D. (1981). Symmetry of the Actiniaria and the significance of symmetry features for the classification of the Anthozoa. *Dokl. (Proc.) Acad. Sci. U.S.S.R., Biol. Sci.* 253, 430-2.

Grimmelikhuijzen, C. J. P. (1983). Coexistence of neuropeptides in *Hydra*. *Neuroscience* 9, 837-45.

Gupta, B. L. and Hall, T. A. (1984). Role of high concentrations of Ca, Cu and Zn in the maturation and discharge *in situ* of sea anemone nematocysts as shown by X-ray microanalysis of cryosections. *Proc. Int. symp. Toxins, drugs and pollutants in marine animals.* pp. 77-95. Springer-Verlag, Berlin.

Hadzi, J. (1963). *The evolution of the Metazoa*. Pergamon, Oxford.

Haeckel, E. (1874). *Natürliche schöpfungsgeschichte* (5th edn). G. Reimer, Berlin.

—— (1899). *Kunst-Formen der natur*. Leipzig und Wien.

Hand, C. (1963). The early worm: a planula. In *The lower Metazoa* (ed. E. C. Dougherty) pp. 33-9. University of California Press, Berkeley, Ca.

—— (1966). On the evolution of Actiniaria. In *The Cnidaria and their evolution* (ed. W. J. Rees). *Symp. zool. Soc. Lond.* 16, 135-46.

Harris, L. G. (1973). Nudibranch associations. *Curr. Top. comp. Pathobiol.* 2, 213-315.

Holstein, T. (1981). The morphogenesis of nematocytes in *Hydra* and *Forskålia*: an ultrastructural study. *J. Ultrastruc. Res.* 75, 276-90.

Holstein, T. and Tardent, P. (1984). An ultrahigh-speed analysis of exocytosis: nematocyst discharge. *Science, N.Y.* 223, 830-3.

Hovasse, R. (1965). Trichocystes, corps trichocystoïdes, cnidocystes et colloblastes. *Protoplasmologica* 3, 1-57.

Hyman, L. H. (1940). *The invertebrates*. Vol. 1. *Protozoa through Ctenophora*. McGraw-Hill, New York.

Karling, T. G. (1966). On nematocysts and similar structures in turbellarians. *Acta zool. fenn.* 116, 1-28.

Lakatos, I. (1978). *The methodology of scientific research programmes*. Vol. 1. Cambridge University Press, Cambridge.

Larsson, R. (1981). Description of *Nosema tractabile* n. sp. (Microspora, Nosematidae), a parasite of the leech *Helobdella stagnalis* (L.) (Hirudinea, Glossiphonidae). *Protistologica* 17, 407-22.

Lenhoff, H. M. (ed.) (1983). *Hydra: research methods*. Plenum Press, New York.

Lom, J. (1964). Notes on the extrusion and some other features of myxo-sporidian spores. *Acta protozool.* 2, 321-7.

—— (1969). Notes on the ultrastructure and sporoblast development in fish parasitizing myxosporidians of the genus *Sphaeromyxa. Z. Zellforsch. mikrosk. Anat.* 97, 416-37.

—— and de Puytorac, P. (1965). Studies on the myxosporidian ultrastructure and polar capsule development. *Protistologica* 1, 53-65.

Lubbock, R. and Amos, W. B. (1981). Removal of bound calcium from nematocyst contents causes discharge. *Nature, Lond.* 290, 500-1.

Lubbock, R., Gupta, B., and Hall, T. A. (1981). Novel role of calcium in exocytosis: mechanism of nematocyst discharge as shown by X-ray microanalysis. *Proc. natn. Acad. Sci. U.S.A.* 78, 3624-8.

Lubbock, R. and Shelton, G. A. B. (1981). Electrical activity following recognition of self and non-self in a sea anemone. *Nature, Lond.* 289, 59-60.

Mackie, G. O. (1970). Neuroid conduction and the evolution of conducting tissues. *Q. Rev. Biol.* 45, 319-32.

Marcum, B. A. and Campbell, R. D. (1978). Development of *Hydra* lacking nerve and interstitial cells. *J. Cell Sci.* 29, 17-33.

Mariscal, R. N. (1974). Nematocysts. In *Coelenterate biology* (eds L. Muscatine and H. M. Lenhoff) pp. 129-78. Academic Press, New York.

Martin, S. M. and Spencer, A. N. (1983). Neurotransmitters in coelenterates. *Comp. Biochem. Physiol.* 74C, 1-14.

Moore, R. C. (ed.) (1956). *Treatise on invertebrate palaeontology. Part F, Coelenterata.* Geological Society of America and University of Kansas Press, Lawrence, Ka.

Pantin, C. F. A. (1951). Organic design. *Advmt Sci., Lond.* 8, 138-50.

—— (1960). Diploblastic animals. *Proc. Linn. Soc.* 171, 1-14.

Petersen, K. W. (1979). Development of coloniality in Hydrozoa. In *Biology and systematics of colonial organisms* (eds G. Larwood and B. Rosen), pp. 105-39. Academic Press, London.

Pugh, P. R. (1983). Benthic siphonophores: a review of the family Rhodaliidae (Siphonophora, Physonectae). *Phil. Trans. R. Soc.* B301, 165-300.

Pulsford, A. and Matthews, R. A. (1982). An ultrastructural study of *Myxobolus exiguus* Thelohan, 1895 (Myxosporea) from grey mullet, *Crenimugil labrosus* (Risso). *J. Fish Dis.* 5, 509-26.

Raikov, I. B. (1978). Ultrastructure du cytoplasme et des nématocystes du cilié *Remanella multinucleata* Kahl (Gymnostomata, Loxodidae). Existence de nématocystes chez les ciliés. *Protistologica* 14, 413-32.

Raup, D. M. and Valentine, J. W. (1983). Multiple origins of life. *Proc. natn. Acad. Sci. U.S.A.* 80, 2981-4.

Robson, E. A. (1975). The nervous system in coelenterates. In *'Simple' nervous systems* (eds P. N. R. Usherwood and D. R. Newth) pp. 169-209. Edward Arnold, London.

Rosen, B. R. (1981). The tropical high diversity enigma—the corals'-eye view. In *Chance, change and challenge. The evolving biosphere* (ed. P. L. Forey) pp. 103-29. British Museum (Natural History) and Cambridge University Press, Cambridge.

Saville-Kent, W. (1893). *The Great Barrier Reef of Australia*. W. H. Allen, London.

Schaller, H. C., Grimmelikhuijzen, C. J. P., and Schmidt, T. (1983). Assay and isolation of substances controlling morphogenesis in *Hydra*. In *Hydra: research methods* (ed. H. M. Lenhoff) pp. 311-24. Plenum Press, New York.

—— Rau, T., and Bode, H. (1980). Epithelial cells in nerve-free *Hydra* produce morphogenetic substances. *Nature, Lond.* 283, 589-91.

Schmidt, H. (1974). On evolution in the Anthozoa. *Proc. Second International Coral Reef Symposium* 1, 533-60.

—— (1981). Die Cnidogenese der Octocorallia (Anthozoa, Cnidaria): I. Sekretion und Differenzierung von Kapsel und Schlauch. *Helgoländer wiss. Meeresunters.* 34, 463-84.

—— and Zissler, D. (1979). Die Spermien der Anthozoen und ihre phylogenetische Bedeutung. *Zoologica, Stuttg.* 44, 129, 1-46.

Scrutton, C. T. (1979). Early fossil cnidarians. In *The origin of major invertebrate groups* (ed. M. R. House). Systematics Association Special Vol. 12, pp. 161-207. Academic Press, London.

Shelton, G. A. B. (ed.) (1982). *Electrical conduction and behaviour in 'simple invertebrates*. Clarendon Press, Oxford.

Shostak, S. (1981). Regeneration in young *Bunodactis verrucosa* Pennant (Actiniaria, Anthozoa). *Wilhalm Roux Arch. EntwMech. Org.* 190, 274-82.

Sleigh, M. A. (1979). Radiation of eukaryote Protista. In *The origin of major invertebrate groups* (ed. M. R. House). Systematics Association Special Vol. 12, pp. 23-54. Academic Press, London.

Smid, I. and Tardent, P. (1982). The influences of ecto- and endoderm in determining the axial polarity of *Hydra attenuata* Pall. (Cnidaria, Hydrozoa). *Wilhelm Roux Arch. EntwMech. Org.* 191, 64-7.

Tardent, P. (1978). Coelenterata, Cnidaria. In *Morphogenese der Tiere* (ed. F. Siedel) Lief. 1: A-1, pp. 69-391. Gustav Fischer, Stuttgart.

—— and Tardent, R. (eds) (1980). *Developmental and cellular biology of coelenterates*. Elsevier-North Holland, Amsterdam.

Totton, A. K. (1965). *A synopsis of the Siphonophora*. British Museum (Natural History), London.

Uspenskaya, A. V. (1977). Extrusomes of invertebrates. In *Motility of non-muscular cells and their components*, pp. 73-100. Nauk, Leningrad (in Russian).

—— (1982). New data on the life cycle and biology of Myxosporidia. *Archs Protistenk.* 126, 309-38.

Vávra, J. (1976). Structure of the Microsporidia. In *Comparative pathobiology*. Vol. 1 *Biology of the Microsporidia* (eds L. A. Bulla and T. C. Cheng), pp. 1-85. Plenum Press, New York.

Weill, R. (1934). *Contribution a l'étude des cnidaires et de leurs nématocystes*. Presses Universitaires de France, Paris.

Werner, B. (1973). New investigations on systematics and evolution of the class Scyphozoa and the phylum Cnidaria. *Publs Seto mar. biol. Lab.* 20, 35-61.

—— (1975). Bau und Lebensgeschichte des Polypen von *Tripedalia cystophora* (Cubozoa, class nov., Carybdeidae) und seine Bedeutung für die Evolution der Cnidaria. *Helgoländer wiss. Meeresunters.* 27, 461-504.

—— (1979). Coloniality in the Scyphozoa: Cnidaria. In *Biology and systematics of colonial organisms* (eds G. Larwood and B. R. Rosen), pp. 81-103. Academic Press, London.

—— (1983). Die Metamorphose des Polypen von *Tripedalia vystophora* (Cubozoa, Carybdeidae) in die Meduse. *Helgoländer wiss. Meeresunters.* 36, 257-76.

Westfall, J. A. (1970). The nematocyte complex in a hydromedusan, *Gionemus vertens. Z. Zellforsch. mikrosk. Anat.* 110, 457-70.

——, Bradbury, P. C., and Townsend, J .W. (1983). Ultrastructure of the dinoflagellate *Polykrikos*. 1. Development of the nematocyst-taeniocyst complex and morphology of the site for extrusion. *J. Cell Sci.* 63, 245-61.

Whittington, H. B. (1979). Early arthropods, their appendages and relationships. In *The origin of major invertebrate groups* (ed. M. R. House). Systematics Association Special Vol. 12, pp. 253-68. Academic Press, London.

Wolpert, L., Hornbruch, A., and Clarke, M. R. B. (1974). Positional information and positional signalling in *Hydra. Am. Zool.* 14, 647-63.

Yanagita, T. M. (1973). The "cnidoblast" as an excitable system. *Publs Seto mar. biol. Lab.* 20, 675-93.

Yochelson, E. L. (1979). Early radiation of Mollusca and mollusc-like groups. In *The origin of major invertebrate groups* (ed. M. R. House). Systematics Association Special Vol. 12, pp. 323-58. Academic Press, London.

Ziman, J. (1968). *Public Knowledge.* Cambridge University Press, Cambridge.

6. On the classification and evolution of the Ctenophora

G. R. HARBISON

Woods Hole Oceanographic Institution, Massachusetts, USA

Abstract

Relationships between ctenophores and other lower metazoans are unclear, and there is no compelling evidence that ctenophores are more closely related to cnidarians than they are to platyhelminths. All of the presently recognized orders of the Ctenophora appear to be monophyletic, except for the Cydippida. The families in this group are redefined, in an attempt to incorporate them into a phylogenetic classification. Although conclusions can be drawn concerning interrelationships between the monophyletic groups, present knowledge is inadequate to determine the direction of important evolutionary changes within the phylum. Two conflicting phylogenetic scenarios are presented, with a preference for the assumption that the common ancestor of the phylum was atentaculate. Although presently accepted classifications of the Ctenophora are clearly not phylogenetic, it seems premature to alter drastically the conventionally recognized taxa until new information is forthcoming. It appears that there is a large undescribed deep-sea fauna, which may clarify our understanding of evolution within the group.

Introduction

Ctenophores form a small, well-defined phylum of lower metazoans. All ctenophores are predators, and all but one order are planktonic. It has become apparent recently that ctenophores are far more widespread and abundant than was previously thought (Harbison, Madin, and Swanberg 1978), and thus constitute a highly successful plan of organization. Most are oceanic; of the planktonic forms, only a handful of species live close to shore in bays and estuaries. Most ctenophores are extremely fragile and cannot be collected in recognizable conditions with nets; this has greatly hampered systematic research. Gentle collection techniques using submersibles have revealed the existence of a diverse and as yet undescribed deep-sea fauna.

Ctenophores are virtually absent from the fossil record (Conway Morris, this volume). A radiographic technique has revealed an organism from the Devonian resembling a ctenophore (Stanley and Stürmer 1983). If their reconstruction is correct, this fossil differs little from recent cydippids. The simplicity of the bodyplan of ctenophores, with a poorly differentiated gelatinous mesogloea and rudimentary organ systems, has led most workers to regard the phylum as ancient (e.g. Hyman 1940). Although I agree with this opinion and consider that they have been a separate group long enough to obscure their relationships with other lower metazoans, I must admit that there is no positive evidence to support it. In general, they are easy to recognize because they have eight rows (comb rows) of fused macrocilia (comb plates), which provide the primary means of propulsion and give the phylum its name ('comb-bearers'). The comb rows are mechanically coupled to an aborally located statocyst. Pole plates, structures of unknown function, lie next to the statocyst. Members of the Platyctenida lose the comb rows as adults, but they are present in the juveniles. All platyctenes that have lost the comb rows retain the statocyst, except *Savangia atentaculata* Dawydoff 1950.

Under each comb row there is a meridional canal which forms part of the gastrovascular system. The gastrovascular system is essentially the same in all ctenophores and can be considered as divided into an axial and a peripheral portion. The axial portion consists of the mouth, the large stomodaeum (pharynx), the infundibulum (gut), the infundibular canal, and the anal (excretory) canals and pores (Fig. 6.1). The infundibular canal is absent in the Beroida. The anal pores do not serve as a true anus, and most undigested material is voided through the mouth. The peripheral portion of the gastrovascular system is connected to the infundibulum, and consists of the perradial, interradial, adradial and meridional canals, the tentacular canals, and the paragastric canals. Any of these canals, except for the adradial and meridional canals, may be missing (Fig. 6.2).

Most ctenophores have two tentacles, that may or may not have side branches (tentilla). Tentacles lacking side branches are referred to as 'simple'. The base of the tentacle (tentacle bulb) is often located deep inside the animal, so that the main filaments exit through tentacle sheaths. The tentacles of most ctenophores are equipped with structures called colloblasts, which cause prey to adhere to the tentacles.

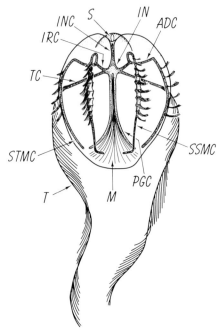

Fig. 6.1. Juvenile *Bolinopsis vitrea* (Lobata), viewed in the tentacular plane. After Chun (1892). Abbreviations for this, and all subsequent figures: ADC, adradial canal; AP, aboral papillae; AU, auricle; IN, infundibulum; INC, infundibular canal; IRC, interradial canal; M, mouth; MC, meridional canal; OL oral lobe; OS, opening of tentacle sheath; PGC, paragastric canal; PP, pole plate; PRC, perradial canal; S, statocyst; SSCR, substomodaeal comb row; SSMC, substomodaeal meridional canal; STCR, subtentacular comb row; STMC, subtentacular meridional canal; T. tentacle; TC, tentacular canal; TS, tentacle sheath.

Tentacles are absent in the Beroida and reduced or absent in two families of the Lobata (Harbison and Madin 1982).

Since the stomodaeum is flattened, it can be regarded as forming a plane (the stomodaeal, sagittal or pharyngeal plane). An imaginary plane at right angles to it passing through the tentacle bulbs is termed the tentacular (transverse) plane. Structures lying closest to the stomodael plane are termed 'substomodaeal', and structures lying closest to the tentacular plane are termed 'subtentacular'. There have been several attempts in the past to provide a uniform nomenclature for the body parts and planes of symmetry within the Ctenophora, but all have failed. Most of the common synonyms are listed by Harbison and Madin (1982) and Hyman (1940).

Most ctenophores are simultaneous hermaphrodites. Gonads are usually located in the walls of the meridional canals, and eggs and sperm are shed through separate gonoducts. External self-fertilization is common in the laboratory. Cleavage is determinate, differs from

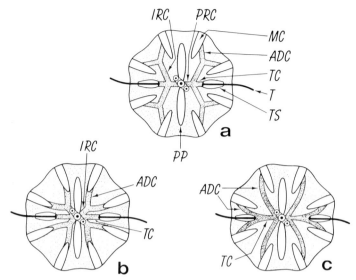

Fig. 6.2. Diagrammatic views from the aboral pole of (a) a pleurobrachiid; (b) a lobate or pleurobrachiid; and (c) a mertensiid, bathyctenid or platyctene, showing how the adradial and tentacular canals can be connected to the infundibulum. The connections of the paragastric canals are not shown.

other groups, and is termed 'biradial' (Pianka 1974). This term is also employed to describe the symmetry of the adult, which is neither truly bilateral nor radial.

Because many Recent species and genera have yet to be described, and the higher taxa are greatly in need of revision, and because radiographic or other techniques may allow us to study fossil ctenophores, many or most of the conclusions reached in this paper should be regarded as only tentative in preparation for the wealth of new information that may soon be forthcoming.

Relationship with other phyla

It has been proposed that ctenophores are descended from ancestors resembling Porifera (Schneider 1902), Hydrozoa (e.g. Chun 1892; Haeckel 1896; Hyman 1940), Scyphozoa (Goette, in Krumbach 1925), Anthozoa (Heider, in Krumbach 1925), and Platyhelminthes (Hadzi 1953). Ctenophores have been regarded as resembling the ancestral forms of polyclad flatworms (Lang 1884; Mortensen 1912; Krumbach 1925) and trochophore larvae (Hatschek 1911). Some have regarded ctenophore-like animals as ancestral to the Bilateria (Haeckel 1896; Bonik, Grasshoff, and Gutmann 1976) and others have considered them evolutionary dead ends (Hyman 1940). The most commonly expressed viewpoint, and still the most widely held opinion today, is that they are more closely allied to the Cnidaria than to other phyla. Until relatively recently, they were united with the Cnidaria in the phylum Coelenterata, and Komai (1963) still regards them as a subphylum. In my view, there is very little evidence to support a close relationship between the Ctenophora and Cnidaria.

Komai (1963) gives the following list of similarities between ctenophores and cnidarians: radial (non-bilateral) symmetry in body shape and organ arrangement, a branching gastrovascular system, a diffuse nervous system, absence of a coelom, and absence of an excretory system and accessory genital apparatus. Other characters in common listed by Hyman (1940) are gelatinous mesogloea, statocyst, and general lack of organ systems. Four of these nine similarities are negative characters, and as such can scarcely be cited as evidence demonstrating a close relationship between the two phyla. In my opinion, the five characters that remain are little better. There are many oceanic animals that are both gelatinous and obviously not closely related to ctenophores, e.g. pseudothecosome pteropods and salps. There are also many organisms other than ctenophores and cnidarians which have statocysts, branching gastrovascular systems or diffuse nervous systems.

No trace of bilateral symmetry occurs in ctenophores, but it is difficult to characterize this non-bilaterality as radial, or to go as far as Hyman (1940), who terms it 'generally tetramerous'. Numerous paired structures create a biradial symmetry: the four-fold symmetry of ctenophores is restricted to the comb row apparatus, statocyst comb row complex, the gastrovascular canals that underlie the comb rows, and to the auricles of the Lobata. This four-fold symmetry is a consequence of the mechanical coupling of the statocyst to the comb rows; it rests on four balancer cilia, and from them four pairs of ciliary tracks connect with the comb rows. These ciliary tracks are asymmetrical (Tamm 1982), and the adjacent comb rows they supply usually have differing lengths and numbers of comb plates. These differences are carried to their greatest extreme in the Cestida, where the subtentacular comb rows are reduced to a few plates, and the substomodaeal comb rows are greatly elongated. Tetramerous symmetry is lacking in other parts of the body: the tentacles (when present) and tentacular canals are always paired, as are the paragastric canals and pole plates, and the stomodaeum and infundibulum are flattened in opposite planes. This non-bilaterality indicates either that the common ancestor of the group was not bilaterally symmetrical, or that ctenophores have been a separate line for so long that all traces of bilaterality have been lost. Even the Platyctenida retain biradial symmetry.

A number of researchers, most notably Lang (1884) and Mortensen (1912), proposed a close relationship between polyclad flatworms and ctenophores. Both Hyman (1951) and Komai (1963) have dismissed this theory, and have argued for a close relationship between the Ctenophora and trachyline medusae. Hyman (1951) gives the following list of similarities between platyctene ctenophores and polyclad flatworms: 'their oval flattened shape with dorso-ventral differentiation, . . . the presence of two dorsal tentacles, . . . creeping upon the entire ventral surface, the centrally located ventral mouth and blindly ending digestive canals; and . . . the radiating, anastomosing nervous system. . . . Many polyclads have a swimming larval stage with eight ciliated lobes which seem comparable to the eight comb rows of ctenophores.' These similarities, which Hyman (1951) and others have considered as being without phylogenetic significance, seem to me to be at least as convincing as the previous list of similarities with cnidarians. It is also worthy of note that two of them (branching gastrovascular system and diffuse nervous system) are on both lists!

It is, to my way of thinking, force of tradition that has kept the Ctenophora and Cnidaria together for so long. The gelatinous, radial nature of their simple bodies has overshadowed the profound differences between the two groups. Most ctenophores are hermaphrodites, and shed sexual products through gonoducts. Early cleavage and subsequent development is strikingly different in the two groups. The statocyst-comb row complex, so characteristic of ctenophores, has no parallel within the Cnidaria. The colloblasts which tentaculate ctenophores use to capture prey are never found in cnidarians, and nematocysts are never found in ctenophores, except as secondarily acquired kleptocnidae. Ctenophores have a mesodermal musculature, never have alternation of generations, and never form colonies.

Three observations have added weight to the hypothesis that ctenophores and cnidarians are more closely allied with one another than with other phyla. The first was the discovery of nematocysts in the tentacles of the ctenophore *Haeckelia rubra* (Gegenbaur, Kölliker, and Müller 1853). Two recent papers (Carré and Carré 1980; Mills and Miller 1984) present convincing evidence that these nematocysts are acquired from their narcomedusan prey. The second was the description of the hydromedusa, *Hydroctena salenskii* Dawydoff 1903. Mortensen (1912) and Hyman (1940) argue convincingly that the similarities between this animal and the Ctenophora are only superficial. The third was the discovery of a planula larva (Komai 1922), ascribed to *Gastrodes parasiticum* Korotneff 1888, presently known as *Lampea pancerina* (Chun 1880). Mills and Miller (1984) express scepticism that this larva is truly homologous to a cnidarian planula. The validity of their analysis depends on whether or not the larva of *L. pancerina* does indeed behave as they hypothesize. I would argue that the behaviour does not matter, since if the larva does indeed exist, it could equally well be regarded as resembling the stereogastrula of a platyhelminth, which also resembles a cnidarian planula (Hyman 1951). Thus, even if *L. pancerina* has a larval stage closely resembling a cnidarian planula, this adds no evidence to support a closer phylogenetic relationship with the Cnidaria than with the Platyhelminthes.

Ctenophora, Cnidaria, and Platyhelminthes have simple bodyplans, with only the most rudimentary of organ systems. This indicates, in my opinion, that all three groups must have arisen in the earliest stages of metazoan evolution. Since all three are monophyletic, they must be related in one of three ways (Fig. 6.3). Given present knowledge, each of the three alternatives can be supported. I regard alternative two as the most likely, because ctenophores share the following traits with the Platyhelminthes that they do not share with the Cnidaria: gonoducts, mesodermal musculature, and determinate cleavage. These three traits do not provide very strong evidence, so it is preferable only to assert that all three groups must have diverged before the fundamental characters of each group were developed. Thus, the common ancestor of ctenophores and cnidarians must have lacked nematocysts, colloblasts and the statocyst-comb row complex, and the common ancestor of ctenophores and platyhelminths must have lacked these structures as well as an excretory or copulatory system. Alternatively some of these characters became lost after the phyla separated. It is perhaps plausible that pelagic ctenophores could have lost excretory and copulatory systems, but the loss of nematocysts or colloblasts is more difficult to rationalize. If the common ancestor of ctenophores and cnidarians captured prey with tentacles, structures other than nematocysts or colloblasts were used to hold the prey fast. Thus, I can see no more reason to assume that the tentacles of ctenophores and cnidarians are homologous than to assume that the tentacles of temnocephalid flatworms are the homologues of the tentacles of ctenophores.

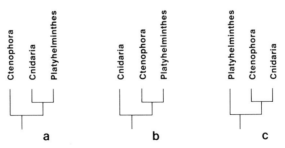

Fig. 6.3. The possible relationships between ctenophores and the two phyla that are generally considered to be their nearest relatives.

Thus, our knowledge of the relationships between ctenophores and other lower metazoans can still be expressed in the same way that Krumbach did in 1925: 'So leicht es in jedem einzelnen Falle ist, zu entscheiden, ob ein gegebenes Tier zu den *Rippenquallen* gehört oder nicht, so unsicher geht das Urteil, wenn es bestimmen soll, wie nahe oder fern die Ctenophoren anderen Tierformen stehen.'

Natural groups within the Ctenophora

Traditionally, the phylum is divided into two classes, the Nuda, containing the single order Beroida, and the Tentaculata, containing the orders Cydippida, Platyctenida, Cestida, Ganeshida, Thalassocalycida and Lobata (Harbison and Madin 1982). In my opinion, all of these orders are monophyletic except for the Cydippida.

Two orders, the Ganeshida Moser 1908 and the Thalassocalycida Madin and Harbison 1978, are monogeneric, and were proposed because neither group could be fitted into the definition of the other orders. Both show a number of affinities with the Lobata, and when a phylogenetic classification is developed, their status will probably be changed.

The Beroida contains two genera, *Beroe* and *Neis*, which are closely related. In this order, tentacles and tentacular canals are absent, the infundibular canal is lacking, and all adradial canals arise directly from the infundibulum (Fig. 6.4). All described species have finger-like projections (aboral papillae) on the margins of the pole plates, and circumoral connections of the meridional and paragastric canals. All, except *Beroe gracilis* Künne 1939, have diverticulae (which sometimes anastomose) on the meridional canals.

The Cestida contains two genera, *Cestum* and *Velamen*, which closely resemble one another. Both are ribbon-shaped with only rudimentary subtentacular comb rows. The oral margin is greatly elongated, extending the entire width of the body, and is fringed with tentillae (Fig. 6.5(a)).

The Lobata contains six families, indicating extensive adaptive radiation. All members of the Lobata, as presently defined, have auricles and oral lobes. Tentacle sheaths are absent in all adult lobates, and in two families, the Kiyohimeidae and the Ocyropsidae, the tentacular apparatus is reduced or absent. Since all lobates have auricles and oral lobes, there is no question that the group is monophyletic.

Komai (1963) considers the Platyctenida as polyphyletic, 'since most of the forms belonging to this group are relatively rare and known from widely separated localities.' However, all of the presently described genera have special chambers in which the eggs develop (brood

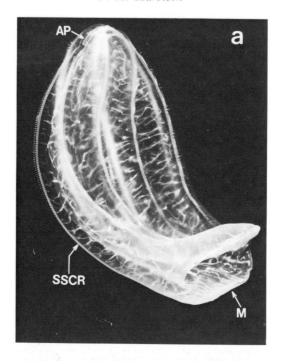

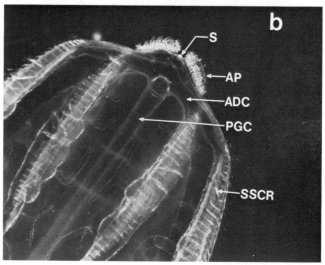

Fig. 6.4. Beroida. (a) *Beroe cucumis*? showing the elaborately anastomosing diverticulae of the meridional canals. The massive stomodaeum occupies most of the body. Length *c.*55 mm. (b) *Beroe* sp. juvenile. The diverticulae of the meridional canals do not form anastomoses. Note the absence of the infundibular canal (Compare Figs 6.6(d) and 6.7(a).) Length *c.*20 mm.

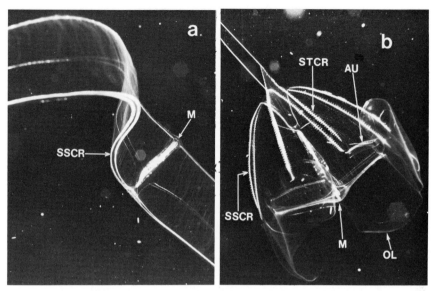

Fig. 6.5. Field pictures of two tentaculate ctenophores, showing *in situ* feeding postures. (a) *Cestum veneris* (Cestida) feeds by capturing prey on tentilla streaming from the oral edge of the animal. Cilia convey the captured prey along the oral edge to the centrally-located mouth. The tentacular apparatus of (b) *Eurhamphaea vexilligera* Gegenbaur 1856 (Lobata) closely resembles that of *C. veneris*. Note the tentilla along the oral margin streaming over the sides of the body. The oral lobes, which are missing in the Cestida, serve as additional surfaces for the capture of prey in lobates.

chambers), and these structures are unique to the Platyctenida. It is difficult to imagine that such specialized structures could have arisen independently in all platyctenes and yet be absent in all other cteno-phores. Further, one can easily reconstruct the evolution of the group from Recent specimens. Species of *Ctenoplana*, which closely resemble pelagic ctenophores, have functional comb rows throughout their lives, but are also able to creep on the surface. Species of *Coeloplana* and *Vallicula*, which have comb rows only as juveniles, actively creep on surfaces, extending the margins of their creeping lobes in the form of temporary 'chimneys'. In species of *Lyrocteis* and *Tjalfiella*, these chimneys have become permanent structures. Several species in different genera are found in association with soft corals and gorgonians and may be ectoparasites. These habits provide additional evidence that the Platyctenida is monophyletic.

The last order, the Cydippida, has characters indicating adaptive radiations as extensive as those of the Lobata. All Cydippida, as presently constituted, have tentacles exiting through tentacle sheaths, and paragastric and meridional canals ending blindly at the oral pole. Several species have been described, which would otherwise be placed in the Cydippida, that have canals which connect at the oral pole (e.g. Dawydoff 1946). Many of these forms could be larval lobates, but some may be legitimate species. It is generally agreed that cydippid ctenophores resemble the ancestors of platyctenids, ganeshids, lobates, and cestids, since developmental stages resembling adult cydippids are found in all four orders. Cydippid ctenophores are sufficiently diverse, however, that the term 'cydippid larva' has little meaning.

Harbison and Madin (1982) recognize five families within the Cydippida. Between the writing of that paper and this contribution, I have studied a number of living and preserved specimens of cydippids, including *Tinerfe cyanea* (Chun 1889) and *T. beehleri* Mayer 1912 (Fig. 6.6(a,b)), and have concluded that these five families must be re-defined. The gelatinous tuberosities on the form described as *T. beehleri* bear no resemblance to the aboral projections of *T. cyanea*, and more closely resemble the aboral papillae of beroids and many platyctenes (Fig. 6.6(c)). In *T. beehleri*, the canals supplying the tentacular apparatus (Fig. 6.6(d)) appear to be identical to those figured by Komai and Tokioka (1942) for *Haeckelia rubra*, and thus Mayer's (1912) *T. beehleri* should be placed in the genus *Haeckelia*. *Tinerfe cyanea* has perradial and interradial canals, and the tentacle sheath opens aborally (Chun 1898); thus it belongs in the Pleurobrachiidae. The form Mayer (1912) describes as *T. lactea* resembles *Lampea pancerina* (Chun 1880) more closely than it does either *T. cyanea* or *H. beehleri*, and thus belongs in the Lampeidae.

I redefine the five families as follows, and also list (in parentheses) after each definition those species I have personally studied:

(i) Haeckeliidae: Body cylindrical. Openings of tentacle sheaths between mouth and oral end of comb rows. Tentacles simple. Mouth widely expandable. Paragastric canals absent in *Haeckelia rubra*, present in others. Tentacle bulbs supplied by canals from substomodaeal adradial canals. Probably two genera, *Haeckelia* and *Dryodora*, although the latter is inadequately described. (*Haeckelia beehleri*).

(ii) Bathyctenidae: Body slightly compressed in stomodaeal axis. Opening of tentacle sheaths between mouth and oral end of comb rows. Tentacles simple. Mouth widely expandable. Paragastric and meridional canals with diverticulae. Tentacle bulbs supplied by per-radial canal from which subtentacular adradial canals arise at tentacle bulb (Fig. 6.2(c)). Two genera, *Bathyctena* and *Aulococtena*. (*Aulococtena acuminata*, Mortensen 1911).

(iii) Lampeidae: Body cylindrical. Openings of the tentacle sheaths between mouth and oral end of comb rows. Tentacles with tentilla. Mouth widely expandable. Paragastric canals present in this and in next two families. Tentacle bulbs supplied by separate tentacular

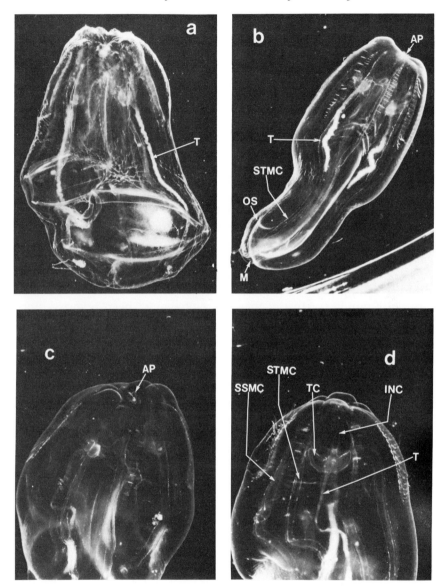

Fig. 6.6. *Haeckelia beehleri* (Haeckeliidae). (a) The two bullet-shaped objects in the stomodaeum are the anterior and posterior nectophores of the siphonophore, *Chelophyes appendiculata*. (b) Another specimen with an empty stomodaeum. Note tentacles are usually retracted into the tentacle sheaths. (c) An aboral view, showing the papillae surrounding the statocyst. (d) The tentacular canals of *H. beehleri* provide connections between adjacent subtentacular meridional canals. Length of animals between 7 and 15 mm.

canals. Adradial canals arise directly from infundibulum. One genus, *Lampea*. (*Lampea pancerina*).

(iv) Mertensiidae: Body strongly compressed in stomodaeal axis, with aborally projecting 'wings'. Openings of tentacle sheaths near aboral pole. Tentacles with side branches. Mouth expandable but not to the extent of the first three families (Fig. 6.7(b)). Tentacle bulbs supplied by perradial canals from which subtentacular adradial canals arise at tentacle bulb, as in Bathyctenidae. Two genera, *Mertensia* and *Callianira*. (*Callianira bialata* Delle Chiaje 1841).

(v) Pleurobrachiidae: Body cylindrical or slightly compressed in stomodaeal axis. Openings of tentacle sheaths near aboral pole. Tentacles with side branches. Mouth as expandable as mertensiids in *Tinerfe* and *Hormiphora* (Fig. 6.7(a)), rigid in *Pleurobrachia* and *Euplokamis*. Tentacle bulbs supplied by separate tentacular canal. Subtentacular and substomodaeal adradial canals united by common interradial canal (Fig. 6.2(a,b)). Four genera. (*Tinerfe cyanea, Hormiphora plumosa* Agassiz 1860, *Hormiphora* spp., *Pleurobrachia bachei* Agassiz 1860).

Other cydippid ctenophores have been described which may not fit into these families. Most of these forms have not been reported since their original descriptions, and some may be larval stages of other species. It is to be expected, however, that as more forms are studied, a redefinition of the families may again be necessary, or new ones may be added.

Harbison and Madin (1982) divided cydippid ctenophores into two large groups: those with tentacles that exit through openings in the tentacle sheaths near the aboral pole (Mertensiidae, Pleurobrachiidae; Fig. 6.7) and those with tentacles that exit through openings near the mouth (Haeckeliidae, Bathyctenidae, Lampeidae). Members of the first group all have to spin in order to ingest prey captured by the tentacles (Connell, in Hardy 1956). Spin ingestion consists of a complex series of movements involving the reversal of adjacent subtentacular comb row cilia, with a coordinated contraction of the tentacle (Tamm 1982). Members of the second group can simply draw prey into the mouth, or if the prey is much larger than the ctenophore, draw themselves onto the prey. The feeding habits of only three species in this group are known, but all eat gelatinous plankton. *Lampea pancerina* eats salps (Harbison and Madin 1982), *Haeckelia rubra* eats narcomedusae (Carré and Carré 1980; Mills and Miller 1984), and *H. beehleri* has been collected three times with the siphonophore *Chelophyes appendiculata* in its stomodaeum (Fig. 6.6(a)). Thus, species with simple tentacles eat tentaculate prey, while *L. pancerina* eats non-tentaculate prey. Paired simple tentacles would appear to be of limited usefulness for the capture of non-tentaculate prey, so it is likely that the Bathyctenidae also eat medusae, siphonophores, or other tentaculate ctenophores. How-

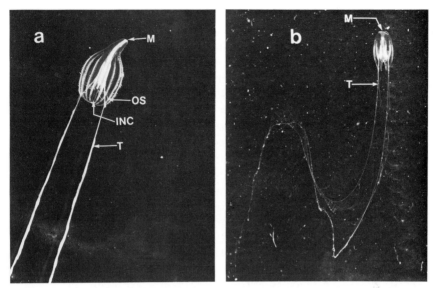

Fig. 6.7. Two spin-ingesting cydippids. (a) *Hormiphora* sp. (Pleurobrachiidae) can make the same movements with its mouth as haeckeliids, but these movements are restricted by the rigid mesogloea that fills most of the body. Length *c*.15 mm. (b) *Callianira bialata* (Mertensiidae) opening its mouth to its widest extent. This fast-moving ctenophore is strongly compressed in the stomodaeal axis. Length *c*.25 mm.

ever, *Aulococtena acuminata* has curious knobs on the ends of its tentacles which may function as lures. Thus, members of the second group resemble beroids, which are known to include salps and other ctenophores in their diet (Hernandes-Nicaise, in Greve, Stockner and Fulton 1976; Harbison, Madin, and Swanberg 1978).

Relationships within the Ctenophora

Members of the Beroida resemble members of the Cestida, Ganeshida, and Lobata in having circumoral connections of the paragastric and meridional canals. Most species of beroids have diverticulae on the meridional canals, as do members of the Platyctenida, the Bathyctenidae, and one family of lobates, the Ocyropsidae. Although Mayer (1912) considers that the circumoral connections of the meridional and paragastric canals indicate a close relationship between the Beroida and the Cestida, I think that both the circumoral connections and the diverticulae are simply means of achieving an improved circulation rather than indicators of a phylogenetic relationship. Small or inactive species and early developmental stages often lack them. For example, while diverticulae are present on the substomodaeal meridional canals of *Ocyropsis maculata*, they are wanting in the much smaller *O. crystallina* (Fig. 6.8). The sluggish *Thalassocalyce inconstans*, which is closely allied with the Lobata, has blindly ending paragastric canals. Therefore, I regard the oral connections of the meridional and paragastric canals, and the presence or absence of diverticulae radiating from these canals, as being of limited taxonomic usefulness.

There are two other similarities between beroids and lobates. One lobate, *Ocyropsis crystallina*, has no trace of a tentacular apparatus as an adult (Fig. 6.8(b)). However, even in adult *O. maculata*, the tentacular canal and reduced tentacle bulb are clearly visible (Fig. 6.8(a)), and the juveniles have functional tentacles. Komai (1963) argues that this shows a tendency within the Ctenophora for the reduction of tentacles, and thus the Beroida should not be placed in a separate class. However, beroids and ocyropsids feed in entirely different ways. Beroids are active predators, seeking out prey, probably using chemotaxis (Swanberg 1974), but ocyropsids are ambush predators which trap animals that blunder into their oral lobes (Harbison, Madin, and Swanberg 1978). Beroids never have functional tentacles, even as juveniles (Agassiz 1874; Chun 1880; Metschnikoff 1885). Chun (1880) and Agassiz (1874) find no trace of tentacles or tentacular rudiments in developing beroids, but Metschnikoff (1885) reports that a cluster of cells homologous to the tentacle primordia of *Callianira bialata* is present in embryos of *Beroe*. The illustration he gives of these cells is not convincing, and he could not follow the fate of these cells, as the embryo became too opaque. Further studies are necessary to decide between these conflicting observations.

Beroids also resemble lobates and cestids in that all are flattened in the tentacular axis. Mayer (1912), among others, regards this (together with similarities in the circumoral connections) 'as an indication that the Beroida and Cestidae are related and derived from a common stock.' However, cestids and beroids also feed in entirely different ways. Cestids move through the water, streaming tentilla over the sides of their bodies (Fig. 6.5(a)), which serve to entrap small organisms, in a manner similar to the lobates (Fig. 6.5(b)). Both are flattened, because both move through the water while feeding. Flattening is one way to develop a streamlined shape. The rapidly-moving mertensiids are also extremely flattened (Fig. 6.7(b)), but in the opposite axis. In slow-moving cydippids, the body is nearly cylindrical. Mertensiids, which spin to ingest prey, must be flattened in the stomodaeal axis. Lobates and cestids, which stream tentilla over the sides of their bodies from the sides of an elongate oral edge, must be flattened in the tentacular axis. Beroids could have become flattened in either axis. That they are flattened in the same axis as cestids is, in my opinion, due to chance rather than an indication that they are closely related. In

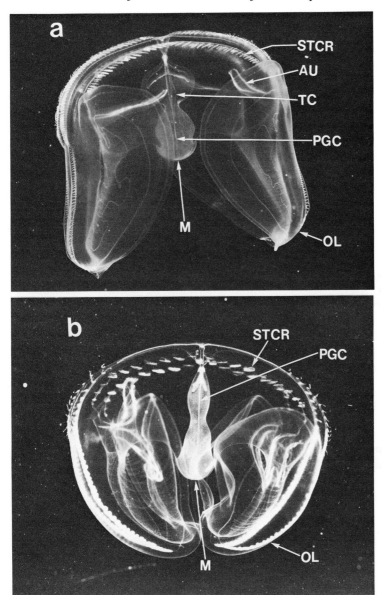

Fig. 6.8. Lobata. (a) *Ocyropsis maculata* (Rang 1828) has rudimentary tentacles near the centre of its hourglass-shaped stomodaeum. The tentacular canal can be clearly seen. Stomodaeal length *c.*45 mm. (b) *Ocyropsis crystallina* (Rang 1828) has no trace of tentacular canals. Stomodaeal length *c.*20 mm. Note also that diverticulae of the meridional canals are seen as faint lines under the comb rows of the meridional canals of *O. maculata*, but are not present in *O. crystallina*.

fact, very young larvae of cestids and lobates are slightly flattened in the stomodaeal axis, which others (e.g. Chun 1880) have taken as evidence for a close relationship with the cydippids.

The animals that most closely resemble beroids are members of the Haeckeliidae, particularly *Haeckelia beehleri* (Fig. 6.6). It makes all of the body movements observed in beroids, even to the extent of turning itself inside out when repeatedly irritated. The aboral papillae bear a striking resemblance to those of *Beroe*. *H. beehleri* resembles a beroid in overall appearance so closely that it can easily be misidentified when superficially examined. Members of the Bathyctenidae also resemble beroids. Their simple tentacles, exiting close to the mouth, would also appear to ally them closely to the Haeckliidae. However, the best-described bathyctenid, *Aulococtena acuminata*, differs in one important way from haeckeliids. In the Haeckeliidae, the tentacle bulb is supplied by canals originating on the subtentacular adradial canals (Fig. 6.6(d)). In *A. acuminata*, the subtentacular adradial canals originate from a common canal that also supplies the tentacle bulb (Fig. 6.2(c); Mortensen 1913). A similar condition is found in *Callianira bialata* (Mertensiidae) and in most platyctenes (e.g. Fricke and Plante 1971). Most platyctenes and all mertensiids differ in two important ways from bathyctenids: their tentacles have side branches, and usually exit aborally from the tentacle sheaths. Some platyctenes have aboral papillae that resemble those of beroids, while mertensiids have pole plates without these papillae. Also, lampeids creep on their widely expanded pharynges, as do platyctenes. Although beroids and haeckeliids also expand their mouths widely, neither exhibits this creeping behaviour. Lampeids differ from platyctenes in that they have a separate tentacular canal, and resemble beroids in that all of the adradial canals arise directly from the infundibulum.

The last family of cydippids, the Pleurobrachiidae, also has a separate tentacular canal, but the adradial canals are united by an interradial canal (Fig. 6.2(a,b)). Interradial canals and separate tentacular canals are found in all members of the Lobata, Ganeshida, Thalassocalycida, and Cestida. I consider this as evidence that these four orders are more closely allied to the Pleurobrachiidae than they are to other families of cydippids. During the development of both lobate and cestid ctenophores, in the earliest stages, tentacles exit aborally, as in the Pleurobrachiidae and Mertensiidae.

Cestids resemble lobates in several ways: compression of the tentacular axis, with a corresponding decrease in the length of the subtentacular comb rows; an orally-located tentacle bulb with an elongate oral groove, fringed by tentilla; and both have the interplate ciliated groove (Chun 1880), which is also found in *Thalassocalyce inconstans* (S. L. Tamm, personal communication). Cestids have no trace of auricles or oral lobes, however, and lobates lack a tentacle sheath.

Both the Ganeshida and Thalassocalycida appear to be more closely allied to the Lobata than to the Cestida. Species of *Ganesha* resemble juvenile lobates so closely that they may be only the developmental stage of a lobate. They differ from any described developmental stage in that the substomodaeal and subtentacular meridional canals are connected by a circumoral canal. *Thalassocalyce inconstans* is also closely allied to the Lobata. It differs from a lobate ctenophore in that the paragastric canals end blindly, and there is a single, medusa-like bell, rather than oral lobes and auricles.

In summary, the various families of cydippids often seem to be more closely allied to members of other orders than to one another. The Haeckeliidae appear to be most closely related to the Beroida and Bathyctenidae. The Bathyctenidae, in turn, show affinities with the Mertensiidae and Platyctenida. The relationships of the Lampeidae are obscure, but they seem closest to the Pleurobrachiidae, yet resemble the Platyctenida in behaviour. The Cestida and Lobata have developmental stages resembling the Pleurobrachiidae. The Ganeshida and Thalassocalycida appear to be the most closely related to the Lobata, and both resemble the developmental stages of some lobates.

Evolution within the Ctenophora

The fundamental problem in determining relationships within the Ctenophora is in establishing directions of evolutionary change. Although this problem is certainly not unique to the Ctenophora, it is complicated by the virtual absence of a fossil record and by our limited knowledge of the physiology and functional morphology of ctenophores. Although development provides many clues to phylogeny, the ontogeny of very few species has been studied. Therefore, the central problem remains as to what characters are primitive and what characters are derived. Such questions as: Did the first ctenophores have tentacles? or Is the absence of side branches on the tentacles a primitive or derived condition in the Haeckeliidae? cannot be answered, given present information. Rather, I present two cladograms that answer those questions in opposite ways (Fig. 6.9), and lean toward the alternative that the first ctenophores had no tentacles (Fig. 6.9(b)).

The simplest evolutionary scenario regards the first ctenophores as organisms resembling present-day beroids; simply ciliated sacks, with modified cilia for grasping and macerating prey. This animal would have fed on other gelatinous planktonic animals. The development of tentacles led to a drastic alteration in behaviour; prey capture became a

passive process. The first tentaculate forms resembled Recent Haeckeliidae, and probably fed on tentaculate prey such as jellyfish. The development of tentilla led to an extensive series of adaptive radiations,
since side branches on the tentacles allowed prey of any size to be captured with greater efficiency. The hypothesis that the first ctenophores
fed on other gelatinous organisms has been suggested by Bonik *et al.*
(1976), and the general outline of this scenario (Fig. 6.9(b)) is in basic
agreement with their theory.

The alternative scenario assumes that the first ctenophores had
tentacles with tentilla (Fig. 6.9(a)). Two lines became established; one
consisting of animals feeding on gelatinous prey and the other consisting of animals eating smaller prey. In the first line, the tentacular
apparatus became progressively reduced and eventually disappeared
altogether in the Beroida. The major difficulty with this scenario, and
the reason I prefer Fig. 6.9(b), is that the chemotactic prey-seeking
behaviour of beroids (Swanberg 1974) has not been reported for other
ctenophores. Prey-seeking demands an effectively coordinated system
of receptors and effectors, and it appears to me that the selective pressure for their development is lacking in the tentaculate feeders. The
tentaculate forms which move through the water to feed (the Lobata
and Cestida) all appear to be non-perceptual feeders; the prey is simply
entrapped when it blunders into the tentilla streaming over the sides of
the animals.

The greatest uncertainty in both schemes is in the placement of the
Lampeidae. In this family, all adradial canals arise directly from the
infundibulum, as in the Beroida and Haeckeliidae, but I have considered the existence of a separate tentacular canal as of greater importance. This leads to the conclusion that spin ingestion must have
evolved twice. A possible solution to this problem could be to place the
Lampeidae between the Haeckeliidae and the Bathyctenidae, and to
assume that simple tentacles arose independently in the two groups.
However, it would then also have to be concluded that a separate
tentacular canal evolved independently in the Lampeidae and the Pleurobrachiidae. I feel that the similarities between the Haeckeliidae,
Bathyctenidae, and Beroida are sufficiently strong that they should be
more closely allied to one another than to the Lampeidae.

The relationships between the Pleurobrachiidae, Thalassocalycida,
Ganeshida, and Lobata are the best established part of both schemes.
The developmental studies of Chun (1880) on *Leucothea multicornis*
(Quoy and Gaimard 1824) show that this animal passes sequentially
through stages resembling the adults of these groups.

Regardless of which of these two phylogenetic scenarios is preferable,
or whether some third alternative is chosen (which is quite likely), it
must be concluded that the Cydippida do not constitute a monophyletic

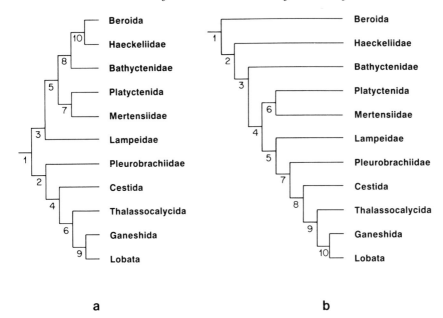

Fig. 6.9. Two attempts at a phylogenetic ordering of monophyletic groups in the Ctenophora. The numbers at the branch points are listed below:

(a) Synapomorphic characters. 1. Comb rows, statocyst, blindly ending meridional canals, tentacles with side branches, colloblasts, pole plates lacking papillae on their borders. 2. Interradial canals, tentacles exiting aborally. 3. Adradial canals arising directly from infundibulum, tentacles exiting orally. 4. Tentilla forming fringe along elongate oral margin, interplate ciliated groove. 5. Subtentacular adradial canals as in Fig. 6.2(c), papillae on margins of pole plates. 6. Enlarged oral capture surface, loss of tentacle sheaths. 7. Tentacles exiting aborally. 8. Loss of tentilla. 9. Oral capture surface divided into two lobes. 10. Adradial canals arising directly from infundibulum.

(b) Synapomorphic characters. 1. Comb rows, statocyst, blind meridional canals, pole plate with papillae on margins, adradial canals arising directly from infundibulum, no infundibular canal. 2. Simple tentacles exiting from sheaths near mouth, tentacular canals as in Fig. 6.6(b), infundibular canal. 3. Subtentacular adradial canals as in Fig. 6.2(c). 4. Tentacles with side branches, colloblasts? 5. Tentacular canal arising from infundibulum. 6. Tentacles exiting from sheaths aborally. 7. Tentacles exiting from sheaths aborally, interradial canals. 8. Tentilla forming fringe along elongate oral margin, interplate ciliated groove. 9. Enlarged oral capture surface, loss of tentacle sheaths. 10. Oral capture surface divided into two lobes.

group. At this stage in our knowledge, however, it seems preferable not to propose a drastic alteration of the presently recognized taxa. Too many groups are too poorly known for any such classification scheme to be very robust. I hope that this attempt at a phylogenetic classification of the Ctenophora will serve as a stimulus for others to falsify or improve it, so that an understanding of evolution within the group will be improved.

Acknowledgements

I thank R. W. Gilmer for his assistance. Figures 6.4(a), 6.7(a,b), 6.8(a,b) (photographs) by Martin Jones. Figure 6.5 (photograph) by John Carlton. This research was supported by the Australian Institute of Marine Science, the Woods Hole Oceanographic Institution, and NSF Grant No. OCE-8209341. This is contribution 5607 from the Woods Hole Oceanographic Institution.

References

Agassiz, A. (1874). Embryology of the ctenophorae. *Mem. Am. Acad. Arts Sci.* 10, 357-98.

Agassiz, L. (1860). *Contributions to the natural history of the United States of America*, Vol. 3. Little, Brown & Co., Boston.

Bonik, K., Grasshoff, M., and Gutmann, W. F. (1976). Die Evolution der Tierkonstrucktionen II. Vielzeller und die Evolution der Gallertoide. *Natur Mus.* 106, 132-43.

Carré, C. and Carré, D. (1980). Les cnidocysts du cténophore *Euchlora rubra* (Kölliker 1853). *Cah. Biol. mar* 21, 221-26.

Chun, C. (1880). Ctenophoren des Golfes von Neapel. *Fauna Flora Golf. Neapel* 1.

—— (1889). Bericht über eine nach den canarischen Inseln im Winter 1887/88 ausgeführte Reise. *Sber. preuss. Akad. Wiss.* for 1889, 519-53.

—— (1892). Die Dissogonie. Eine neue Form der geschlechtlichen Zeugung. *Separat-abdruck aus der Festschrift zum 70st Geburtstage Rudolf Leuckarts.* W. Englemann, Leipzig.

—— (1898). Die Ctenophoren der Plankton-Expedition. *Ergebnisse der Plankton-Expedition* 2. Lipsius und Tischer, Kiel.

Dawydoff, C. (1903). *Hydroctena salenskii* (Etude morphologique sur un nouveau coelentére pélagique). *Zap. imp. Akad. Nauk* (Classe phys.-math.) (8ème Sér.) 14, 1-16.

—— (1946). Contribution à la connaissance des cténophores pélagiques des eaux de l'Indochine. *Bull. Biol. Fr. Belg.* 80, 113-70.

—— (1950). La nouvelle forme de cténophores planarisés sessiles provenant de la Mer de Chine Méridionale (*Savangia atentaculata* nov. gen. nov. spec.). *C.r. hebd. Séanc. Acad. Sci., Paris* 231, 814-16.

Delle Chiaje, S. (1841). *Descrizioni e notomia degli animali invertebrati della Sicilia citeriore*, Vol. 4. Stamperia Societá Tipografica, Napoli.

Fricke, H. W. and Plante, R. (1971). Contribution a l'étude des ctenophores platycténides de Madagascar: *Ctenoplana* (*Diploctena* n.s. gen.) *neritica* n. sp. et *Coeloplana* (*Benthoplana* n.s. gen.) *meteoris* (Thiel 1968). *Cah. Biol. mar.* 12, 57-75.

Gegenbaur, C. (1856). Studien über Organisation und Systematik der Cteno-phoren. *Arch. Naturgesch.* 22, 163-205.

—— Kölliker, A., and Müller, H. (1853). Bericht über einige im Herbste 1852 in Messina angestellte vergleichend-anatomische Untersuchungen. *Z. wiss. Zool.* 4, 299-370.

Greve, W., Stockner, J., and Fulton, J. (1976). Towards a theory of speciation in *Beroe*. In *Coelenterate ecology and behavior* (ed. G. O. Mackie) pp. 251-58. Plenum Press, New York.

Hadzi, J. (1953). An attempt to reconstruct the system of animal classification. *Syst. Zool.* 2, 145-54.

Haeckel, E. (1896). *Systematische Phylogenie der wirbellosen Thiere* (*Invertebrata*). G. Reimer, Berlin.

Harbison, G. R. and Madin, L. P. (1982). Ctenophora. In *Synopsis and classification of living organisms.* (ed. S. P. Parker) Vol. 1, pp. 707-15. McGraw-Hill, New York.

—— and Swanberg, N. R. (1978). On the natural history and distribution of oceanic ctenophores. *Deep Sea Res.* 25, 233-56.

Hardy, A. C. (1956). *The open sea, its natural history.* Houghton Mifflin, Boston.

Hatschek, B. (1911). *Das neue zoologische System.* W. Engelmann, Leipzig.

Hyman, L. H. (1940). *The invertebrates.* Vol. 1. *Protozoa through Ctenophora.* McGraw-Hill, New York.

—— (1951). *The invertebrates.* Vol. 2. *Platyhelminthes and Rhyncocoela. The acoelomate Bilateria.* McGraw-Hill, New York.

Komai, T. (1922). *Studies on two aberrant ctenophores, Coeloplana and Gastrodes.* Published by the author, Kyoto.

—— (1963). A note on the phylogeny of the Ctenophora. In *The lower metazoa: comparative biology and phylogeny* (ed. E. C. Dougherty) pp. 181-88. University of California Press, Berkeley, Ca.

—— and Tokioka, T. (1942). Three remarkable ctenophores from the Japanese seas. *Annot. zool. jap.* 21, 144-51.

Korotneff, A. (1888). *Cunoctantha* und *Gastrodes. Z. wiss. Zool.* 47, 650-7.

Krumbach, T. (1925). Erste und einzige Klasse der Acnidaria. Vierte Klasse des Stammes der Coelenterata. Ctenophora. *Handbuch Zool.* 1, 905-95.

Künne, C. (1939). Die *Beroe* der Südlichen Nordsee, *Beroe gracilis* n. sp. *Zool. Anz.* 127, 172-4.

Lang, A. (1884). Die Polycladen (Seeplanarien) des Golfes von Neapel und der angrenzenden Meerabschnitte. *Fauna Flora Golf. Neapel* 11, 1-688.

Madin, L. P. and Harbison, G. R. (1978). *Thalassocalyce inconstans*, new genus and species, an enigmatic ctenophore representing a new family and order. *Bull. mar. Sci.* 28, 680-7.

Mayer, A. G. (1912). *Ctenophores of the Atlantic coast of North America.* Carnegie Institute Publications, Washington, D.C., No. 162.

Metschnikoff, E. (1885). Vergleichend-embryologische Studien. 4. Über die Gastrulation und Mesodermbildung der Ctenophoren. *Z. wiss. Zool.* 42, 648-56.

Mills, C. E. and Miller, R. L. (1984). Ingestion of a medusa (*Aegina citrea*) by the nematocyst-containing ctenophore *Haeckelia rubra* (formerly *Euchlora rubra*): phylogenetic implications. *Mar. Biol.* 78, 215-21.

Mortensen, T. (1912). Ctenophora. *Danish Ingolf-Exped.* Vol. 5, Part 2, H. Hagerup, Copenhagen.

—— (1913). Ctenophora. *Rep. Sci. Results Michael Sars Exped.*, 1910 Vol. 3, Part 2, 1-9.

Moser, F. (1908). Cténophores de la Baie d'Amboine. *Rev. suisse Zool.* 16, 1-26.

Pianka, H. D. (1974). Ctenophora. In *Reproduction of marine invertebrates.* (eds A. C. Giese and J. S. Pearse) pp. 201-65. Academic Press, New York.

Quoy, J. R. C. and Gaimard, J. P. (1824). Zoologie, deuxième partie (pp. 409-61). In *Voyage autour du monde, fait par ordre du Roi, sur les corvettes de S.M. D'Uranie et la Physicienne, pendant les années* 1817, 1818, 1819 *et* 1820, (author L. de Freycinet). Pillet Aine, Paris.

Rang, M. (1828). Établissement de la famille des Béroïdes dans l'ordre des Acalèphes libres, et description de deux genres nouveaux qui lui appartiennent. *Mém. Soc. Hist. nat. Paris* 4, 166-73.

Schneider, K. C. (1902). *Lehrbuch der vergleichenden Histologie der Tiere.* G. Fischer, Jena.

Stanley, G. D. and Stürmer, W. (1983). The first fossil ctenophore from the Lower Devonian of West Germany. *Nature, Lond.* 303, 518-20.

Swanberg, N. (1974). The feeding behavior of *Beroe ovata. Mar. Biol.* 24, 69-74.

Tamm, S. L. (1982). Ctenophora. In *Electrical conduction and behaviour in 'simple' invertebrates* (ed. G. G. Shelton) pp. 266-358. Clarendon Press, Oxford.

7. The phylogenetic status of the acoelomate organization within the Bilateria: a histological perspective

REINHARD M. RIEGER

Department of Biology and Curriculum of Marine Sciences, University of North Carolina, Chapel Hill, North Carolina, USA

Introduction

Despite the diverse views about the origin of the Bilateria, their ancestral organization is currently viewed either as acoelomate or coelomate. Acoelomate concepts of bilaterian origin predict endomesoderm of a mesenchymal origin with the muscles and connective tissue which fill the space between epidermis and gut differentiating from this mesenchyme. In contrast, coelomate concepts argue for an epithelial origin of the endomesoderm from the myoepithelium in gastric pockets of an ancestral radiate. The impact of electron microscopy on our understanding of the histological organization of the mesoderm and its derivatives in adults is recent but already significant, and has produced new evidence for a coelomate origin of the acoelomate Bilateria (Platyhelminthes, Nemertea, Gnathostomulida). A review of recent histological and ultrastructural studies of the parenchyma, tissues filling the body cavity between body wall and organ systems, suggests a significant heterogeneity. As long as we lack a unifying pattern for the structure and organization of the parenchyma in these supposedly plesiomorphic acoelomate phyla, convergence in that character remains a probability. This is corroborated by the occurrence of acoelomate-like organization in the forms derived from coelomate groups (e.g. Hirudinea, interstitial Annelida, Enteropneusta), because these groups illustrate a similar picture of convergent evolution of acoelomate organization. If the acoelomate phyla are envisaged as being derived in convergent lines of evolution from a coelomate stock, their origin from larval and juvenile stages of coelomates through progenesis would appear to be a possibility. This paper discusses the significance of parenchymal tissues and their origins in the different evolutionary lines leading to acoelomates.

Historical review

The foundation of our present understanding of the basic design of bilaterian (triploblastic) phyla was laid in the second half of the last century through the advent of light histology (Hertwig and Hertwig 1882). During this century, a major contribution to the issue was Hyman's (1951) concise synthesis of the literature and her clear recognition of three basic types of body cavity within the Bilateria: the acoelomate, pseudocoelomate, and coelomate conditions.

In spite of the bewildering diversity of views on the origin of the Bilateria, only two alternative body cavity organizations are discussed today as possible ancestral conditions: either an acoelomate planuloid or gallertoid, or an oligomerous coelomate (Clark 1979; Gutmann 1981). Gutmann's (1981) gallertoid proposal was primarily derived from biomechanical considerations: as such it represents a valid systematic hypothesis about the plesiomorphic characters of Bilateria. However, for a rejuvenation of Lang's theory (the ctenophore origin of the Bilateria) new intermediate forms would be needed to provide homologies between the ctenophores and the lower Bilateria: see Hyman (1951) for discussion of Lang's theory. For the remaining two alternatives of the bilaterian stem group, Clark (1964) has convincingly demonstrated from a biomechanical analysis that only the acoelomate planuloid can be easily transformed into a coelomate bilaterian. A biomechanical analysis of the oligomerous coelomate theory usually involves the discrepancy of the evolution of large fluid-filled cavities in small organisms moving by ciliary swimming or ciliary gliding (Clark 1979).

Recent advances in our understanding of the histological organization of extant acoelomates clearly widens the gap between acoelomate flatworms and the cnidarian planula, both of which are seen as important links in the transformation sequence from the radiate to the bilaterian organization in hypotheses concerning acoelomate origins (Smith and Tyler, this volume). On the other hand, the discovery of myoepithelial organization in part of the mesoderm in various coelomates (e.g. Gardiner and Rieger 1980; Wood and Cavey 1981), as well as the recognition of acoelomate conditions derived from the coelomate organization in interstitial Annelida (Fransen 1980), strengthen the hypothesis that a coelomate construction is the plesiomorphic condition in Bilateria.

These recent advances are primarily a function of the higher resolving power of the electron microscope. This allows the identification of many new features and, even more importantly for the present issue, the unambiguous identification of special relationships between the different cells and their matrices. The lack of such detail, together with the scarcity of intermediate forms at the level of phyla, was certainly the main cause for the heated phylogenetic rebuttals between leading systematists (such as Hyman, Remane, Marcus, Steinbock, Hadzi, and Hanson) during the late fifties and early sixties. Today, the lack of understanding of the methodological relationship between the search for the extent of character homologies (cladistic analysis) and the limitations set by functional design and material properties (functional morphology) hinder further advances in this broad area. The two separate approaches could possibly be used as a reciprocal procedure for the same data set (Rieger and Tyler 1979) and could thus enable us to test cladistic analysis with functional morphology For it is clear that ancestral species must have been functional organisms just as all functional (e.g. biomechanical) models must satisfy comparative morphological criteria.

In order to understand clearly the controversy between the schools arguing for the oligomerous organization of the bilaterian stem group and those claiming an acoelomate origin of the Bilateria, it is necessary to identify the profound histological differences in tissues derived from the endomesoderm (true mesoderm) as postulated by the two theories (Fig. 7.1(a)). While the ectomesoderm (= ectomesenchyme, Starck and Siewing 1980) generally retains a simple connective tissue type organization, the endomesoderm differentiates into both extremes of tissue types within the Bilateria: as a connective tissue of cells aggregated randomly with respect to their own polarity and embedded in a matrix; or as an epithelial tissue of cells aggregated with the individual axis of polarity being aligned and pointing in the same direction and with the matrix (if present) restricted to the apical (cuticles) and basal (basement membrane) sides of the cell (see also Beklemischev 1969). With a coelomate origin of the Bilateria one postulates an epithelial organization of the primitive endomesoderm (derived from a myoepithelium of gastric pockets in the Radiata), but with an acoelomate origin one argues for a connective tissue organization as the plesiomorphic condition of the bilaterian endomesoderm.

The main corollary of this controversy is the following question: Is the endomesodermal muscle tissue in the original Bilateria organized as a myoepithelium or is it in the form of fibre-type muscle cells embedded within the matrix of connective tissue? Both conditions can be found in the plesiomorph outgroup, the Radiata, and therefore we cannot decide on the outgroup principal (Remane 1952) which of the two is more

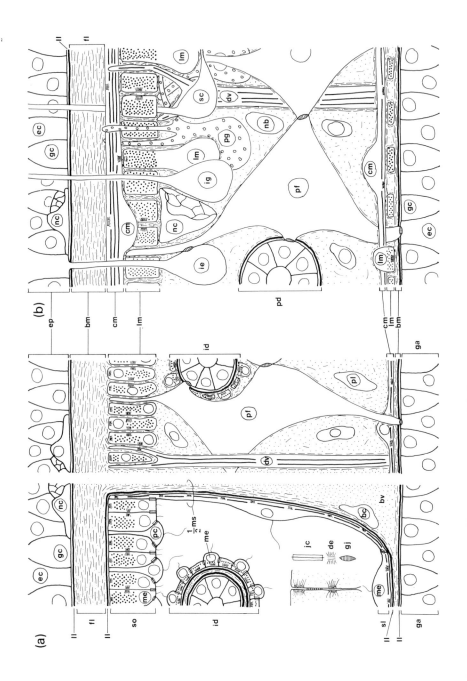

Fig. 7.1. (a) Schematic cross section of body cavity in mid-body region, according to coelomate (left) or acoelomate (right) origin of the Bilateria. Inset at lower left illustrates graphical symbols used for identification of different cell junctions. (b) Model of matrix configuration and common cell types as seen in body cavity of many Platyhelminthes in the polyclad-neoophoran line of evolution. Abbreviations: bc, blood cell; bm, basement membrane; bv, blood vessel; cm, circular muscle cells; de, desmosomes (spot configuration); dv, dorsoventral muscle cell; ec, regular epithelial cell ep. epidermis; fl, fibrillar layer of basement membrane; ga, gastrodermis; gc, gland cells; gj, gap junctions (spot configuration); id, invagin-ated tubular ducts (e.g. from excretory system, reproductive system or as tubular glands); ie, invaginated epidermal cells; ig, invaginated gland cells; jc, junctional complex (belt morphology, usually zonula adhaerentes and septate junction in invertebrates); ll, limiting layer of basement membrane; lm, longitudinal muscle cells; me, myoepithelial cells; ms, mesentery; nb, neoblasts; nc, nerve cells; pc, peritoneal cells; pd, protonephridial duct; pf, fixed parenchyma cells; pg, pigment cells. pl, free parenchyma cells; sc, primary sensory cell; sl, slanchnopleura. so, somatopleura.

primitive. The ctenophores exhibit a connective tissue type muscle organization, the Cnidaria possess myoepithelia (Hertwig and Hertwig 1882). One possible future test could be the finding of structural and functional differences between the muscles derived from ectoderm (epidermis of Cnidaria, muscles of Ctenophora, some muscles of protostome larvae derived from ectomesenchyme) and muscles derived via the endoderm (most bilaterian musculature, cnidarian endodermal muscles).

The basic concept of acoelomate connective tissue

Since Hyman (1951), the acoelomate body cavity has traditionally been associated with three bilaterian phyla: Platyhelminthes, Nemertea, and Gnathostomulida. This paper neither addresses the protracted debate about the acoelomate/coelomate origin of the Mollusca nor does it attempt to review critically the phylogenetic relationship of the above-listed acoelomate phyla with the pseudocoelomates, which show in several cases an acoelomate-like organization (See Salvini-Plawen 1980).

In the Platyhelminthes, three main lines of evolution are distinguishable: the nemertodermatid-acoel line, the catenulid line, and the line leading to the remaining orders (Rieger 1981a). The latter line shows a clear division into the haplopharyngid-macrostomid and polyclad-neoophoran subgroups. As demonstrated by Ehlers (this volume), trematodes and cestodes are side branches within the polyclad-neoorphoran line of free-living Turbellaria. Recognizing these three evolutionary lines in the Platyhelminthes is important in the discussion of the variations in acoelomate organization because the apomorphies employed to link the lines are not securely based and are open to controversy. It is also important to remember that we do not yet know from other evidence on morphology, life cycles, etc. any intermediates that link the organization of the Nemertea and Gnathostomulida with that of the three platyhelminth lines. Therefore, it appears that the acoelomate body plan represents an important unifying character set for these phyla. The question then is: can one detect a unifying principle in the body cavity organization which could be interpreted as homologous among these acoelomate groups?

The acoelomate organization has traditionally been conceived as a connective tissue (parenchyma or mesenchyme) containing two classes of cells: fixed parenchyma cells and free parenchyma cells (Prenant 1922; see also Fig. 7.1(b)). This tissue exhibits all gradations from a matrix-free cellular connective tissue (e.g. some Acoela) to a more vertebrate-like connective tissue with ample amounts of intercellular matrix (e.g. certain polyclads and Nemertea; Pedersen 1961, 1964, 1966, 1968). The intercellular matrix generally consists of two components: the ground substance which is largely composed of aminated, sulphated, carboxylated, or phosphorylated polysaccharides attached to proteins; and the fibrillar system of 80-100 nm thin filaments, which are probably collagenous (Curtis, Cowden, Moore, and Robertson 1983). Pedersen (1961, 1966, 1968) and others have clearly shown that the fibrillar system is continuous with the fibrillar layer of the body wall basement membrane and, if present, the basement membrane on the gut. The limiting layer of the basement membrane is chemically similar to the ground substance of the parenchyma in that both are polysaccharide-protein complexes (e.g. Pedersen 1966).

The fixed parenchyma cells are usually large cells provided with numerous cell processes contacting the gut, body wall, and other organ systems. Two main trends in cytological differentiation of this cell type are seen. In many cases their cytoplasm appears homogeneous, fluid-like, and often contains large amounts of glycogen. These cells we here designate as the typical parenchyma cells. In many parenchyma cells extremely large intracellular vacuoles occur. These cells are identified as chordoid cells and might be another cytological differentiation of the fixed parenchyma cell type (Rieger 1981a). A derivative of such vacuolated cells could possibly be the calcareous corpuscle cells in Cestoda (Hess 1980).

That free parenchyma cells form a set of morphologically and histologically diversified cells is evident from cell dissociations reported in the older literature (Prenant 1922). One of these types, clearly identifiable with EM, is a small, round basophilic cell with a high N/C ratio, and with a cytoplasm containing numerous free ribosomes and chromatoid bodies. Endoplasmic reticulum is characteristically lacking in this cell type (Pedersen 1972; Hay and Coward 1975). Hay and Coward claim that a large number of similar basophilic cell types identified by maceration techniques and light histological procedures may actually be the plasmatic portion of muscle cells and certain gland cells: see also Pedersen (1976) for discussion on the neoblast concept.

Although controversies exist as to the name, function, and origin of these cells, it is safe to say that they are important both for regular tissue growth and for regeneration within triclad flatworms. Because of priority, they will here be called neoblasts. Some evidence suggests that the typical parenchymal cell can dedifferentiate into neoblast cells (Bowen, Ryder, and Dark 1976). Morphological evidence shows that neoblasts can differentiate into a variety of tissues (e.g. muscle cells: Pedersen 1972; Hori 1978; epidermal cells: Pedersen 1972; Hori 1978).

Besides the fixed and free parenchyma cells already distinguished in the technically excellent study by Prenant (1922), two more common

cell types must be considered as general features of the acoelomate connective tissue. These are parenchyma pigment cells, usually associated
with the body wall musculature and basement membrane (their origin
from fixed parenchyma cells is unlikely), and the invaginated portions
of epidermal and gastrodermal cells (Rieger 1981a). Although this last
category is not derived from mesoderm, it needs to be considered because
such cells can quantitatively dominate in filling certain body regions in
some groups (see below).

Comparison of acoelomate connective tissue organization

Our understanding of the fine structure of this tissue is very incomplete,
despite the work of Pedersen (1961, 1964, 1966, 1968) and others. The
present summary is given to help test specific hypotheses about the
organization of connective tissues, through careful three-dimensional
reconstructions of each of the major acoelomate groups.

1. The nemertodermatid–acoel line

A unique feature in the body cavity construction of these Platyhelminthes is the lack of an intercellular matrix, even at the interface of
epidermis and body wall musculature (location of basement membrane
in other forms). The report of matrix remnants between muscle cells
and epidermal cells and between muscle fibres and muscle cells in the
body wall of some Nemertodermatida (Smith and Tyler, this volume)
is therefore of considerable phylogenetic interest. The matrix structure
is in this case fibrogranular, but appears to lack filaments larger than
80 nm.

  Within the Acoela an intercellular matrix is only present around the
statocyst. Recent evidence suggests that some (possibly all?) parenchyma cells filling the space between body wall and central digestive
parenchyma can be regarded functionally and phylogenetically as part
of the central digestive parenchyma, which is equivalent to the epithelial gut in other Turbellaria (Smith and Tyler, this volume). While
the complete lack of a separate peripheral parenchyma is likely in
Diopisthoporus and *Paratomella*, other genera show typical fixed parenchyma cells (e.g. *Convoluta*, *Kuma*), or chordoid tissue (e.g. *Oligochoerus*,
Convoluta). The fixed parenchyma cells are often filled with glycogen or
appear empty in transmission electron microscope (TEM) sections.
These cells intimately interdigitate with the epidermis (Fig. 7.2(a)).

  Although the existence of neoblasts in Acoela is suggested by Pedersen's (1964) study on *Convoluta convoluta* as well as by Mamkaev and
Markosova's (1978) study on *Oxyposthia praedator*, such cells have not

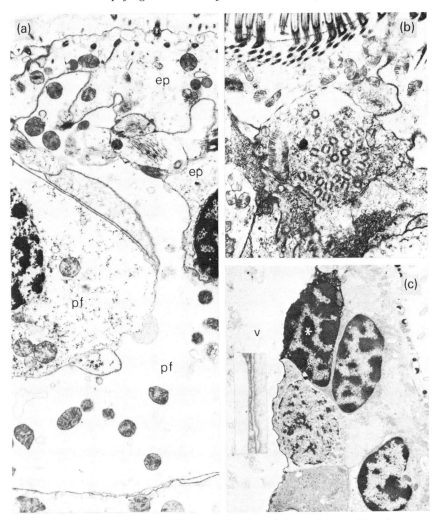

Fig. 7.2. Special features of parenchyma in Acoela and Catenulida. (a) *Solenofilomorpha* sp., cross section: observe interdigitation of parenchyma cell with epidermal cell; × 5800. (b) *Retronectes atypica*, longitudinal section: observe epidermal replacement cell with numerous centrioles at base of pharyngeal epithelium; × 11 700 (courtesy Dr D. A. Doe). (c) *Paracatenula* sp., cross section: observe large vacuole surrounded by thin processes of special parenchyma cell (*); × 6500. Inset: enlargement of thin processes surrounding vacuole. × 28 500. Abbreviations: ep, epidermal cell; pf, parenchyma cell; v, vacuole.

been reported by other investigators. In these species, the cells also appear different from typical neoblast cells described in the triclad literature because they lack chromatoid bodies. On the other hand, unique epidermal replacement cells are found in almost all acoel families. It is still uncertain whether these peculiar cells actually replace epidermal cells or whether they are dedifferentiating epidermal cells (Tyler 1984). 'Parenchyma gland cells' are also reported in certain acoels (e.g. Solenofilomorphidae: Crezee 1975). They are, however, most probably extremely elongated mucous glands, which one would expect to originate in the epidermis or in the pharyngeal epithelium.

A cellular connective tissue as described above for some Acoela has not yet been confirmed with TEM for the Nemertodermatida (Smith and Tyler, this volume). Except for the gonads, the only tissue filling the spaces between cellular gut and epidermis is muscle tissue.

2. The catenulid line

Within the order Catenulida, which is the only order in the phyletic line, a connective tissue as described earlier appears to be present only in the family Catenulidae (Moraczewski 1981). Older light histological studies mention the existence of various free parenchyma cell types (see summary in Borkott 1970). Generally, the lack of connective tissue like that of other flatworms has been stressed by light microscopists (e.g. Borkott 1970). In the Stenostomidae especially the space between body wall and gut appears to be filled with a fluid-like intercellular matrix, found either between radial muscle cells (Borkott 1970) or between coelom-like cavities (Reisinger 1976). Large fluid-filled spaces, possibly of an intercellular nature, are seen in *Rhynchoscolex* (Reisinger 1924) and in *Paracatenula* (Fig. 7.2(c)). Ultrastructural studies show the existence of neoblast-like cells and epidermal replacement cells in the family Catenulida (Moraczewski 1981; see Fig. 7.2(b)). Within the marine family Retronectidae, peculiar 'parenchymal strands' have been described by Sterrer and Rieger (1974). Preliminary TEM investigations of this tissue have revealed elongated tubular cells filled with what appear to be secretory granules and empty intercellular spaces between these cells (Doe and Rieger 1977). No connection to the gastrodermis or epidermis has yet been established, but it is possible that these are also 'parenchyma gland cells' similar to those of Acoela.

Besides the fluid-like intercellular matrix, intercellular matrix is restricted to the infrequent occurrence of very thin basement membranes at the epidermis (Rieger 1981a) and possibly on other epithelial tissues such as gut and excretory organs.

3. The remaining orders

It has been emphasized (Rieger 1981a) that invaginated gland cell bodies appear to be the most common filling tissue in the body cavity of the macrostomid-haplopharyngid line. Only in some forms does one find reports of a significant elaboration of parenchyma cells, which are possibly derivations from the basic fixed parenchyma cell type. In some members of the Dolichomacrostomidae these cells are very large and vacuolated, the vacuoles being intracellular (Rieger 1971; supported by recent unpublished TEM studies). In a new species of *Paramyozonaria*, large 'parenchyma cells' occur. These cells can be seen as unique epidermal gland cells, because they penetrate the epidermis with thin processes (Fig. 7.3(a)). Possibly similar cells, previously identified as fluid-filled parenchyma cells, are found in *Myozona* (Rieger 1981a). Finally, it should be mentioned that epidermal replacement cells with numerous centrioles (as in the Catenulida and some Tricladida) have also been reported from the Haplopharyngida (Rieger 1981a).

An intercellular matrix is present between all cells of the filling tissue. In some cases the matrix is lacking in places between epidermal cells and muscle cells (Doe 1981; Tyler 1984). The matrix is usually electron-dense with a fine granular appearance, sometimes with filaments typical of the size range of the fibrillar component in the matrix of other forms (Fig. 7.3(b)). A distinct limiting layer (*lamina densa*) at the epidermis-muscle interface is either missing or only very weakly developed.

The connective tissue filling the body cavity in the polyclad-neoophoran line, including the Trematoda, Cestoda and Temnocephalida, in general corresponds well with the configuration described above as the basic plan (Fig.7.1(b)). Variations of this arrangement are differences in the amount of intercellular matrix and the presence or absence of typical neoblasts. While the Polycladida and the Lecithoepitheliata generally show an especially well developed matrix, two main trends are discernible in the other free-living flatworm orders. In the Proseriata and Tricladida the matrix is often less well developed but is usually continuous from the muscle cells to the basement membrane on the gut or other internal organs. Usually the fibrillar system is particularly well developed around the subepidermal muscle layer. On the other hand, in most Prolecithophora and Rhabdocoela the matrix is restricted to the basement membrane of the epidermis and to thinner basement membrane on the gut or other internal organs.

In the endoparasitic flatworms, the intercellular matrix (ground substance and fibrillar system) is usually well developed and may be very fluid-like in some Trematoda and Cestoda (Threadgold and Gallagher 1966; Reissig and Colucci 1968; Rohde 1971, 1974; Strong and Bogitsch 1973; Threadgold and Arme 1974a; Matricon-Gondran 1977; Wittrock 1978), whereas the ectoparasitic Temnocephala (Williams 1979) appear to have matrix conditions more similar to the rhabdocoel type described above.

Neoblasts or similar germinal cells are commonly found in the Tricladida and Cestoda (Pedersen 1972; Bonsdorff, Forssten, Gustafsson,

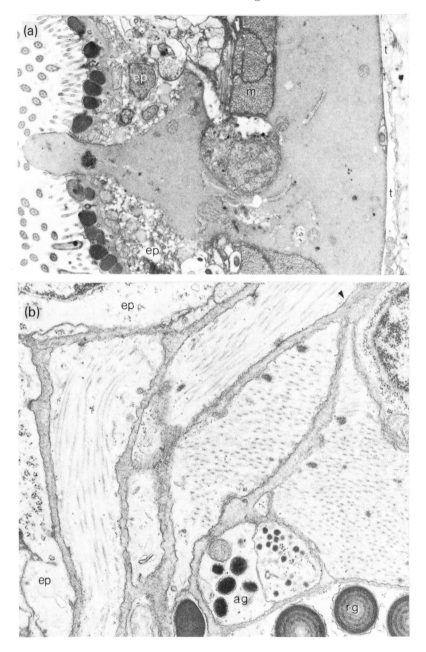

Fig. 7.3. Special features of the parenchyma in the Macrostomida. (a) *Paramyozonaria* sp., cross section: 'parenchyma gland cell' with homogeneous cytoplasm; × 10 545. (b) *Paromalostomum* sp., cross section: intercellular matrix; note fibrillar system (arrow) within intercellular matrix; × 19 950 (courtesy Dr S. Tyler). Abbreviations: ag, adhesive glands; ep, epidermis; m, body wall muscles; rg, rhabdite glands; t, testis lining cell.

and Wikgren 1971; Hess 1980). Pedersen (1966) describes a cell type from polyclad connective tissue which resembles the neoblasts of Tricladida; the homology is, however, uncertain because of a lack of understanding of cytological variations found within this cell type. Typical neoblasts have not yet been reported from Lecithoepitheliata, Rhabdocoela, Temnocephalida, or Trematoda. In the latter group, stem cells with a high N/C ratio and numerous free ribosomes have been described from the excretory ducts (Bennett 1977) and the vitellaria (Irwin and Threadgold 1970). Stem cells are also expected within the subtegumental layer for epidermal replacement, because replacement of the larval epidermis by special cells from the parenchyma is also well established for the Trematoda (Lyons 1977). Similar cells with rhabdoid inclusions are known from growth and regeneration areas of the epidermis in the Tricladida. These 'rhabditogen cells' are probably epidermal replacement cells (Tyler 1984). Finally, it should be mentioned here that MacKinnon, Burt, and Pike (1981) report an epidermal replacement cell in adults of the rhabdocoel *Paravortex*, as well as differentiating 'neoblasts' in the embryo of this species.

Other trends of tissue specialization include the occurrence of fixed parenchyma cells, which are mostly filled with fluid-like cytoplasm. In some forms such cells occupy most of the space between gut and body wall (Fig. 7.4(a)), often contain varying amounts of glycogen (Threadgold and Arme 1974a; Bennett 1977) or appear almost empty in TEM images (Fig. 7.4(b)).

(a) Gnathostomulida

Light histological studies emphasized the general lack of a typical connective tissue in this group (Sterrer 1969); no comparative study is yet available at the EM level. However, Mainitz (1979) states that a parenchyma is mostly lacking in these animals and preliminary studies on *Haplognathia* c.f. *simplex* and on *Semaeognathia sterreri* strongly support this notion (Fig. 7.5). In most regions the large gut cells fill the entire space to the body wall (Fig. 7.5(a,c)). Occasionally, cells which could be interpreted as parenchyma cells are seen at the base of the gut (Fig. 7.5(b)), but serial sections are needed to establish that these are not basal processes of gut cells.

The matrix in Gnathostomulida is restricted to a thin but well developed fibrogranular basement membrane on the epidermis and the foregut and around the reproductive system. In some forms (e.g. *Saemeognathia sterreri*) a similar electron-dense matrix can be observed in intracellular lacunae between the gut and muscle cells (Fig. 7.5(c)).

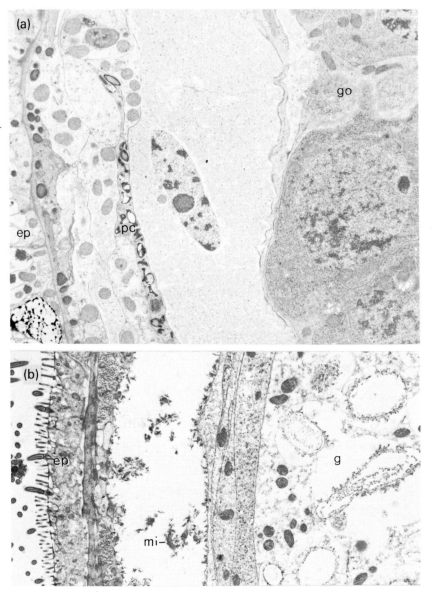

Fig. 7.4. Specialization of fixed parenchyma cells for hydrostatic functions within neoophoran flatworms. (a) *Archimonotresis* c.f. *limophila*, cross section: observe large parenchyma cell with homogeneous cytoplasm; × 6400 (courtesy Dr. S. Tyler). (b) *Neoschizorhynchus parvorostro*, cross section: note empty parenchyma cell; × 6000. Abbreviations: ep, epidermis; g, gut; go, female gonads; mi, mitochondria; pc, pigment cell.

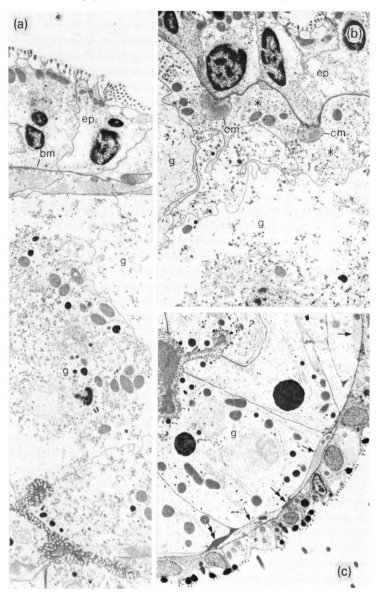

Fig. 7.5. Body cavity organization in mid-body regions of the Gnathostomulida. (a) *Haplognathia* c.f. *simplex*, longitudinal section; × 4675. (b) *Haplognathia* c.f. *simplex*, longitudinal section: note presence of cell body (*) between epidermis and gut and profiles of circular muscle cells; × 4675. (c) *Semaeognathia sterreri*, cross section: again observe the lack of parenchyma cells between gut and body wall musculature (arrows). Note electron-dense matrix (double arrows) filling spaces between gut and body wall; × 3825 (courtesy Dr M. Mainitz). Abbreviations: bm, basement membrane; cm, circular muscle cells; ep, epidermis; g, gut.

(b) Nemertea

According to Hyman (1951) nemerteans represent the culmination of the acoelomate construction. The only TEM study on the connective tissue of the group, by Pedersen (1968), suggests that it is similar to that of Polycladida with respect to the amount of intercellular matrix. The three groups of connective tissue cells distinguished by Pedersen (1968) are difficult to compare with connective tissue cells within Platyhelminthes, because of the polymorphism of nemertean connective tissue cells. The absence of neoblast-like cells from the connective tissue is significant, but similar interstitial cells are found in some species within the epidermis (Pedersen 1968). Recently, Turbeville and Ruppert (1983) have reported the occurrence of 'epidermal basal cells' in the body wall musculature. Should basal cells be derived from interstitial cells, then this would be the first report of the type of cell below basement membrane. Turbeville and Ruppert's study on *Carinoma tremaphoros* also suggests that the intercellular matrix is only weakly developed in this species. Further investigations are therefore likely to yield similar variations in connective tissue structures (cell types, matrix elaboration) as have been described for the Platyhelminthes.

Discussion

The present review indicates that a typical parenchyma cell is lacking at least at the origin in the nemertoderm-acoel line and possibly in the line leading to the Gnathostomulida. The space between gut and epidermis is here primarily filled by muscle cells, a matrix being found only in the Gnathostomulida and in some Nemertodermatida between epidermis and subepidermal musculature.

In the macrostomid-haplopharyngid subgroup it is suggested that the space between gut and body wall is primarily occupied by the invaginated portions of gland cells of the epidermis, foregut, and genital organs. The epidermal replacement cells filled with centrioles in *Haplopharynx* have their counterpart in the epidermal replacement cells of Catenulida and some Tricladida (Smith and Tyler, this volume). While the haplopharyngid-mascrostomid subgroup exhibits a matrix with fibrillar components possibly derived from the basic connective tissue plan, the Catenulida are peculiar because of their tendency to a pseudocoel organization. So far, what would traditionally be considered typical parenchyma of acoelomates appear to be restricted to the polycladid-neoophoran group and to the Nemertea. Even between these two lines of homology of the connective tissue has not yet been demonstrated (Pedersen 1968). Based on the organization of the connective tissue it is therefore difficult at present to link the three main lines of the Platyhelminthes with each other or with the Nemertea and Gnathostomulida. Convincing intermediates are lacking and precise three-dimensional reconstructions of this spatially complicated tissue have yet to be provided.

Within the Platyhelminthes, the findings of this review underscore the existence of three main evolutionary lines as outlined earlier. At least for this phylum it can also be said that size of organisms is related to tissue organization (Rieger 1981a). Smaller forms usually have less matrix than larger ones, although significant exceptions do exist. The fibrillar component of the intercellular matrix appears as an especially important structural element, based on its particular position in the basement membrane and around muscle cells. That the parenchyma constitute a hydrostatic skeleton in the acoelomate organization (Clark 1964) is also corrobarated by the finding of very large cells with essentially fluid-like cytoplasm. Glycogen metabolism (Threadgold and Arme 1974b) and cellular communication (Gallagher and Threadgold 1967; Baron 1968; Matricon-Gondran 1977; Quick and Johnson 1977) seem to be important functions of the fixed parenchyma cell. Some kind of totipotent neoblast cell or specialized stem cell is of widespread occurrence throughout the phylum. The origin of these cells, though not completely clear, is most probably from dedifferentiating somatic or germ cells, as well as from totipotent cells derived from the embryo (Gremigni 1981). Dedifferentiating gastrodermal cells have been described as possible neoblast precursors (Coward 1979; see Hay and Coward 1975 for older literature) and thus endodermàl cells could possibly give rise to epidermal replacement cells; the equipotency of neoblasts has, however, been seriously questioned (Hay and Coward 1975). The migration of endodermal cells into the epidermis is described for several cnidarians (Summers and Haynes 1969; Campbell 1974; Chapman 1974; Fennhoff 1980). Similar migrations of cells from endomesoderm into the epidermis are suspected in some coelomates (Rieger 1981b). Whether there is a common basis for such patterns of cell migrations and the replacement of epidermal cells from the parenchyma in Platyhelminthes must be clarified by future investigations.

As long as a clear homology of the connective tissue in the acoelomate phyla discussed here cannot be established, the possibility will remain that they have reached the acoelomate condition along convergent lines of evolution. Fransen (1980) and Rieger (1980) have produced new evidence that the acoelomate organization can be seen as a derivative of the coelomate body plan. A survey of the occurrence of 'functional acoelomates' (organisms with a body cavity filled with cells and their matrix) in otherwise pseudocoelomate or coelomate phyla is impressive

(Ax 1963): most free-living nematodes lack any large cavity between body wall and gut, and the Nematomorpha-Gordioidea have a paren-chyma-like tissue completely filling the body cavity (Eakin and Brand-enburger 1974). Gastrotricha are functional acoelomates, as has been shown by Rieger, Ruppert, Rieger, and Schoepfer-Sterrer (1974), and entoprocts are often referred to as acoelomates (Salvini-Plawen 1980). However, in all these cases it remains controversial whether the acoelomate condition is primitive or derived.

Within the Annelida the interstitial families (Fransen 1980) and the Hirudinea (Sawyer and Fitzgerald 1981) are good examples of the transformation of the coelomate body cavity into an acoelomate organ-ization. Fransen (1980) demonstrated that small body size and ciliary gliding are closely correlated with the occurrence of the acoelomate condition in interstitial Annelida. Finally, Hyman (1959) stresses the obliteration of coelomic spaces in some Enteropneusta by muscle tissue. This is interesting in light of the above-mentioned possibility that muscle tissue is the only tissue found between epidermis and organ systems in the nemertodermatid flatworms and the Gnathostomulida.

The variety of cellular organization of these functional acoelomates provides examples of how the various tissues derived from endomeso-derm in the true acoelomate phyla could have developed from a coelo-mate stock. Moreover, the mesoderm in larvae and young juveniles of coelomates in the spiralian line is usually described as mesenchymal bands (Starck and Siewing 1980). It has also been suggested (Rieger 1980) that the acoelomates have originated from larvae or juveniles of macrofauna through early maturation of gonads (progenesis). Should this hypothesis be substantiated, it would appear that the mesenchymal tissue organization of the mesoderm in these larval forms would allow one to envisage the origin of acoelomates without having to postulate the reduction of the coelom from the coelomate ancestor.

Acknowledgements

This work has been supported through NSF grant No. DEB 8119652. For critical reviews of various versions of this paper I thank my wife, Dr Gunde E. Rieger, Dr Seth Tyler, Dr Julian Smith III, Dr Wolfgang Sterrer, and Dr Julian Lombardi. Part of the material presented here has been obtained during the tenure of an Alexander v. Humboldt Fel-lowship at the II Zoologisches Institut, Universität Göttingen, FRG.

References

Ax, P. (1963). Relationships and phylogeny of the Turbellaria. In *The lower Metazoa* (ed. E. C. Dougherty) pp. 191-224. University of California Press, Berkeley, Ca.

Baron, P. J. (1968). On the histology and ultrastructure of *Cysticercus longicollis*, the cysticercus of *Taenia crassiceps* Zeder 1800 (Cestoda: Cyclophyllidae). *Parasitology* 58, 497-513.

Beklemishev, V. N. (1969). *Principles of comparative anatomy of invertebrates.* Vol. 2. *Organology.* University of Chicago Press, Chicago, Il.

Bennett, C. E. (1977). *Fasciola hepatica* development of excretory and parenchymal systems during migration in the mouse. *Expl Parasit.* 41, 43-53.

Bonsdorff, C. H., Forssten, T., Gustafsson, M. K. S., and Wikgren (1971). Cellular composition of plerocercoids of *Diphyllobothrium dendriticum* (Cestoda). *Acta zool. fenn.* 132, 1-20.

Borkott, H. (1970). Geschlechtliche Organisation, Fortpflanzungsverhalten und Ursachen der sexuellen Vermehrung von *Stenostomum sthenum* nov. spec. (Turbellaria Catenulida). *Z. Morph. Tiere* 67, 183-262.

Bowen, I. D., Ryder, T. A., and Dark, C. (1976). The effects of starvation on the planarian worm *Polycelis tenuis* Iijima. *Cell Tissue Res.* 169, 193-209.

Campbell, R. D. (1974). Cnidaria. In *Reproduction of marine invertebrates* (eds A. C. Giese and J. S. Pearse) Vol. 1, pp. 133-99. Academic Press, New York.

Chapman, G. (1974). The skeletal system. In *Coelenterate biology* (eds L. Muscatine and M. Lenhoff). Academic Press, New York.

Clark, R. B. (1964). *Dynamics in metazoan evolution: the origin of the coelom and segments.* Clarendon Press, Oxford.

—— (1979). Radiation of the Metazoa. In *The origin of major invertebrate groups* (ed. M. R. House), Systematics Association Special Vol. 12, pp. 55-102. Academic Press, London.

Coward, S. J. (1979). On the occurrence and significance of annulate lamellae in gastrodermal cells of regenerating planarians. *Cell. Biol. Int. Rep.* 3, 101-6.

Crezee, M. (1975). Monograph of the Solenofilomorphidae (Turbellaria, Acoela). *Int. Rev. Hydrobiol.* 60, 769-845.

Curtis, S. K., Cowden, R. R., Moore, J. D., and Robertson, J. L. (1983). Histochemical and ultrastructural features of the epidermis of the land planarian *Bipalium adventitium*. *J. Morph.* 175, 171-94.

Doe, D. A. (1981). Comparative ultrastructure of the pharynx simplex in Turbellaria. *Zoomorphologie* 97, 133-93.

—— and Rieger, R. M. (1977). A new species of the genus *Retronectes* (Turbellaria, Catenulida) from the coast of North Carolina, USA. *Mikrofauna Meeresbod.* 66, 1-10.

Eakin, R. M. and Brandenburger, J. L. (1974). Ultrastructural features of a Gordian worm (Nematomorpha). *J. Ultrastruct. Res.* 46, 351-74.

Fennhoff, F. I. (1980). Embryonic development of *Tubularia corocea* Agassiz, with special reference to the formation of interstitial cells. In *Developmental and cellular biology of coelenterates* (eds P. and P. Tardent) pp. 127-31. Elsevier-North Holland, Amsterdam.

Fransen, M. E. (1980). Ultrastructure of coelomic organization in annelids. I. Archiannelida and other small polychaetes. *Zoomorphologie* 95, 235-49.

Gallagher, S. S. E. and Threadgold, L. T. (1967). Electron microscope studies of *Fasciola hepatica* II. The interrelationship of the parenchyma with other organ systems. *Parasitology* 57, 627-32.

Gardiner, S. L. and Rieger, R. M. (1980). Rudimentary cilia in muscle cells of annelids and echinoderms. *Cell Tissue Res.* 213, 247-52.

Gremigni, V. (1981). The problem of cell totipotency, dedifferentiation and transdifferentiation in Turbellaria. *Hydrobiologia* 84, 171-9.

Gutmann, W. F. (1981). Relationships between invertebrate phyla based on functional-mechanical analysis of the hydrostatic skeleton. *Am. Zool.* 21, 63-81.

Hay, E. D. and Coward, S. J. (1975). Fine structure studies on the planarian *Dugesia*. I. Nature on the "neoblast" and other cell types in non-injured worms. *J. Ultrastruct. Res.* 50, 1-21.

Hertwig, O. and Hertwig, R. (1882). Die Coelomtheorie. Versuch einer Erklaerung des mittleren Keimblattes. *Jena. Z. Naturw.* 15, 1-150.

Hess, E. (1980). Ultrastructure study of the Tethridium of *Mesocestoides corti*; tegument and parenchyma. *Z. ParasitKde* 61, 135-60.

Hori, I. (1978). Possible role of rhabdite-forming cells in cellular succession of the planarian epidermis. *J. Electron Microsc.* 27, 89-102.

Hyman, L. H. (1951). *The invertebrates: Platyhelminthes and Rhynchocoela. The acoelomate Bilateria.* Vol. 2 McGraw-Hill, New York.

—— (1959). *The invertebrates: smaller coelomate groups. Chaetognatha, Hemichordata, Pogonophora, Phoronida, Ectoprocta, Brachiopoda, Sipunculida. The coelomate Bilateria* Vol. 5. McGraw-Hill, New York.

Irwin, S. W. B. and Threadgold, L. T. (1970). Electron-microscope studies on *Fasciola hepatica* VII. The development of vitelline cells. *Expl Parasit.* 28, 399-411.

Lyons, K. M. (1977). Epidermal adaptations of parasitic Platyhelminthes. *Symp. zool. Soc. Lond.* 39, 97-144.

MacKinnon, B. M., Burt, M. D. B., and Pike, A. W. (1981). Ultrastructure of the epidermis of adult and embryonic *Paravortex* species (Turbellaria, Eulecithophora). *Hydrobiologia* 84, 241-52.

Mainitz, M. (1979). The fine structure of gnathostomulid reproductive organs I. New characters in the male copulatory organ of *Scleroporalia*. *Zoomorphologie* 92, 241-72.

Mamkaev, Y. V. and Markosova, T. G. (1978). Electron-microscopic study of the parenchyma in Acoela. *Trudy zool. Inst., Leningr.* 84, 7-12 (in Russian).

Matricon-Gondran, M. (1977). Diversité de jonctions "gap" dans le parenchyme d'un trematode Digenien. *Biol. Cellulaire* 30, 16a.

Moraczewski, J. (1981). Fine structure of some Catenulida (Turbellaria, Archoophora). *Zoologica Pol.* 28, 367-415.

Pedersen, K. J. (1961). Studies on the nature of planarian connective tissue. *Z. Zellforsch. mikrosk. Anat.* 53, 569-608.

—— (1964). The cellular organization of *Convoluta convoluta*, an acoel turbellarian: a cytological, histochemical and fine structural study. *Z. Zellforsch. mikrosk. Anat.* 64, 655-87.

—— (1966). The organization of the connective tissue of *Discocoelides langi* (Turbellaria, Polycladida). *Z. Zellforsch. mikrosk. Anat.* 71, 94-117.

—— (1968). Some morphological and histochemical aspects of nemertean connective tissue. *Z. Zellforsch. mikrosk. Anat.* 90, 570-95.

—— (1972). Studies on regeneration blastemas of the planarian *Dugesia tigrina*, with special reference to differentiation of the muscle-connective tissue filament system. *Wilhelm Roux Arch. EntwMech. Org.* 169, 134-69.

—— (1976). Scanning electron microscopical observations on epidermal healing in the planarian *Dugesia tigrina*. *Wilhelm Roux Arch. EntwMech. Org.* 179, 251-73.

Prenant, M. (1922). Recherches sur le parenchyme de Plathelminthes. *Archs Morph. gén. exp.* 5, 1-174.

Quick, D. C. and Johnson, R. G. (1977). Gap junctions and rhombic particle arrays in Planaria. *J. Ultrastruct. Res.* 60, 348-361.

Reisinger, E. (1924). Die Gattung *Rhynchoscolex*. *Z. Morph. Ökol. Tiere* 1, 1-37.

—— (1976). Zur Evolution des stomatogastrischen Nervensystems bei den Plathelminthen. *Z. zool. Syst. Evolutionsforsch.* 14, 241-53.

Reissig, M. and Colucci, A. (1968). Localization of glycogen in the cestode, *Hymenolepis diminuta. J. Cell Biol.* 39, 754-63.

Remane, A. (1952). *Die Grundlagen des natuerlichen systems, der Vergleichenden anatomie und der Phylogenetik. Theoretische Morphologie und Systematik. I. Akademische verlagsges.* Geest and Portig, Leipzig.

Rieger, R. (1971). Die Turbellarienfamilie Dolichomacrostomidae nov. fam. (Macrostomida). I. Vorbemerkungen und Karlingiinae nov. subfam. I. *Zool. Jb. Abt. Syst.* 98, 236-42.

—— (1980). A new group of interstitial worms, Lobatocerebridae nov. fam. (Annelida) and its significance for metazoan phylogeny. *Zoomorphologie* 95, 41-84.

—— (1981a). Morphology of the Turbellaria at the ultrastructural level. *Hydrobiologia* 84, 213-29.

—— (1981b). Fine structure of the body wall, nervous system, and digestive tract in the Lobatocerebridae Rieger and the organization of the gliointerstitial system in Annelida. *J. Morph.* 167, 139-65.

—— Ruppert, E. E., Rieger, G. E., and Schoepfer-Sterrer, C. (1974). On the fine structure of gastrotrichs, with the description of *Chordodasys antennatus* sp. n. *Zool. Scr.* 3, 219-37.

—— and Tyler, S. (1979). The homology theorem in ultrastructural research. *Am. Zool.* 19, 655-64.

Rohde, K. (1971). Untersuchungen an *Multicotyle purvisi* Dawes 1941 (Trematoda: Aspidogastraea). IV. Ultrastruktur des Integuments der geschlechtsreifen Form und der freien Larve. *Zool. Jb. Abt. Anat.* 88, 365-86.

—— (1974). Light- and electron-microscopic studies of the posterior pharynx and the anterior and posterior glands of *Polystomoides* (Monogenea: Polystomidae). *Zool. Jb. Abt. Anat.* 92, 1-17.

Salvini-Plawen, L. von (1980). Phylogenetischer Status and Bedeutung der mesenchymaten Bilateria. *Zool. Jb. Abt. Anat.* 103, 354-73.

Sawyer, R. T. and Fitzgerald, S. (1981). Leech circulatory system. In *Invertebrate blood cells* (eds N. A. Ratcliffe and A. F. Rowley) pp. 141-59. Academic Press, London.

Starck, D. and Siewing, R. (1980). Zur Diskussion der Begriffe Mesenchym und Mesoderm. *Zool. Jb. Abt. Anat.* 103, 374-88.

Sterrer, W. (1969). Beiträge zur Kenntnis der Gnathostomulida I. Anatomie und Morphologie des Genus *Pterognathia* Sterrer. *Ark. Zool.* 22, 1-125.

—— and Rieger, R. M. (1974). Retronectidae—a new cosmopolitan marine family of Catenulida (Turbellaria). In *Biology of the Turbellaria* (eds N. W. Riser and M. P. Morse) pp. 63-92. McGraw-Hill, New York.

Strong, P. A. and Bogitsch, B. J. (1973). Ultrastructure of the lymph system of the trematode *Megalodiscus temperatus. Trans Am. microsc. Soc.* 92, 570-8.

Summers, R. G. and Haynes, F. (1969). The ontogeny of interstitial cells in *Pennaria triaxella. J. Morph.* 129, 81-8.

Threadgold, L. T. and Arme, C. (1974a). *Hymenolepis diminuta*: an electron microscopic study of ion absorption. *Expl Parasit.* 35, 475-91.

—— —— (1974b). Electron microscopic studies of *Fasciola hepatica* Part II: autophagy and parenchymal cell function. *Expl Parasit.* 35, 389-405.

Threadgold, L. T. and Gallagher, S. S. E. (1966). Electron microscope studies of *Fasciola hepatica* I. The ultrastructure and interrelationship of the parenchymal cells. *Parasitology* 56, 299-304.

Turbeville, J. M. and Ruppert, E. E. (1983). Epidermal muscles and peristaltic burrowing in *Carinoma tremaphoros* (Nemertini): correlates of effective burrowing without segmentation. *Zoomorphology* 103, 103-20.

Tyler, S. (1984). In *Biology of the integument* (eds J. Bereiter-Hahn, A. G. Matoltsy and K. S. Richards) Vol. 1. Springer, Heidelberg.

Williams, J. B. (1979). Studies on the epidermis of Temnocephala IV. Observations on the ultrastructure of the epidermis of *Temnocephala dendyi*, with notes on the role of the Golgi apparatus in autolysis. *Aust. J. Zool.* 26, 217-24.

Wittrock, D. D. (1978). Ultrastructure of the ventral papillae of *Quinqueseralis quinqueseralis* (Trematoda: Notocotylidae). *Z. Parasitkde* 57, 145-54.

Wood, R. L. and Cavey, M. J. (1981). Ultrastructure of the coelomic lining in the podium of the starfish *Stylasterias forreri. Cell Tissue Res.* 218, 449-73.

8. The acoel turbellarians: kingpins of metazoan evolution or a specialized offshoot?

JULIAN SMITH III and *SETH TYLER*

Department of Zoology, University of Maine, Orono, Maine, USA

Abstract

Because acoel turbellarians are anatomically relatively simple, they have been pivotal in several theories on the origins of lower metazoans. To some this simplicity indicates that acoels are like the most primitive metazoans, being models for the link between ciliated protozoans and the Metazoa; for others this simplicity is like that of the coelenterate planula larva, so that acoels are invoked as models for the link between the coelenterates and the Bilateria; for yet others the simplicity is seen as secondary reduction from a coelomate ancestor so that acoels themselves have no bearing on relationships of groups outside the Turbellaria. Discussions of these theories have been based largely on light microscopy and so have dealt perforce with a paucity of characters. To update the standing of these theories we have assembled information on ultrastructural characters and searched among them for homologies that relate acoels to other lower animals. Considering ultrastructural characters of the body wall, parenchyma, and digestive tract, it appears that acoels are derived, and cannot logically be fitted into any of the current theories of interphyletic relationships, either as models of primitiveness or as reduced coelomates. Instead, the Nemertodermatida, sister group to the Acoela, and the Catenulida appear more primitive, and consideration of these groups for phylogenetic schemes is more likely to bear fruit.

Introduction

A logical starting point for a phylogenetic tree of lower invertebrates is a taxonomic group with simple morphology from which other groups can be derived by progressive development of more complicated structures. The apparent simplicity of the acoel turbellarians, particularly their lack of digestive and body cavities, has attracted attention in this regard for two prominent theories on metazoan phylogeny: von Graff's (1891, 1908) planula-acoel theory (see also Beklemischev 1963, 1969), in which acoels are seen as neotenous descendants of a cnidarian planula (see Salvini-Plawen 1978), and Hadzi's (1963) ciliate-acoel theory in which they would be derived by cellularization of a ciliated protozoan (Hanson 1977). For both of these theories the acoels stand as ancestors to all other bilaterians. In contrast, according to the archicoelomate-enterocoel theory the acoels' simplicity can be viewed as a secondary reduction from a coelomate bilaterian ancestor (Ax 1963; Siewing 1980).

Discussion of these theories, even by modern proponents, has revolved around relatively simplistic concepts of acoel morphology; for the most part only the relatively few characters discernible by light microscopy have been considered. More recent ultrastructural studies have produced a significant body of information, much of which has not yet been brought to bear on these theories. Our aim is to show how ultrastructural characters of acoels and their relatives affect the standing of these theories.

The question posed by the title of this paper is really twofold: first, From what group have the acoels descended? and second, Which groups, if any, are to be regarded as the acoels' descendants? It is logical to look first at the relationships the Acoela has with the other turbellarian orders. Unless the Acoela is the most primitive turbellarian order, acoels cannot be the most primitive bilaterians.

Ultrastructure of the parenchyma

The most important aspect of acoelan simplicity from a phylogenetic standpoint is the nature of their parenchyma. As Clark (1979) has pointed out, a major difference in the two most widely discussed theories on the origin of the Bilateria, the archicoelomate-enterocoel theory (see Siewing 1980) and the planula-acoel theory (see Salvini-Plawen 1978) lies in the prediction that each makes about the primitive state of the mesoderm in bilaterians. The archicoelomate-enterocoel theory holds a coelomate ancestor to be more primitive than the Turbellaria;

thus, parenchyma in the most primitive turbellarians should show evidence of being a reduced coelom, perhaps resembling other known cases of coelomic reduction in which the body cavity is secondarily filled by mesenchyme (see Rieger, this volume). On the other hand, the planula-acoel theory requires that mesoderm in the Turbellaria be the most primitive occurrence of mesoderm in the Bilateria, and thus the arrangement of the mesodermal derivatives in all other bilaterians should have had its origin in the parenchyma of acoel turbellarians (Hyman 1951; Beklemischev 1963, 1969).

Light-microscopic studies of acoel turbellarians lead to the assumption that their parenchyma consists of two parts: a central, or digestive, parenchyma that corresponds to the digestive epithelium of non-acoel turbellarians, and a peripheral parenchyma that fills space between the central parenchyma and body wall (e.g. Westblad 1948). As with the parenchyma of non-acoel turbellarians, the peripheral parenchyma of acoels has been defined negatively as being all cells that are not part of the digestive tract, body wall, gonads or sex organs (see Pedersen 1964, for definitions and review of literature). At the electron microscopic level, the parenchyma in acoels can be seen to be more diverse than light microscopists have assumed.

1. Central parenchyma

Klima (1967) and Kozloff (1972) published ultrastructural evidence of the presence of both syncytial and cellular elements in the digestive parenchyma of acoels, corroborating some earlier conclusions based on light microscopy (Boguta and Mamkaev 1972). A recent ultrastructural study of central parenchyma in six species of acoels (Smith 1981b) concluded that the digestive tract in Acoela most commonly appears as a central, membrane-bounded syncytial mass separated from other tissues by one or more types of specialized cells (Figs 8.1 and 8.2). Possession of a digestive tract with a syncytial element is characteristic of many acoels, having been documented by EM in representatives of eight of the fifteen families of acoels (Table 8.1). In some cases, e.g. *Paedomecynostomum* sp. and *Philactinoposthia* sp., the syncytium is apparently formed only after ingestion of food and is shed when digestion is complete, leaving a large central cavity (Mamkaev and Markosova 1979; Smith 1981b).

The specialized cells bordering the syncytium appear in a variety of forms in different species. In a relatively simple arrangement, these are large, pale-staining cells that wholly (in *Diopisthoporus* cf. *longitubus*, *Hesiolicium inops*) or partly (in *Kuma* sp., *Diopisthoporus* sp.) isolate the syncytium from the rest of the body (Fig. 8.1). In a more complex

Table 8.1. Distribution of the central syncytium in Turbellaria Acoela

Family	Species	Reference
Diopisthoporidae	*Diopisthoporus* cf. *longitubus*	Smith 1981b
	Diopisthoporus sp.	Smith, unpublished
Paratomellidae	*Hesiolicium inops*	Smith, unpublished
Convolutidae	*Convoluta convoluta*	Mamkaev and Markosova 1979
	Oxyposthia praedator	Mamkaev and Markosova 1979
	Oligochoerus limnophilus	Klima 1967
	Convoluta sp. nov.	Smith 1981a,b
Mecynostomidae	*Paedomecynostomum* sp.	Smith 1981b
	Paramecynostomum diversicolor	Ivanov and Mamkaev 1977
Anaperidae	*Anaperus biaculeatus*	Mamkaev and Markosova 1979
	Anaperus sp.	Smith 1981b
Haploposthiidae	*Kuma* sp.	Smith 1981b
Childiidae	*Philactinoposthia* sp.	Smith 1981b
Otocelidae	*Otocelis luteola*	Kozloff 1972

arrangement, such as that of *Convoluta* sp. nov., there are three distinct types of highly branched wrapping cells lying next to the syncytium and isolating it from the rest of the body (Fig. 8.2).

It is evident that the syncytium and wrapping cells are a functional unit, i.e. together they constitute the digestive parenchyma, and that the syncytium is formed and periodically renewed by fusion of the bordering cells. Morphological evidence for this comes from a study of *Convoluta* sp. nov. examined at several stages in its life cycle (Smith 1981a,b). In all of the stages studied, from prehatchling to adult, the wrapping cells appear intimately associated with the syncytium, forming with it a clearly recognizable morphological unit. The syncytium is fully formed before hatching (Smith 1981a) and at this stage has a relatively high level of synthetic activity as evidenced by the amount of rough endoplasmic reticulum and Golgi apparatus present. As the animal matures, this synthetic activity drops off markedly (Smith 1981b). Synthetic activity remains high throughout the life cycle in one of the wrapping cell types, and these are the only cells in the body that appear to be making the 200-400 nm diameter vesicles that are characteristically contained in the syncytium of all life-history stages. These vesicles are not produced in the syncytium of older individuals, and it is concluded that they reach the syncytium by fusion of this wrapping cell type with the syncytium (Smith 1981b). Functions of the other wrapping cell types are uncertain.

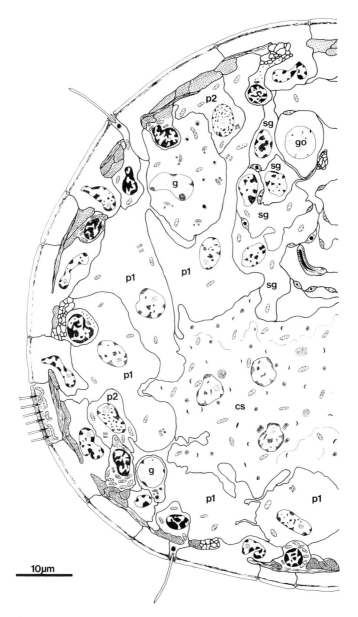

Fig. 8.1. *Diopisthoporus* c.f. *longitubus*: Schematic cross section through half the gonadal region, based on information from TEM. Cellular interdigitation greatly reduced for clarity. Epidermal cilia, mitochondria, epitheliosomes, and infraciliature shown for only one cell. Abbreviations: cs, central syncytium; g, epidermal gland cells; go, gonial cells; p1, p2, types 1 and 2 wrapping cells; sg, stromal cells of the gonad.

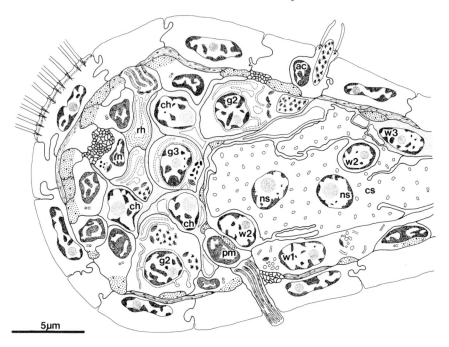

Fig. 8.2. *Convoluta* sp. nov.: Schematic half cross section anterior to the mouth of a juvenile specimen, based on TEM information. Cellular interdigitation greatly reduced for clarity. Epidermal cilia, infraciliature, and terminal web shown for only one cell. Abbreviations: ac, accessory cell; ch, chordoid cells; cs, central syncytium; g2, putative adhesive gland cells; g3, type 3 epidermal gland cells; n, neuron; ns, nuclei of central syncytium; pm, parenchymal muscle; rh, rhabdoid gland cells; w1, w2, w3, types 1, 2 and 3 wrapping cells.

Similar arguments on the relationship between the syncytium and its bordering cells can be made for other species. For example, in *Diopisthoporus* cf. *longitubus* (Fig. 8.1), the central parenchyma is regarded as comprising the syncytium plus the wrapping cells: one of two types of wrapping cell apparently is the progenitor of the second type, which in turn fuses with the syncytium (Smith 1981b).

Experimental evidence for the formation and renewal of the syncytium by fusion of cells lying peripheral to it comes from autoradiographic studies tracing the origin of nuclei in the syncytium and from histochemical studies of digestive enzymes in the syncytium and bordering cells (Markosova 1976; Ivanov and Mamkaev 1977; Mamkaev 1979; Mamkaev and Markosova 1979). While these reports state that the fusing peripheral cells belong to the peripheral parenchyma, it is better to regard cells intimately involved with the digestive process and synthesizing digestive enzymes as being part of the central parenchyma (Smith 1981a).

A quite different form of central parenchyma has been found in *Paratomella rubra* (Smith 1981b). Here, the centre of the body is filled with processes of large, parenchymally arranged cells and with small, membrane-bounded cytoplasmic bodies that appear to arise from the parenchymal cells (Fig. 8.3). Nuclei of this region occur individually in membrane-bounded masses of cytoplasm, i.e. not in a syncytium. At least some of the parenchymal cells bordering this central region extend to the body wall. Other parenchymally arranged cells (Fig. 8.3), similar in appearance to those in the central region, form what would certainly be considered the peripheral parenchyma on the basis of light microscopy (cf. Crezee 1978). The parenchymal cells in the central region differ from those elsewhere in showing synthetic activity (Fig. 8.4). Whether this difference indicates the presence of two types of parenchymal cell, one bounding the central region (a digestive parenchymal cell), and another not reaching the central region (a peripheral parenchymal cell) cannot be decided at present; if there are two types, they are quite similar in morphology. Digestion in *Paratomella* apparently proceeds extracellularly by digestive enzymes released through necrotic breakdown of the innermost parenchymal cells.

2. Peripheral parenchyma

With the central parenchyma identified, all remaining cells of the parenchyma can be considered as peripheral parenchyma, three varieties of which have so far been found.

(a) 'Dark cells.'

Found in *Kuma* sp., these cells have homogeneous dark cytoplasm, and contain myofilaments and sarcoplasmic reticulum. They are thus ultrastructurally similar to modified muscle cells in other invertebrates, especially gastrotrichs, in which such modified muscle cells have recently been shown to contain myoglobin (Ruppert 1978, his Fig. 13F; Ruppert and Travis 1982).

(b) Fixed parenchymal cells.

Found in *Convoluta* sp. nov. and *Kuma* sp. (Smith 1981b) and in *Convoluta convoluta* (Pedersen 1964). These are large, highly branched cells with pale-staining cytoplasm. The cells denoted as fixed parenchymal cells in *Hesiolicium inops* by Crezee and Tyler (1976) occupy the anatomical position of wrapping cells, i.e. next to the syncytium, and it is thus best to term them wrapping cells. These wrapping cells lie against cells of the body wall.

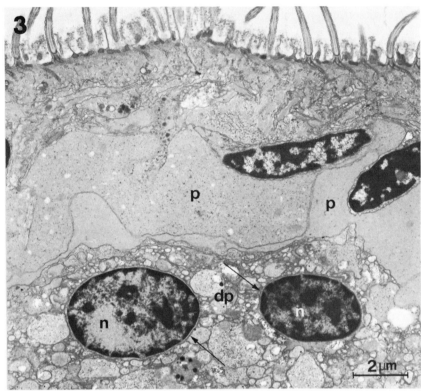

Fig. 8.3. *Paratomella rubra*: ventral body wall, parenchymal cells and digestive parenchyma. Nuclei in digestive parenchyma are each surrounded by a thin layer of cytoplasm (arrowed).

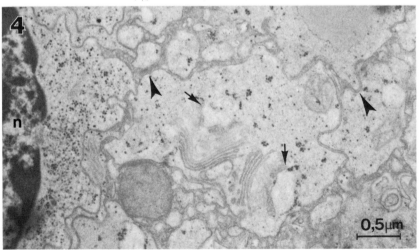

Fig. 8.4. *P. rubra*: central parenchyma, with nucleus, processes and cytoplasmic bodies
of the parenchymal cells. Golgi bodies appear to be producing vesicles with flocculent
contents (arrowed). Note dense extracellular material in the digestive cavity (arrow-
heads). Abbreviations: dp, digestive parenchyma; n, nucleus; p, parenchymal cells.

(c) *Chordoid cells.*

Found in *Convoluta* sp. nov. as well as in a variety of other acoels (Ax 1966; Dörjes 1968). These are branching cells characterized by a greatly vacuolated cytoplasm, usually with one large central vacuole and many smaller vacuoles in their processes. Ax (1966) suggested on theoretical grounds that the chordoid spaces of acoels were intracellular vacuoles. Klima (1967), on the other hand, stated that the spaces observed with TEM in *Oligochoerus limnophilus* were extracellular, although he did not document this; similarly, Mamkaev (1979) and Siewing (1980) compared the chordoid spaces to a schizocoel. However, for *Convoluta* sp. nov. and other acoels we have studied it appears from both TEM and light-microscopical observations that the chordoid spaces are intracellular vacuoles. The vacuolated processes of the chordoid cells branch extensively between the cells of the body wall (Fig. 8.2), penetrate between cells of the testes, and surround developing eggs (Smith, unpublished).

There are some acoels that appear to lack peripheral parenchyma, namely *Diopisthoporus* cf. *longitubus* (Fig. 8.2), a new *Diopisthoporus* species from North Carolina, *Hesiolicium inops* (see above), and *Paratomella rubra*.

Phylogenetic implications of parenchymal structure in the Acoela

The form of the parenchyma in the Acoela, particularly of the digestive tract, is unique. Outside the Acoela, in other turbellarians as well as other lower invertebrates, the digestive tract is epithelial. Despite claims to the contrary (Mamkaev 1979), published TEM results show that the digestive tract in non-acoel turbellarians is epithelial, e.g. Haplopharyngida and Macrostomida (Doe 1981), Tricladida (Bowen, Ryder, and Thompson 1974; Bowen 1980), Rhabdocoela (Holt and Mettrick 1975), Polycladida (Smith, unpublished results on *Notoplana atomata*), Catenulida (Moraczewski 1981). If the Acoela is to occupy a pivotal position in phylogenetic speculations, the transition between parenchymal, syncytial, and epithelial digestive tracts must be explained, and the direction of that transition must be determined. Also critical is the structure of the peripheral parenchyma because it should be the homologue of mesodermal derivatives in other invertebrates. It is appropriate to begin with a consideration of the group most closely related to the Acoela, its sister group, the Nemertodermatida (Karling 1974; Tyler and Rieger 1977; Ehlers, this volume).

The nature of the digestive tract in nemertodermatids has been a point of heated debate in the past: cf. Steinböck (1938) and Mamkaev (1979) who argue, respectively, that it is syncytial or parenchymal, with Westblad (1937, 1949) and Karling (1967) who propose that it is epithelial. However, recent EM studies of *Nemertoderma* and *Flagellophora* make it clear that the digestive tract is not merely cellular (Tyler and Rieger 1977), but epithelial, with phagocytes and gland cells extending between the muscle cells to the base of the epidermis, and belt junctions delimiting an occluded lumen (Smith 1981b).

The close relationship between the Nemertodermatida and the Acoela allows the construction of a character transformation series for the digestive tract (Fig. 8.5). At one end of this sequence is the epithelial digestive tract of Nemertodermatida; at the other is the parenchymal and syncytial digestive tract seen in the majority of acoels; and standing as a likely intermediate between these extremes is the parenchymal digestive tract of *Paratomella*. There can be little doubt of the homology of these three states.

Establishing the direction of this transformation series is a matter of debate. Ivanov and Mamkaev (1977) propose that a non-epithelial stage like that shown here for *Paratomella* (Fig. 8.5, state 2) is primitive for the acoels and for Bilateria in general. Ivanov and Mamkaev (1977) state that the digestive parenchyma in *Oxyposthia praedator*, among others, exhibits this state. Subsequent studies on this species (Mamkaev and Markosova 1979) show that the digestive parenchyma is essentially similar to that in *Kuma* sp. Also, Mamkaev (1979) does not consider the digestive tract in *Nemertoderma* to be epithelial, so he does not deal with a condition like state 1 in Fig. 8.5. According to Mamkaev (1979), the development of central syncytium occurs within the Acoela, and from the original parenchymal condition, epithelial formation and differentiation of gland cells supposedly occurred in several independent lines, leading to the digestive tract of other turbellarians and bilaterians.

We feel the evidence is stronger for regarding the condition of *Nemertoderma* as most primitive and have constructed a cladogram accordingly (Fig. 8.6). One justification is that the epithelial intestine of the nemertodermatids is more like that found in other turbellarian groups, including the Catenulida and Haplopharyngida-Macrostomida, which are potentially primitive to the Nemertodermatida (Karling 1974; Ax 1963), as well as in all other metazoan groups that could be chosen for outgroup comparison, than is the parenchymal or syncytial digestive tract of the Acoela. Another compelling reason is that the uniflagellate spermatozoan of the Nemertodermatida is more likely to be a primitive feature than the biflagellate sperm of acoels (Tyler and Rieger 1975, 1977; Hendelberg 1977). It can also be argued that the structure of the epidermal ciliary rootlet system and presence of extracellular matrix

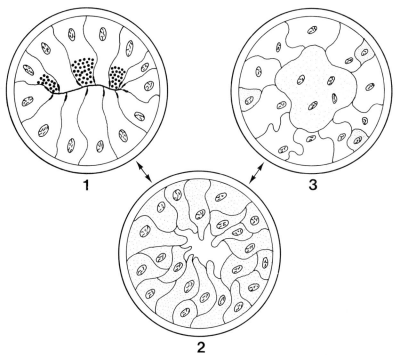

Fig. 8.5. Morphological transformation sequence for the digestive tract in nemertodermatids and acoels. The states 1-3 depicted schematically may be characterized as follows.

State 1: (Order Nemertodermatida). Ciliation perhaps absent (but note cilium in Tyler and Rieger 1977, Fig. 8; see also Karling 1967). Gland cells present (presence of digestive gland cells in *Flagellophora* has not been verified by TEM, but see Faubel and Dörjes 1978). Lumen occluded, but defined by septate and adhaerens-like junctions and by distal finger-like processes of the phagocytes.

State 2: (*Paratomella rubra*). Gland cells absent. An irregular, central extracellular space between parenchymally arranged digestive cells serves as a 'lumen'.

State 3: (*Diopisthoporus* spp., *Hesiolicium inops*). Two types of digestive parenchymal cells (second type, stippled, so far known only from *Diopisthoporus* cf *longitubus*) and central syncytium compose the digestive parenchyma.

are indicators of a more primitive position for the Nemertodermatida (see below).

If the nemertodermatids have a more primitive digestive tract than acoels, the digestive tract in the turbellarian ancestor must have been more like that seen in other lower invertebrates: an epithelial tract enclosing a lumen and incorporating gland cells (state 0 in Fig. 8.6). Loss of the lumen and gland cells in the line leading to acoels may be functionally correlated; it is unlikely that glands producing enzymes for extracellular digestion would be found in the absence of a junction-delimited digestive lumen. While intracellular digestion is the predominant mode for the Turbellaria as a whole (Jennings 1974), the acoels have apparently specialized in such digestion. As Mamkaev and Seravin (1963) suggested, the central syncytium may have evolved to allow the intracellular digestion of food items too large to be phago-cytosed by a single cell.

Evolution of peripheral parenchyma must of course follow the same cladogram (Fig. 8.6). The more primitive members of this cladogram are those lacking a peripheral parenchyma, *Nemertoderma* (Smith 1981b), *Paratomella, Hesiolicium*, and *Diopisthoporus*, and it must be con-cluded that its possession (by members of state 4) is an advanced feature. This implies that turbellarians are not reduced coelomates.

Other features of phylogenetic significance

Most other ultrastructural features that have been discussed in the literature regarding the phylogenetic position of the Acoela indicate simply that they stand unique in the Turbellaria and, for that matter, among metazoans in general. Particularly characteristic of the acoels is the structure of their body wall, especially the form of epidermal ciliation (Tyler 1979). Their cilia have a distinctive pattern of axonemal micro-tubules in their tips and a rootlet system that interconnects adjacent cilia in a way different from any other turbellarian group. Only the nemertodermatids have a ciliary ultrastructure resembling this, evi-dence of a close relationship between acoels and nemertodermatids (Tyler and Rieger 1977). The structure of the ciliation is so complex and distinctive in the nemertodermatid-acoel line (N-A line) that the transition to or from ciliation of other turbellarians is difficult to envisage, even though it is clear that these ciliation patterns must be related. Our attempts to identify homologies between component root-lets in acoels and other metazoans have proved fruitless. Nevertheless, the course of development of the cilia may hold keys to explaining dif-ferences in the infraciliature. Cilia in the developing epidermis of an acoel embryo arise from basal bodies (centrioles) produced solely by a kinetosomal mode of genesis, and the pattern of ciliary interconnec-tions in the rootlets seems to arise from the orientation of newly forming procentrioles to parts of already established cilia (Tyler 1984a). In a macrostomid embryo, in contrast, procentrioles arise through *de novo* and kinetosomal origins, and the centrioles thus formed move indi-vidually to the cell surface to grow an axoneme (Tyler 1981). Evidence of similar *de novo* centriologenesis has been seen in microstomids (M. Reuter, personal communication), and in adults of several turbellarians: Catenulida (Moraczewski 1977, 1981), Haplopharyngida (Doe 1981; Tyler, unpublished), and Tricladida (Hori 1978).

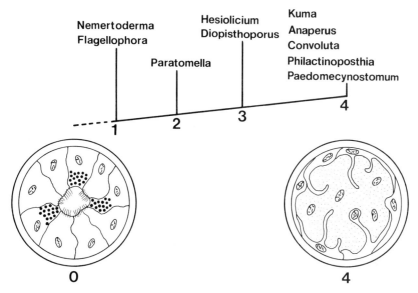

Fig. 8.6. Proposed phylogenetic transformation series for the digestive tract in the N-A line. States 1-3 from Fig. 5, states 0 and 4 characterized as follows.

. State 0: Most probable primitive condition in the Turbellaria; one-layered epithelial digestive tract with ciliated cells and digestive gland cells (see Karling 1974; Ax 1963, this volume). Among archoophorans, this state is found in many Catenulida (Retronectidae: Sterrer and Rieger 1974; Stenostomidae: Marcus 1945); in Haplopharyngida and Macrostomida (Doe 1981); and in many Polycladida (Lang 1881; Bock 1913). The digestive tract in Polycladida Acotylea is said to unciliated (Jennings 1974) but observations on *Notoplana atomata* (Smith, unpublished) show that the phagocytes of this species are ciliated.

State 4: (Many acoels; only genera studied by us are listed). At least one type of flattened, highly branched wrapping cell and a central syncytium constitute the digestive parenchyma.

Interconnections of rootlets in acoels and nemertodermatids may be a consequence of a more primitive mode of producing multiciliated cells, one requiring that newly forming basal bodies are oriented to existing cilia, whereas the non-interconnected pattern of higher turbellarians may have arisen through the acquisition of a new developmental capability to produce basal bodies en masse by *de novo* genesis. This difference in ciliogenesis may also explain the contrast in putative epidermal replacement cells in the two groups: 'pulsatile bodies,' containing fully formed cilia and infraciliature (kinetosomal mode genesis) in the N-A line (Dorey 1965), and centriole-containing 'stem cells of the epidermis' (*de novo* mode genesis) in the other turbellarians (cf. references above). Thus, the phylogenetic link between ciliation in the N-A line and that in the rest of the Turbellaria may lie in the plesiomorphy of kinetosomal ciliogenesis.

The acoel body wall lacks any sort of intercellular matrix, including basement membrane. In other turbellarian groups such matrices are usually well developed (Rieger 1981; Tyler 1984b). For the Turbellaria as a whole one can identify a broad spectrum of matrix development, ranging from complete absence in acoels, through states with an irregularly arranged matrix between all subepidermal cells, to discrete basement membrane plus dermal connective matrices (Tyler 1984b). Although it is tempting to arrange this series in such a way that lack of a matrix would be considered primitive while possession of well-organized matrices would be considered advanced, this is probably not legitimate. Other lower metazoans, including coelenterates and sponges, have an intercellular matrix, and other turbellarian groups that could potentially be considered primitive to the acoel line, i.e. Catenulida and Haplopharyngida, have such a matrix. Particularly noteworthy is the presence of a matrix in at least some nemertodermatids (Figs 8.7-8.9). In view of other evidence pointing to nemertodermatids being more primitive than acoels, this lack of matrix in acoels must be viewed as derived (see also Rieger, this volume).

There are several other recently discovered characters besides the structure of epidermal ciliation that set the N-A line apart from other turbellarians. Rhabdoids and epitheliosomes (ultrarhabdites) are composed of mucoid material in the N-A line and of basic protein in higher turbellarians (Smith, Tyler, Thomas, and Rieger 1982). Glandular adhesive organs show comparable cytochemical differences. What appear to be such organs in nemertodermatids and in a variety of acoels are composed of a single mucus gland cell enwrapped by a collar-like, cilia-bearing epidermal cell (Fig. 8.10 and unpublished personal observations). In the higher turbellarians glandular adhesive organs are composed of two glands, one secreting basic protein, emerging through a modified epidermal cell which usually bears only modified microvilli (Tyler 1976). Epidermal septate junctions of members of the N-A line are exceedingly difficult to see in standard TEM preparations, and although they appear in negatively-stained preparations (Fig. 8.11) to be basically similar to the pleated type found in polyclads and triclads, preliminary data indicate that the septa of acoel junctions tend to be anastomosed, unlike those of the higher turbellarians studied so far (cf. Green 1981; Green and Bergquist 1982).

Conclusion

The structure of the digestive tract, sperm, and body wall show the Nemertodermatida is primitive to the Acoela, and on balance other ultrastructural characters do not contradict this conclusion. Whether the Nemertodermatida is primitive to other turbellarians or derived from them, the Acoela stands as a specialized branch of the turbellarian phylogenetic tree. Acoels cannot be considered ancestral to other turbellarian groups, let alone other lower metazoans.

Having the Nemertodermatida primitive to the Acoela does not simply mean that they are to be substituted for acoels in present theories on the origin of lower metazoans. The ciliate-acoel theory relied on the untenable supposition that syncytiality in the Acoela is primitive. Furthermore, as has been recognized even on the basis of light microscopic characters (Ax 1963) there are no homologies between acoels and ciliates. The existence of closed mitosis in an acoel has been cited as one possible homology (Hanson 1977). Our observations (unpublished) of dividing cells in *Paratomella rubra, Diopisthoporus* cf. *longitubus*, and *Archaphanostoma* cf. *agile* show that mitosis is open; i.e. the nuclear membrane does not remain during mitosis. The planula-acoel theory rests on the similarity in body form between the planula and the acoel *Diopisthoporus* (Beklemischev 1963; Salvini-Plawin 1978), and ultrastructural studies show that these similarities are superficial. The position of the mouth in the two is analogous, and the structure of the digestive tract in *Diopisthoporus* is syncytial rather than parenchymal or epithelial. Furthermore, the epidermal ciliation in *Diopisthoporus*, as well as in the rest of the N-A line, is derived and cannot be considered a bridge between the monociliated condition in *Trichoplax* and the Cnidaria on one hand and the monociliated and multiciliated conditions in the other lower metazoans on the other (see Rieger 1976).

The archicoelomate-enterocoel theory does not deal directly with the position of the Acoela, but by considering coelomate phyla as more primitive leaves the Acoela as a secondarily reduced and derived group. Evidence contrary to this position, however, is the discovery that peripheral parenchyma is absent from more primitive members of the N-A line and developed only in more advanced members. Even other orders that have been considered potentially the most primitive turbellarians, i.e. the Haplopharyngida-Macrostomida (Ax 1963) or the Catenulida (Karling 1974), apparently lack parenchyma: cf. Rieger (1981) for Haplopharyngida-Macrostomida; Moraczewski (1981) for Catenulida; (see also Rieger, this volume). In the absence of other cases of reduced coeloms morphologically similar to the condition in primitive turbellarians (cf. Rieger, this volume) it is equally possible that the lack of parenchyma in these archetypical acoelomates, which resembles more closely the condition of the body cavity in some pseudocoelomates, is a trait more primitive than the annelid coelom.

None of the three prominent theories on metazoan origins, therefore, can satisfactorily account for the ultrastructural features we find in acoels and their relatives. The ciliate-acoel and planula-acoel theories

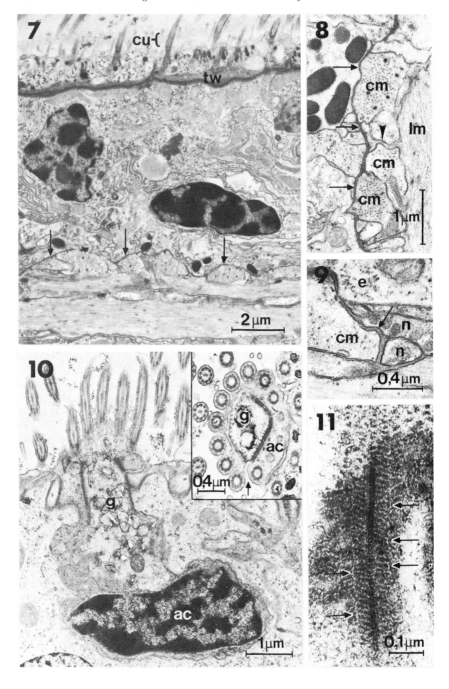

Fig. 8.7   Fig. 8.8   Fig. 8.9   Fig. 8.10   Fig.8.11

cannot legitimately use acoels as models for the ancestral bilaterian. Yet the remaining archicoelomate-enterocoel theory fails to explain the origin of the Platyhelminthes, whatever its success in explaining the existence of other lower invertebrate phyla. The Platyhelminthes remains problematic. Its members are attractive to phylogeneticists for their relatively simple morphology and for their tremendous variety of structure, and yet the Acoela, ostensibly its simplest members, are surprisingly derived in comparison with other lower metazoans.

Acknowledgements

Thanks are due especially to R. M. Rieger for the free and open exchange of ideas, and to Eddie Meisner for typing the manuscript. This research was supported by NSF grants GB42211 (R. M. Rieger, P.I.) and BSR-8116894 (S. Tyler, P.I.).

References

Ax, P. (1963). Relationships and phylogeny of the Turbellaria. In *The lower Metazoa* (ed. E. C. Dougherty) pp. 191-224. University of California Press, Berkeley, Ca.

—— (1966). Das chordoide gewebe als histologisches Lebensformmerkmal der Sandlükenfauna des Meeres. *Naturw. Rdsch., Stuttgart.* 19, 282-9.

Beklemischev, V. N. (1963). On the relationship of the Turbellaria to other groups of the animal kingdom. In *The lower Metazoa* (ed. E. C. Dougherty) pp. 234-44. University of California Press, Berkeley, Ca.

Fig. 8.7. New genus (Nemertodermatida): body wall. Note thick, bilayered terminal web, thin mucous cuticle, and extracellular matrix (arrowed).

Fig. 8.8. New genus (Nemertodermatida): base of epidermis (left) and body wall circular and longitudinal muscles, with electron-dense extracellular matrix between (arrowed). Processes from longitudinal muscles reach the matrix (arrowhead).

Fig. 8.9. *Flagellophora* sp. (Nemertodermatida): Circular muscles of body wall, base of epidermal cell and nerve cell processes, with fibrogranular electron-dense matrix between (arrowed).

Fig. 8.10. *Convoluta* sp. nov.: neck of putative adhesive gland cell in epidermis, enwrapped by accessory cell bearing modified cilia. Inset: cross-section of gland neck and accessory cell. The seam (arrowed) where the accessory cell closes on itself is oriented posteriorly.

Fig. 8.11. *Convoluta* sp. nov.: tricellular region of epidermal septate junction, lanthanum impregnated. Epidermal-epidermal septate junction sectioned perpendicularly, epidermal-rhabdoid gland neck septate junctions sectioned tangentially. Note pleated septal structure, and additional perpendicularly oriented strands (arrowed). Abbreviations: ac, accessory cell; cm, body wall circular muscles; cu, mucous cuticle; e, epidermal cell; g, gland cell; lm, body wall longitudinal muscles; n, nerve cell process; tw, terminal web.

—— (1969). *Principles of comparative anatomy of invertebrates*. Vol. 1. University of Chicago Press, Chicago, Il.

Bock, S. (1913). Studien über Polycladen. *Zool. Bidr. Upps.* 2, 31-344.

Boguta, K. K. and Mamkaev, Yu. V. (1972). Structure of the parenchyma of acoelous turbellarians. *Vestn. Leningr. Univ., Biol.* 27, 15-29 [in Russian].

Bowen, I. D. (1980). Phagocytosis in *Polycelis tenuis*. In *Nutrition in the lower Metazoa* (eds D. C. Smith and Y. Tiffon) pp. 1-14. Pergamon Press, Oxford.

——, Ryder, T. A., and Thompson, J. A. (1974). The fine structure of the planarian *Polycelis tenuis* (Iijima). II. The intestine and gastrodermal phagocytosis. *Protoplasma* 79, 1-17.

Clark, R. B. (1979). Radiation of the Metazoa. In *The origin of major invertebrate groups* (ed. M. R. House), Systematics Association Special Vol. 12, pp. 55-102. Academic Press, London.

Crezee, M. (1978). *Paratomella rubra* Rieger and Ott, an amphiatlantic acoel turbellarian. *Cah. Biol. mar.* 19, 1-9.

—— and Tyler, S. (1976). *Hesiolicium* gen. n. (Turbellaria Acoela) and observations on its ultrastructure. *Zool. Scr.* 5, 207-16.

Doe, D. A. (1981). Comparative ultrastructure of the pharynx simplex in Turbellaria. *Zoomorphologie* 97, 133-93.

Dorey, A. E. (1965). The organization and replacement of the epidermis in acoelous turbellarians. *Q. J. microsc. Sci.* 106, 147-72.

Dörjes, J. (1968). Die Acoela (Turbellaria) der deutschen Nordseeküste und ein neues System der Ordnung. *Z. zool. Syst. Evolutionsforsch.* 6, 56-452.

Faubel, A. and Dörjes, J. (1978). *Flagellophora apelti* gen. n. sp. n.: a remarcable [sic] representative of the Order Nemertodermatida (Turbellaria: Archoophora). *Senckenberg. marit.* 10, 1-13.

Graff, L. von (1891). *Die organisation der Turbellaria Acoela*. von Wilhelm Engelman, Leipzig.

—— (1908). Acoela und Rhabdocoelida. In Dr H. G. Bronn's *Klassen und Ordnungen des Tierreichs* (ed. H. G. Bronn) Vol. 4, pp. 1733-2599. C. F. Winter'sche, Leipzig.

Green, C. R. (1981). A clarification of the two types of invertebrate pleated septate junction. *Tissue Cell* 13, 173-88.

—— and Bergquist, P. R. (1982). Phylogenetic relationships within the Invertebrata in relation to the structure of septate junctions and the development of 'occluding' junctional types. *J. Cell Sci.* 53, 279-305.

Hadzi, J. (1963). *The evolution of the Metazoa*. Macmillan, New York.

Hanson, E. D. (1977). *The origin and early evolution of animals*. Wesleyan University Press, Middletown, Ct.

Hendelberg, J. (1977). Comparative morphology of turbellarian spermatozoa studied by electron microscopy. *Acta zool. fenn.* 154, 149-62.

Holt, P. A. and Mettrick, D. F. (1975). Ultrastructural studies of the epidermis and gastrodermis of *Syndesmis franciscana* (Turbellaria: Rhabdocoela). *Can. J. Zool.* 53. 530-49.

Hori, I. (1978). Possible role of rhabdite-forming cells in cellular succession of the planarian epidermis. *J. Electron Microsc.* 27, 89-102.

Hyman, L. H. (1951). *The invertebrates*. Vol. 2. Platyhelminthes and Rhynchocoela. McGraw-Hill, New York.

Ivanov, A. V. and Mamkaev, Yu. V. (1977). Über die Struktur des Digestions-parenchym bei Turbellaria Acoela. *Acta zool. fenn.* 154. 59-61.

Jennings, J. B. (1974). Digestive physiology of the Turbellaria. In *Biology of the Turbellaria* (eds N. W. Riser and M. P. Morse) pp. 173-97. McGraw-Hill, New York.

Karling, T. (1967). Zur Frage von dem systematischen Wert der Kategorien Archoophora und Neoophora (Turbellaria). *Commentat. biol.* 30, 1-11.

—— (1974). On the anatomy and affinities of the turbellarian orders. In *Biology of the Turbellaria* (eds N. W. Riser and M. P. Morse) pp. 1-16. McGraw-Hill, New York.

Klima, J. (1967). Zur Feinstruktur des acoelen Susswasserturbellars *Oligochoerus limnophilus* (Ax und Dörjes). *Ber. Naturw.-med. Ver. Innsbruck* 55, 107-24.

Kozloff, E. (1972). Selection of food, feeding, and physical aspects of digestion in the acoel turbellarian *Otocelis luteola*. *Trans. Am. microsc. Soc.* 91, 556-65.

Lang, A. (1881). Die Polycladen (Seeplanarien) des Golfes von Neapel und der angrenzenden Meeresabschnitte. *Fauna Flora Golf. Neapel* 11, 1-688.

Mamkaev, Yu. V. (1979). On the histological organization of the turbellarian digestive system. *Trudy zool. Inst., Leningr.* 84, 13-24 (in Russian).

—— and Markosova, T. G. (1979). Electron-microscopic study of the parenchyma in Acoela. *Trudy zool. Inst., Leningr.* 84, 7-12 (in Russian).

—— and Seravin, L. N. (1963). Feeding of the acoelous turbellarian *Convoluta convoluta* (Abildgaard). *Zool. Zh.* 47, 197-205 (cited from Boguta and Mamkaev 1972).

Marcus, E. (1945). Sobre Catenulida Brasilieros. *Bolm Fac. Filos. Ciênc. Univ. S. Paulo, Ser. Zool.* 10, 3-133.

Markosova, T. G. (1976). Acid phosphatase activity during digestion in the anintestinal Turbellaria *Convoluta convoluta*. *Zh. Evol. Biokhim. fiziol.* 12, 183-4.

Moraczewski, J. (1977). Asexual reproduction and regeneration of *Catenula* (Turbellaria, Archoophora). *Zoomorphologie* 88, 65-80.

—— (1981). Fine structure of some Catenulida (Turbellaria, Archoophora). *Zoologica Pol.* 28, 367-415.

Pedersen, K. J. (1964). The cellular organization of *Convoluta convoluta*, an acoel turbellarian. A cytological, histochemical, and fine-structural study. *Z. Zellforsch. mikrosk. Anat.* 64, 655-87.

Rieger, R. M. (1976). Monociliated epidermal cells in Gastrotricha: significance for early metazoan evolution. *Z. zool. Syst. Evolutionsforsch.* 14, 198-226.

—— (1981). Morphology of the Turbellaria at the ultrastructural level. *Hydrobiologia* 84, 213-29.

Ruppert, E. E. (1978). The reproductive system of gastrotrichs III. The genital organs of the Thaumastodermatinae subfam. n. and Diplodasyinae subfam. n. with discussion of reproduction in Macrodasyida. *Zool. Scr.* 7, 93-114.

—— and Travis, P. B. (1982). Hemoglobin-containing cells of *Neodasys* (Gastrotricha, Chaetonotida). I. Morphology and ultrastructure. *J. Morph.* 175, 57-64.

Salvini-Plawen, L. von (1978). On the origin and evolution of the lower Metazoa. *Z. zool. Syst. Evolutionsforsch.* 16, 40-88.

Siewing, R. (1980). Das Archicoelomatenkonzept. *Zool. Jb. Abt. Syst.* 103, 439-82.

Smith, J. P. S. (1981a) Fine structural observations on the central parenchyma in *Convoluta* sp. *Hydrobiologia* 84, 259-65.

—— (1981b). Fine-structural anatomy of the parenchyma in the Acoela and Nemertodermatida (Turbellaria). Ph.D. Thesis, University of North Carolina, Chapel Hill, NC.

——, Tyler, S., Thomas, M. B., and Rieger, R. M. (1982). The morphology of turbellarian rhabdites: phylogenetic implications. *Trans. Am. microsc. Soc.* 101, 209-28.

Steinböck, O. (1938). Über die stellung der Gattung *Nemertoderma* Steinböck im System der Turbellarien. *Acta Soc. Fauna Flora fenn.* 62, 1-28.

Sterrer, W. and Rieger, R. (1974). Retronectidae—a new cosmopolitan family of Catenulida (Turbellaria). In *Biology of the Turbellaria* (eds N. W. Riser and M. P. Morse) pp. 63-92. McGraw-Hill, New York.

Tyler, S. (1976). Comparative ultrastructure of adhesive systems in the Turbellaria. *Zoomorphologie* 84, 1-76.

—— (1979). Distinctive features of cilia in metazoans and their significance for systematics. *Tissue Cell* 11, 385-400.

—— (1981). Development of cilia in embryos of the turbellarian *Macrostomum*. *Hydrobiologia* 84, 231-9.

—— (1984a). Ciliogenesis in embryos of the acoel turbellarian *Archaphanostoma*. *Trans. Am. microsc. Soc.* 103, 1-15.

—— (1984b). Turbellarian platyhelminths. In *Biology of the integument* (eds J. Bereiter-Hahn, A. G. Matoltsky and K. S. Richards) Vol. 1. pp. 112-31. Springer, Heidelberg.

—— and Rieger, R. (1975). Uniflagellate spermatozoa in *Nemertoderma* (Turbellaria) and their phylogenetic significance. *Science, N. Y.* 188, 730-2.

—— (1977). Ultrastructural evidence for the systematic position of the Nemertodermatida (Turbellaria). *Acta zool. fenn.* 154, 193-207.

Westblad, E. (1937). Die Turbellarien-Gattung *Nemertoderma* Steinböck. *Acta Soc. Fauna Flora fenn.* 60, 45-89.

—— (1948). Studien über skandinavische Turbellaria Acoela V. *Ark. Zool.* 41, 1-82.

—— (1949). On *Meara stichopi* (Bock) Westblad, a new representative of Turbellaria Archoophora. *Ark. Zool.* Ser. 2 1, 43-57.

9. Phylogenetic relationships within the Platyhelminthes

ULRICH EHLERS

*II Zoologisches Institut und Museum der Universität,
Göttingen, Federal Republic of Germany*

Abstract

Several characteristics found within free-living and parasitic Platyhelminthes are discussed using the methods and principles of phylogenetic systematics. A hypothesis of the phylogenetic relationships among the higher taxa is presented. Evaluating all valid characteristics it is demonstrated that most of the known higher taxa represent natural entities. The autapomorphies of the known and of new monophyletic groups (e.g. Neodermata, Trepaxonemata, Rhabditophora) and the characteristics of the most recent common stem species of the Platyhelminthes are identified. These characteristics are either inherited from the stem species of the Bilateria and Platyhelminthomorpha respectively, or are evolved in the stem-lineage of the taxon Platyhelminthes.

Introduction

Discussions about the taxonomic position of the Platyhelminthes in general or of specific sub-groups such as the Acoela or Polycladida, as well as their importance in the evolution of the Metazoa are long-standing and have led to several quite different conclusions. A major problem has been that no recent analysis of the phylogenetic relationships within all the Platyhelminthes has considered the many morphological features distinguishable at the electron microscope level.

New information on comparative morphology based on light and electron microscopy, from development, reproductive biology, nutrition, locomotion etc. provides valuable pointers in elucidating phylogenetic relationships.

Only a few noteworthy characteristics can be mentioned here and a more comprehensive analysis of the phylogenetic relationships of the Platyhelminthes is given elsewhere (Ehlers 1985; see also Ax 1984 and this volume).

The phylogenetic system presented here commences with the well-known main groups, often called orders in textbooks (e.g. Hyman 1951; de Beauchamp 1961; Kaestner 1984; Barnes 1980) and in the phylogenetic system of the free-living Platyhelminthes given by Karling (1974). Two groups, the Lecithoepitheliata and the Prolecithophora, both of which lack detailed EM studies, are omitted. It is demonstrated that most of the known main groups represent monophyletic taxa (with the possible exception of the 'Typhloplanoida' and 'Dalyellioida'); a few derived characteristics (autapomorphies) of these taxa will be referred to at the end of this paper.

On the premise that the system presented here (which follows the principles and methods of phylogenetic systematics, see e.g. Hennig 1966, Wiley 1981, and Ax 1984) reflects the phylogenetic relationships between the taxa, it is possible to characterize the ancestral or stem species of the Platyhelminthes. Only the characteristics of the most recent common stem species of the Platyhelminthes are significant when discussing the position of the Platyhelminthes within the Metazoa.

Weighting of several characteristic features

1. Neodermis

The bodies of all 'Turbellaria' are covered by an epidermis which is monolayered and normally ciliated and cellular, but may sometimes be aciliated or partially or completely syncytial. Often, the epidermal nuclei are positioned distally (Fig. 9.2(d)), but in species with a weak basal lamina or lacking this intercellular matrix the cell body occasionally lies more proximally below the body musculature (many Acoela, a few Catenulida and Macrostomida, miracidium larva of *Schistosoma*), and in at least several species of Seriata and Lecithoepitheliata such a situation arises by secondarily sunken nuclei (Giesa 1966; Reisinger, Cichocki, Erlach, and Szyskowitz 1974).

The tegument of well-known parasitic taxa, i.e. Trematoda, Monogenea, and Cestoda, is quite distinct from that of free-living species. Without exception, a peripheral syncytium exists with, at least in adults, many small cytoplasmic connections to nucleated regions below the basal lamina (often called pericarya, cytons).

How can these differences be explained? In order to find an answer we have to understand the development of the peripheral epithelium in the different taxa in question. Firstly, it is important to realize that in principle the somatic cells in the Platyhelminthes do not show a mitosis: the epidermal cells are unable to divide when differentiated from blastomeres during embryonic development. But when the embryo or the young worm hatched from the egg continues to grow, the body surface area increases and additional cell material is needed. In free-living Platyhelminthes (including the parasitic 'Dalyellioida') this epidermal material is supplied by differentiation from a reservoir of omnipotent cells, the so-called stem cells or neoblasts, which remain capable of mitosis. When differentiating into an epidermal cell, a cytoplasmic process of a stem cell first penetrates the basal lamina, then the cytoplasm of the differentiating cell occupies an area between the older epidermal cells, and finally the nucleus of the differentiating epidermal cell also normally penetrates the basal lamina. As long as the prospective epidermal cell is situated beneath the basal lamina, the cell is characterized by large amounts of free ribosomes and often clusters of centrioles, but when penetrating the epidermis a rapid differentiation into a typical epidermal cell with all the organelles etc. characteristic for each species in question occurs (unpublished observations).

The differentiation of the tegument in parasitic Platyhelminthes, Trematoda, Monogenea, and Cestoda, is known from several investigations. As an example, I will summarize the observations of Lyons (1973) on the monogenean species *Entobdella solea*: see also Fournier (1976, 1979) for other species of the Monogenea. The oncomiracidium larva of *Entobdella* has a partial covering of ciliated cells arranged in distinct areas of the body, whilst the interciliary cytoplasmic layer is probably a syncytium. Both the ciliated cells and the interciliary cytoplasm are nucleated at an early phase of embryonic development (Fig. 9.1(a)). In a later stage, when the ciliated cells have flattened, the nuclei of these cells degenerate whereas the nuclei of the interciliary cytoplasm project clear of the surface and may be lost (Fig. 9.1(b)). The basal lamina begins to differentiate and beneath this lamina stem cells occur. The epidermal cells of the hatched larva lack nuclei, and beneath these ciliated cells the stem cells develop cytoplasmic processes which form the presumptive layer of the worm (Fig. 9.1(c)). These processes seem to make contact with the now anucleate interciliary cytoplasm. At the end of the free-swimming larval stage the anucleate ciliary cells are cast off and the discontinuous cytoplasm of the definitive tegument spreads out to form a continuous syncytium (Fig. 9.1(d)). Subsequently the nucleated regions beneath the basal lamina establish additional connections with the increasing peripheral cytoplasm. In all Monogenea shedding of the ciliated epidermal cells occurs when the larvae infect the host, and all the post-larval stages possess a newly differentiated syncytial dermis or neodermis.

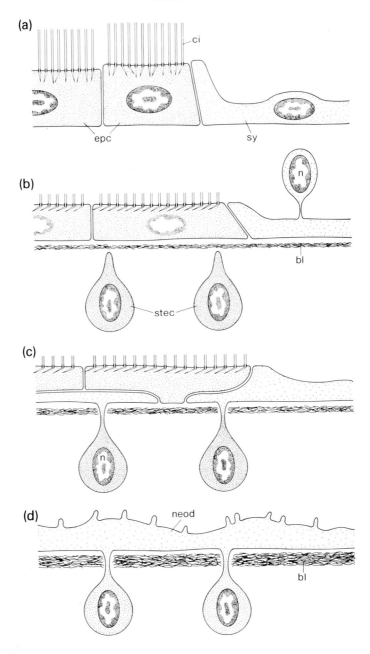

Fig. 9.1. Shedding of the ciliated epidermal cells and differentiation of a neodermis in the oncomiracidium larva of Monogenea (after Lyons 1973). See text for explanations. Abbreviations: bl, basal lamina; ci, cilium; epc, epidermal cells; n, nucleus; neod, neodermis; stec, stem cells; sy, syncytium.

In the Trematoda (Aspidobothrii and Digenea), the same phenomenon as found in the Monogenea has been observed (Køie, Christensen, and Nansen 1976; Fredericksen 1978; Meuleman, Lyaruu, Khan, Holzmann, and Sminia 1978; Coil 1981). The miracidium larvae of Digenea have ciliated cells separated from their neighbours by material which is often called 'ridge cytoplasm'. This material represents part of the definitive epithelium, the neodermis. The epidermal cells are shed at the end of the larval stage, normally when entering the host. The interciliary ('ridge') cytoplasm spreads out on the surface to cover the post-larvae, with further cytoplasmic material being added to the neodermis by cells differentiating from stem cells.

The lycophore and oncosphere larvae of the Cestoda are covered by a nucleated syncytial epidermis, with or without cilia, normally called the embryophore (Grammeltvedt 1973; Fairweather and Threadgold 1981; Rohde and Georgi 1983). This embryophore, often internally strengthened with structural proteins, is shed when the host is infected, and the post-larval stages always have a typical neodermis.

Differentiation of the typical neodermis in Trematoda, Monogenea, and Cestoda, together with the shedding of the epidermis, is unique amongst invertebrates and does not occur in other platyhelminth taxa.

There has been no discussion about the likelihood that the loss of the epidermis and differentiation of the neodermis represent secondarily evolved features. Evidence which supports the homology of this feature between these parasitic taxa is discussed elsewhere (Ehlers 1985).

The likelihood of homology is also supported by the observation that the loss of the epidermis and differentiation of a neodermis is not a prerequisite of parasitism within the Platyhelminthes. For example, in the digestive gland of the common cockle *Cerastoderma edule*, post-larval stages of digeneans may occur which are always covered with a neodermis, whereas *Paravortex cardii*, a parasitic species of the 'Dalyellioida', also living in the digestive gland, possesses a ciliated, cellular epidermis (MacKinnon, Burt, and Pike 1981). As another example, the larva of *Austramphilina elongata* (Cestoda) sheds its ciliated epidermis when penetrating the cuticle of the host, a crayfish or shrimp (Rohde and Georgi 1983), whilst the larva of *Kronborgia amphipodicola* ('Dalyellioida') when penetrating the cuticle of an amphipod does not shed its ciliated epidermis (Køie and Bresciani 1973), and adult *Kronborgia* still have such an epidermis.

The most parsimonious explanation is to accept the existence of a common ancestor of the Trematoda, Monogenea, and Cestoda, so that these taxa together constitute the monophyletic taxon Neodermata. The character 'neodermis' is evolved in the stem-lineage of this natural group Neodermata, the most recent common stem species of the Neodermata possessing such a neodermis, an autapomorphy of the taxon

Neodermata, and passing on this autapomorphy to its descendants to form a synapomorphy of the Trematoda and Cercomeromorpha (Monogenea and Cestoda).

2. Heterocellular female gonad

In the Catenulida, Nemertodermatida, Acoela, Macrostomida, and Polycladida homocellular female ovaries occur, with endolecithal eggs. On the other hand, the taxa Seriata and Rhabdocoela (including Neo-dermata) and the taxa not shown in Fig. 9.3, the Prolecithophora and perhaps also the Lecithoepitheliata, exhibit heterocellular female gonads with ovocytes and yolk cells, the former being produced in layers called germaria, and the latter in the vitellaria. In all the taxa with heterocellular female gonads the eggs are ectolecithal, i.e. each egg is composed of different cells, normally one ovocyte (zygote) and several yolk cells.

The homocellular female ovary is plesiomorphic. Heterocellular female gonads are derived and constitute a synapomorphy of the taxa Seriata and Rhabdocoela (including the Neodermata). The interpret-tation that the apomorphous conditions of the female gonads in the taxa Seriata, Rhabdocoela and at least the Prolecithophora are homo-logous is corroborated by new findings on the fine structure of the gonad cells, especially on the genesis of the eggshell droplets in the vitellocytes (unpublished observations; see also Bunke 1981, 1982). Thus, the monophyly of the taxon Neoophora (including the Seriata, Rhabdocoela, Prolecithophora, and perhaps also the Lecithoepitheliata) seems warranted.

3. 9 + '1' pattern of the axoneme in the spermatozoa

Platyhelminthes have strongly divergent spermatozoa, depending on the special mode of propagation in connection with internal fertilization (Henley 1974; Hendelberg 1977).

In the spermatozoon of the Nemertodermatida only one cilium oc-curs with a 9 + 2 axonemal pattern. The Acoela also possess biciliary spermatozoa with the typical 9 + 2 pattern, although sometimes one or both central microtubules are reduced, so that a 9 + 1 or 9 + 0 pattern occurs. The 9 + 1 pattern of the Acoela does not correspond with the 9 + '1' pattern described below. The Catenulida and the Macrostomida have aciliary spermatozoa.

In Polycladida, Seriata, 'Typhloplanoida' (including Kalyptorhyn-chia), 'Dalyellioida', and all Neodermata, which normally possess biciliary spermatozoa, the pattern of axonemal fine structure is unique. Instead of the two central singlets, an unpaired central core occurs which is not homologous with a single microtubule. In these taxa this axial unit always shows a central electron-dense element, an intermediary transparent zone, and a peripheral electron-dense sheath, with nine spoke-like radiations passing from this sheath to the A-subtubules of the peripheral doublets (Fig. 9.2(a)). In tangential sections (Fig. 9.2(b)) or in negatively stained 9 + '1' axonemes several spirally coiled longitudinal ribbons resembling a helix can be seen in the peripheral electron-dense sheath of the axial unit (Thomas 1975).

This axial unit is without question a highly apomorphic feature and indicates the close relationships between the Polycladida, Seriata, and all Rhabdocoela. These taxa together constitute a monophyletic taxon, the Trepaxonemata, with a common ancestral species.

4. Protonephridium

The only type of excretory organs which exist within the Platyhelminthes are protonephridia, which must be inherited from the most recent common stem species of the Bilateria and the Platyhelminhomorpha respectively (Ax, this volume). The basic characteristic protonephridium shows different structures in distinct taxa of the Platyhelminthes and can therefore be used for clarifying phylogenetic relationships.

In many taxa, the proximal parts of the protonephridia, the cyrtocytes or flame bulbs, are composite structures formed from the terminal cell and the neighbouring tubule cell, and showing two rows of cytoplasmic projections. The inner rows project from the terminal cell, which also bears a tuft of cilia, while the outer rows lead from the canal or tubule cell (Ax, this volume, Fig. 11.5(e,f)). More distally, this tubule cell 'is in the form of a single sheet wrapped around to form a tube, with a desmosome running along the length of the contiguous surfaces' (Wilson and Webster 1974).

Such a two-cell weir is found in the Macrostomida, Seriata, Aspidobothrii, Digenea, Monogenea, and Cestoda. These taxa, as well as many other Bilateria, possess a paired protonephridial system, primarily with two pores. On the other hand, the Catenulida always have an unpaired system, which represents a specific derived feature. In all Catenulida examined so far the terminal cell always bears two cilia, the long rootlets of which project along the narrow side strengthening the weir (Ax, this volume, Fig. 11.5(c,d)). Such a structure, which is unknown in other Bilateria, represents an autapomorphy of the taxon Catenulida. In the Catenulida the weir is formed by the terminal cell only. This is a plesiomorphic feature which can also be found, for example, in the Gnathostomulida, and has been inherited from the most recent common stem species of the Platyhelminthes (for further

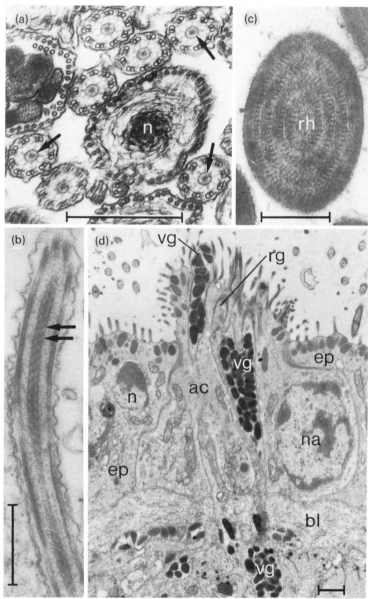

Fig. 9.2. (a and b), 9 + '1' pattern (arrows) of the ciliary axoneme in the spermatozoa of Trepaxonemata. (a) *Notocaryoplanella glandulosa*, Seriata, cross-section, × 70 000. (b) *Kataplana mesopharynx,* Seriata, longitudinal section, × 48 000. (c) Lamellated rhabdite of Rhabditophora (*Paromalostomum fusculum*, Macrostomida, cross-section, × 43 000). (d) Duo-gland adhesive system of Rhabditophora (*Coelogynopora axi*, Seriata, cross section, × 15 000). Abbreviations: ac, anchor cell; bl, basal lamina; ep, epidermis; n, nucleus; na, nucleus of anchor cell; rg, releasing gland neck; rh, lamellated rhabdite; vg, viscid gland and viscid gland neck. Scale bars in each represent 0.5 μm.

details see Ax, this volume). This statement also implies that the composite structure of the weir in Macrostomida, Seriata, and the different taxa of Rhabdocoela is derived. According to the principle of parsimony this two-cell weir has not independently evolved several times but is a common derived feature. Together these taxa constitute a monophyletic taxon called Rhabditophora. Within the Rhabditophora this two-cell weir seems to be secondarily modified in distinct triclads, in species of the 'Dalyellioida' and in the Lecithoepitheliata.

The Nemertodermatida and Acoela do not possess protonephridia. Since there is clear evidence that excretory organs are present in the most recent common stem species of the Platyhelminthes (Ax, this volume), the absence of such a system must be secondarily acquired. The loss of the protonephridia is one of several features that links the Nemertodermatida with the Acoela (Tyler and Rieger 1977; Smith and Tyler, this volume) and again, according to the principle of parsimony, I regard this derived characteristic as a common feature reflecting relationships either as a synapomorphy of these two taxa or an autapomorphy of the Acoelomorpha.

5. Lamellated rhabdites

Epidermal glandular cells can be found in all platyhelminth taxa. The granules of these cells differ from each other with regard to their morphology and histochemistry. Within distinct taxa, a special kind of granule exists, called 'true rhabdites' by Smith, Tyler, Thomas, and Rieger (1982). These authors define rhabdites as 'rod-shaped secretions . . . with one to several concentric striated lamellae constituting its cortex, and with a concentrically lamellated, granulated, or homogeneous medulla'. Such lamellated rhabdites, which are unknown in the Metazoa outside the Platyhelminthes, are found in all the Macrostomida (Fig. 9.2(c)) and Polycladida and in several species of Seriata, 'Typhloplanoida' and 'Dalyellioida'. Lamellated rhabdites do not occur in Catenulida, Nemertodermatida, Acoela, and Neodermata.

There is evidence that the lamellated rhabdites of Macrostomida, Polycladida, Seriata, 'Typhloplanoida' and 'Dalyellioida' are homologous (Smith, Tyler, Thomas, and Rieger 1982). These granules are undoubtedly apomorphic features and according to parsimony represent a common derived characteristic of the above-named taxa, an autapomorphy of the taxon Rhabditophora.

The lack of lamellated rhabdites in the Catenulida, Nemertodermatida, and Acoela represents a plesiomorphous characteristic, whereas the lack of the lamellated rhabdites in the Neodermata occurred secondarily and perhaps represents another autapomorphy of this taxon.

6. Duo-gland adhesive system

Many species of free-living Platyhelminthes are able to adhere to surfaces by means of distinct parts of the epidermis. These prominent structures, often called adhesive papillae, were studied by Tyler (1976) at the EM level. Adhesive structures exist, which are always composed of three different cell types (Fig. 9.2(c)):

(a) one (or several) modified epidermal cell(s) or a part of an epidermal syncytium bearing straight microvilli with fibrous cores;

(b) a gland cell type with large dense membrane-bound secretion granules; and

(c) a second gland cell type characterized by small and less dense membrane-bound secretion granules.

The large secretion granules are of an adhesive substance, which attaches the modified microvilli of the epidermis to surfaces. Tyler (1976) called this type of gland the viscid gland, and the specialized epidermal cell an anchor cell. The other type of gland, with its small less dense granules, is called the releasing gland, its secretion apparently releasing the anchor cells from surfaces.

Such a duo-gland adhesive system occurs in the Macrostomida, Polycladida, Seriata, and 'Typhloplanoida' (including the Kalyptorhynchia). There is clear evidence concerning the homology of the systems between these taxa. Duo-gland adhesive systems can also be found in other metazoan taxa, namely Gastrotricha, and in a few species of Nematoda and Annelida, but the systems within these taxa differ from those of Platyhelminthes (Rieger and Tyler 1979; Tyler and Melanson 1979).

The platyhelminth duo-gland adhesive system appears to be a feature evolved within this taxon. Because the Catenulida, Nemertodermatida, and Acoela lack this characteristic, the system is evolved in the stem-lineage of the Rhabditophora, i.e. the system is another autapomorphy of this taxon.

The 'Dalyellioida' and Neodermata do not possess the duo-gland adhesive system: in these taxa the absence must be secondarily acquired. The Neodermata lose their epidermis, and thus the ability to differentiate epidermal anchor cells, at the end of the larval phase. The majority of the 'Dalyellioida' live in the phytal or on mud: this might hold true for a common stem species of this taxon or of the Doliopharyngiophora, respectively. In either habitat, an adhesive system is of limited use and could thus be reduced. The absence of the system might be a synapomorphy of the 'Dalyellioida' and Neodermata, an autapomorphy of the Doliopharyngiophora.

Characterization of the most recent common stem species of the taxon Platyhelminthes and of the subordinate taxa

On the basis of many previously known characteristics it seems possible to explain the phylogenetic relationships between the higher monophyletic taxa of the Platyhelminthes (Fig. 9.3). The phylogenetic system presented here corresponds in many aspects to the system of free-living taxa published by Karling (1974) and to the ideas of Bychowsky (1937, 1957) and Llewellyn (1965, 1970) concerning the interrelationships of parasitic taxa (see also Brooks 1982).

In the system two names often used in systematics of the Platyhelminthes, are lacking. First, the name 'Archoophora'. The Catenulida, Nemertodermatida, Acoela, Macrostomida, and Polycladida are classified under this name (e.g. Barnes 1980); the specific features of the 'Archoophora' are the homocellular female gonad and the endolecithal eggs. Both features are plesiomorphies, the 'Archoophora' lacking any autapomorphy. The taxon 'Archoophora' is thus a paraphyletic group and not monophyletic. The 'Archoophora' have a common stem species, the common stem species of the Platyhelminthes, but not all of the descendants of this stem species belong to the 'Archoophora', the Neoophora being set aside.

The second name not used here is 'Turbellaria'. Here, we have the same situation as in the 'Archoophora'. All the characteristic features of the 'Turbellaria' (free-living, body covered by a ciliated epidermis, etc.) are plesiomorphies; the 'Turbellaria' without doubt represent a paraphyletic group.

All the plesiomorphies like 'homocellular female gonad' or 'body covered by a ciliary epidermis' must be present in the most recent common stem species of the Platyhelminthes.

By means of a comprehensive analysis of all the characteristics studied one can characterize this most recent common stem species. I can only mention several of these features here, a detailed discussion appearing elsewhere (Ax 1984, this volume; Ehlers 1985):

1. Characteristics inherited from the most recent common stem species of the Bilateria: bilateral symmetry; small body size; round profile of the body; no external cuticle; one-layered cellular epidermis, entirely ciliated; basal lamina lacking or very thin; ciliary locomotion; nervous system infraepidermal-subepidermal; concentration of nerve cells at the anterior end forming a cerebrum; monociliary epidermal sensory cells, no multiciliary sensory cells; no statocyst; mainly or exclusively smooth musculature, weakly differentiated; without circulatory system; without coelom, primary body cavity with only a few

cells; one pair of protonephridia and one pair of pores, formation of the weir by the terminal cell only; simple mouth pore, originating from the blastoporus; intestine with neither proctodaeum nor anus; homocellular female gonad (ovary); one (male) genital pore present (no female pore); endolecithal eggs; spiral (quartet) cleavage; without a larva (no pelago-benthic life cycle); bottom dweller.

2. Characteristics inherited from the most recent common stem species of the Platyhelminthomorpha: hermaphroditism; direct sperm transfer; internal fertilization; modified spermatozoa.

3. Characteristics evolved in the stem-lineage of the Platyhelminthes (autapomorphies of the taxon Platyhelminthes): multiciliary epidermis (0.2-1.8 cilia/μm^2 epidermis); epidermal cilia without accessory centrioles; terminal cell of the protonephridia with two cilia (without accessory centrioles) and without the primitive eight microvilli-rods (Ax, this volume); no mitosis of the epidermal cells and other somatic cells.

Several autapomorphic characteristics of the subordinate taxa which are named in Fig. 9.3:

(Black square 1) Catenulida von Graff 1905: unpaired protonephridium; special organization of the cyrtocyte; dorsally located male genital porus; aciliary spermatoza.

(2) Euplatyhelminthes Bresslau and Reisinger 1928: increase in ciliary density of the epidermis (3-6 cilia/μm^2 epidermis); frontal glands.

(3) Acoelomorpha new monophyletic taxon: epidermal cilia with a shelf-like termination; rostral rootlet of the epidermal cilia with a knee-like bend; posterior rootlet of the epidermal cilia with two fibre bundles; reduction of the protonephridia.

(4) Nemertodermatida Steinböck 1931: statocyst with two statoliths.

(5) Acoela Uljanin 1870: rostral rootlet of the epidermal cilia with two lateral rootlets; statocyst with tubular body, etc.; biciliary spermatozoa; duet cleavage.

(6) Rhabditophora new monophyletic taxon: lamellated rhabdites; duo-gland adhesive system; duo-cell weir of the protonephridia; multiciliary terminal cells of the protonephridia.

(7) Macrostomida (+ Haplopharyngida) von Graff 1882: all glands of the duo-gland adhesive system emerge in one collar of modified microvilli; aciliary spermatozoa.

(8) Trepaxonemata new monophyletic taxon: biciliary spermatozoa; ciliary axoneme of the spermatozoa with a dense core (9 + '1' pattern).

(9) Polycladida Lang 1881: intestine with many highly branched diverticula; resorption of certain blastomeres (macromeres 4A-4D and micromeres 4a-4c).

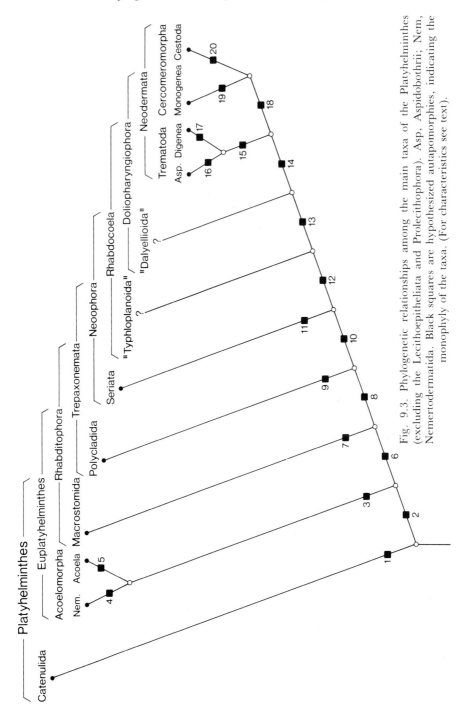

Fig. 9.3. Phylogenetic relationships among the main taxa of the Platyhelminthes (excluding the Lecithoepitheliata and Prolecithophora). Asp, Aspidobothrii; Nem, Nemertodermatida. Black squares are hypothesized autapomorphies, indicating the monophyly of the taxa. (For characteristics see text).

(10) Neoophora Westblad 1948: heterocellular female gonad with ovocytes and vitellocytes; ectolecithal eggs.

(11) Seriata Bresslau 1928-33: pharynx tubiformis; serial arrangement of follicular gonads (testes, vitellaria).

(12) Rhabdocoela Ehrenberg, 1831: unspecialized pharynx bulbosus.

(?) 'Typhloplanoida' (+ Kalyptorhynchia) von Graff 1905: no autapomorphy known.

(13) Doliopharyngiophora new monophyletic taxon: pharynx doliiformis with a terminal-subterminal mouth-opening; reduction of the duo-gland adhesive system.

(?) 'Dalyellioida' (+ Temnocephalida) Meixner 1924: no autapomorphy known.

(14) Neodermata new monophyletic taxon: shedding of the epidermis (cells) at the end of a larval phase; post-larval stages with a syncytial neodermis; cilia of the (larval) epidermis with only one (rostral) rootlet; epithelial sensory cells with EM-dense collars.

(15) Trematoda Rudolphi 1808: epidermal cells of the larvae separated from their neighbour by material of the neodermis; parasites of molluscs.

(16) Aspidobothrii Burmeister 1856: cotylocidium larva with a posterior-ventral disc; development of this disc to a huge ventral sucker with alveoli in the adult; neodermis with specialized microvilli in form of microtubercles; oviducts divided into chambers by septa.

(17) Digenea van Beneden 1858: tiers of epidermis cells in the miracidium larva; cercariae and adult with a ventral sucker; early developing stages (miracidium, sporocyst) without a digestive system; complex life cycle (changes of stages of development and of the hosts).

(18) Cercomeromorpha Bychowsky 1937: larvae and (at least the younger) post-larval stages with 16 marginal hooks.

(19) Monogenea van Beneden 1858: oncomiracidium larva bears three zones of ciliated epidermal cells (one in the middle and one at each end); four (pigmented or unpigmented) rhabdomeric eyespots.

(20) Cestoda Gegenbaur 1859: 10 marginal hooks in the larvae and the post-larval stages (existent in the Gyrocotylidea and Amphilinidea; six hooks only in the Cestoidea, i.e. the Caryophyllidea + Eucestoda); all stages of development and the adult without a digestive system.

References

Ax, P. (1984). *Das phylogenetische System. Systematisierung der lebenden Natur aufgrund ihrer Phylogenese.* G. Fischer, Stuttgart.

Barnes, R. D. (1980). *Invertebrate zoology* (4th edn). Saunders College, Philadelphia.

Beauchamp, P. de (1961). Classe des Turbellariés. In *Traité de zoologie* (ed. P. P. Grassé) Vol. 4, 1, pp. 35-212. Masson et Cie, Paris.

Brooks, D. R. (1982). Higher level classification of parasitic Platyhelminthes and fundamentals of cestode classification. In *Parasites—their world and ours* (ed. D. F. Mettrick and S. S. Desser) pp. 189-93. Elsevier Biomedical Press, Amsterdam.

Bunke, D. (1981). Ultrastruktur- Untersuchungen an Vitellocyten von *Microdalyellia fairchildi* (Turbellaria, Neorhabdocoela). *Zoomorphologie* 99, 71-86.

—— (1982). Ultrastruktur- Untersuchungen zur Eischalenbildung bei *Microdalyellia fairchildi* (Turbellaria). *Zoomorphologie* 101, 61-70.

Bychowsky, B. E. (1937). Ontogenese und phylogenetische Beziehungen der parasitischen Plathelminthes. *Izv. Akad. Nauk SSSR, Biol.* 4, 1353-83 (in Russian).

—— (1957). *Monogenetic trematodes: their systematics and phylogeny.* (in Russian). Akad, Nauk. SSSR, Moscow. (English translation, 1961: ed. W. J. Hargis. American Institute of Biological Science, Washington, D.C.).

Coil, W. H. (1981). Miracidial penetration in *Fascioloides magna* (Trematoda). *Z. Parasitkde* 65, 299-307.

Ehlers, U. (1985). *Das phylogenetische System der Plathelminthes.* Akademie der Wissenschaften und der Literatur, Mainz and G. Fischer, Stuttgart.

Fairweather, I. and Threadgold, L. T. (1981). *Hymenolepis nana*: the fine structure of the embryonic envelopes. *Parasitology* 82, 429-43.

Fournier, A. (1976). Le tégument d'*Euzetrema knoepffleri*, monogène parasite d'amphibien: ultrastructure et évolution au cours du cycle biologique. *Annls Parasit.* 51, 15-26.

—— (1979). Evolution du tégument des *Polystoma* (Mongènes Polystomatidae) au cours du cycle. *Z. ParasitKde* 59, 169-85.

Fredericksen, D. W. (1978). The fine structure and phylogenetic position of the cotylocidium larva of *Cotylogaster occidentalis* Nickerson 1902 (Trematoda: Aspidogastridae). *J. Parasit.* 64, 961-76.

Giesa, S. (1966). Die Embryonalentwicklung von *Monocelis fusca* Oersted (Turbellaria, Proseriata). *Z. Morph. Ökol. Tiere* 57, 137-230.

Grammeltvedt, A. F. (1973). Differentiation of the tegument and associated structures in *Diphyllobothrium dendriticum* Nitsch (1824) (Cestoda: Pseudophyllidea). An electron microscopical study. *Int. J. Parasit.* 3, 321-7.

Hendelberg, J. (1977). Comparative morphology of turbellarian spermatozoa studied by electron microscopy. *Acta zool. fenn.* 154, 149-62.

Henley, C. (1974). Platyhelminthes (Turbellaria). In *Reproduction of marine invertebrates* (eds A. C. Giese and J. S. Pearse) Vol. 1, pp. 267-343. Academic Press, New York.

Hennig, W. (1966). *Phylogenetic systematics.* University of Illinois Press, Urbana, Il.

Hyman, L. H. (1951). *The invertebrates.* Vol. 2. *Platyhelminthes and Rhynchocoela, the acoelomate Bilateria.* McGraw-Hill, New York.

Kaestner, A. (1984). *Lehrbuch der Speziellen Zoologie* Vol. 1, (4th edn). G. Fischer, Stuttgart.

Karling, T. G. (1974). On the anatomy and affinities of the turbellarian orders. In *Biology of the Turbellaria* (eds N. W. Riser and M. P. Morse) pp. 1-16. McGraw-Hill, New York.

158   *Ulrich Ehlers*

Køie, M. and Bresciani, J. (1973). On the ultrastructure of the larva of *Kronborgia amphipodicola* Christensen and Kanneworff, 1964 (Turbellaria, Neorhabdocoela). *Ophelia* 12, 171-203.

—— Christensen, N. Ø. and Nansen, P. (1976). Stereoscan studies of eggs, free-swimming and penetrating miracidia and early sporocysts of *Fasciola hepatica*. *Z. ParasitKde* 51, 79-90.

Llewellyn, J. (1965). The evolution of parasitic platyhelminths. In *Evolution of parasites* (ed. A. E. R. Taylor) (*Symp. British Soc. Parasitol.* Vol. 3), pp. 47-78. Blackwell Scientific Publications, Oxford.

—— (1970). Taxonomy, genetics and evolution of parasites—Monogenea. *J. Parasit.* 56, 493-504.

Lyons, K. M. (1973). Epidermal fine structure and development in the oncomiracidium larva of *Entobdella solea* (Monogenea). *Parasitology* 66, 321-33.

MacKinnon, B. M., Burt, M. D. B., and Pike, A. W. (1981). Ultrastructure of the epidermis of adult and embryonic *Paravortex* species (Turbellaria, Eulecithophora). *Hydrobiologia* 84, 241-52.

Meuleman, E. A., Lyaruu, D. M., Khan, M. A., Holzmann, P. J. and Sminia, T. (1978). Ultrastructural changes in the body wall of *Schistosoma mansoni* during the transformation of the miracidium into the mother sporocyst in the snail host *Biomphalaria pfeifferi*. *Z. ParasitKde* 56, 227-42.

Reisinger, E., Cichocki, J., Erlach, R., and Szyskowitz, T. (1974). Ontogenetische Studien an Turbellarien: ein Beitrag zur Evolution der Dotterverarbeitung im ektolecithalen Ei, part I and II. *Z. zool. Syst. Evolutionsforsch.* 12, 161-95; 241-78.

Rieger, R. and Tyler, S. (1979). The homology theorem in ultrastructural research. *Am. Zool.* 19, 655-64.

Rohde, K. and Georgi, M. (1983). Structure and development of *Austramphilina elongata* Johnston, 1931 (Cestodaria: Amphilinidea). *Int. J. Parasit.* 13, 273-87.

Smith, J., Tyler, S., Thomas, M. B., and Rieger, R. M. (1982). The morphology of turbellarian rhabdites: phylogenetic implications. *Trans. Am. microsc. Soc.* 101, 209-28.

Thomas, M. B. (1975). The structure of the 9 + 1 axonemal core as revealed by treatment with trypsin. *J. Ultrastruct. Res.* 52, 409-22.

Tyler, S. (1976). Comparative ultrastructure of adhesive systems in the Turbellaria. *Zoomorphologie* 84, 1-76.

—— and Melanson, L. (1979). Cytochemistry of adhesive organs in lower metazoans. *Am. Zool.* 19, 985.

—— and Rieger, R. M. (1977). Ultrastructural evidence for the systematic position of the Nemertodermatida (Turbellaria). *Acta zool. fenn.* 154, 193-207.

Wiley, E. O. (1981). *Phylogenetics. The theory and practice of phylogenetic systematics.* John Wiley, New York.

Wilson, R. A. and Webster, A. (1974). Protonephridia. *Biol. Rev.* 49, 127-60.

10. The phylogenetic relationships within the Seriata (Platyhelminthes)

BEATE SOPOTT-EHLERS

II Zoologisches Institut und Museum der Universität, Göttingen, Federal Republic of Germany

Abstract

The phylogenetic relationships among the main taxa of the Seriata are discussed on the basis of a number of characters distinguishable by light and electron microscopy. The common stem species of the Seriata is characterized and a tentative scheme for the phylogenetic relationships within the Seriata is given.

Introduction

Within the free-living Platyhelminthes the triclads or planarians probably represent the best known group. Many ecological, karyological, and morphological studies, as well as intensive research concerning regeneration and cell differentiation, have been carried out on these platyhelminths, which occur in marine, limnetic, and terrestrial environments (Benazzi and Benazzi Lentati 1976; Ball and Reynoldson 1981; Benazzi and Gremigni 1982). The systematic relationships of different triclad groups have been discussed by Ball (1981).

The Tricladida belong to the taxon Seriata. The position of the Seriata within the Platyhelminthes is discussed by Ehlers (this volume). Besides the Tricladida, other less well known groups belong to the Seriata. These groups are predominantly marine (Sopott 1972) and are called Proseriata (Meixner 1938; Ax 1963; Karling 1974; Barnes 1980; Crezée 1982) or, less commonly, Protricladida (de Beauchamp 1961).

To the Proseriata belong the Monocelididae von Graff 1904-08, Monotoplanidae Ax 1958, Nematoplanidae Meixner 1938, Polystyliphoridae Ax 1958, Coelogynoporidae Steinböck 1924, Otoplanidae Hallez 1894, and two monotypic taxa known from the limnetic environment, the Otomesostomatidae Midelburg 1908 and the Bothrioplanidae von Graff 1904-08. Recent EM studies have revealed several new characteristics (epithelial sensory cells and glandular cells) and have provided additional information on previously known features (e.g. statocysts). This is the first attempt to provide a theory on the phylogenetic interrelationships of these proseriate taxa and of the position of the Tricladida within the Seriata.

Discussion of the phylogenetic relationships

1. Seriata Bresslau 1928-33 (syn. Metamerata Reisinger 1924)

The Seriata share certain common characteristics in their digestive and reproductive systems which must have been present in the stem species of this taxon (Karling 1974). One of these attributes is an unbranched rod-shaped and more or less lobed intestine. A pharynx plicatus leads into this intestine, runs primarily in a horizontal direction parallel to the longitudinal axis of the body and has a posteriorly directed opening, the so-called pharynx tubiformis.

Male and female gonads of the Seriata show a typical follicular development and serial arrangement. The follicular testes form one or two longitudinal rows. The female gonads are always separated into germ- and yolk-cell producing regions. The yolk cells or vitellaria normally form lateral cords of follicles, while the germaria are as a rule developed as one pair of compact organs which lie near the pharynx.

2. Lithophora Steinböck 1925 (syn. Crossocoela von Graff 1904–08)

This common organization of the Seriata is primarily found in all representatives of the Lithophora, but the pharynx plicatus secondarily varies in some species, e.g. within the Otoplanidae (Ax 1956). The Lithophora (Monocelididae, Coelogynoporidae, Otoplanidae, Monotoplanidae, Otomesostomatidae) differ from all other Seriata in having a special feature, a statocyst which is invariably situated ventrocranially to the brain. The structure of this organ has been investigated by previously unpublished EM studies of the statocysts in 12 different species of Lithophora. The statocyst has an external capsule formed by intercellular material, which separates the cells within the organ from the surrounding tissue. In all Lithophora studied the arrangement of these internal cells correspond to each other. The cell nuclei have an identical position in all the statocysts:

(a) The nuclei of two cells, more rarely of one, are situated dorsally just beneath the capsule. Ventrally there are the nuclei of two more cells. In some species there may be the nuclei of one or two additional cells. The cytoplasm of these cells, which supposedly produce the capsule of the statocyst, almost completely border the central part of the statocyst.

(b) The centre of the statocyst is mainly occupied by a large cell, which forms the statolith. The nucleus of this cell is strongly lobed and dorsoventrally flattened. The unpaired statolith shows concentric layers and is secreted into a vacuole in the middle of the statolith forming cell.

(c) A further four to eight smaller cells occur in the central part, their nuclei being called 'Nebensteinchen' by earlier light microscopists. These nuclei form two groups of two or four nuclei located ventro-frontally to laterally to the statolith. Between these 'Nebensteinchen' cells, the statolith-forming cell and the peripheral statocyst cells respectively, a more or less voluminous intercellular space is developed. From the 'Nebensteinchen' cells fine cytoplasmic processes lead to the dorsal part of the statocyst, where they communicate with processes of extra-capsular nerve cells entering the statocyst from the dorsal border. The 'Nebensteinchen' cells supposedly play an important role in transmitting stimuli initiated by the statolith to nerve cells, which lie outside the statocyst, i.e. the 'Nebensteinchen' cells should be regarded as neurons.

A statocyst of this type is not known from other Platyhelminthes. The statocysts of some Catenulida, as well as the Nemertodermatida and Acoela, show a different construction. The lithophoran statocysts are also quite different from those of other Metazoa, such as the Cnidaria (Horridge 1969; Singla 1975) or Ctenophora (Krisch 1973; Aronova 1974; Vinnikov, Aronova, Khakeevich, Tsirulis, Lavrova, and Natochin 1981).

The statocyst of the Lithophora thus represents a specific feature of the Monocelididae, Coelogynoporidae, Otoplanidae, Monotoplanidae, and Otomesostomatidae, upon which the close relationship of these taxa is based. There are no indications of secondary reduction of this organ in other groups of Seriata, which do not possess a statocyst.

3. Unguiphora new monophyletic taxon

Like the Lithophora, the Nematoplanidae and Polystyliphoridae also possess serially arranged follicular testes and vitellaria. In contrast to the single pair in the Lithophora, there are two to six pairs of germaria in the nematoplanids and polystyliphorids. The germaria also form follicles and have a serial arrangement: for the Nematoplanidae and Polystyliphoridae see Ax (1966, 1958) respectively. The Nematoplanidae and Polystyliphoridae are further characterized by the existence of a claw-shaped stylet in the male copulatory organ; such a stylet, which differentiates in a very specific manner, is unknown from other Seriata.

These criteria, i.e. the two to six pairs of follicular germaria and a claw-shaped stylet demonstrate the close relationship between the Nematoplanidae and the Polystyliphoridae, which together form the new taxon Unguiphora. The multiplication of the claw-shaped stylet by way of serially arranged prostatoid organs is a characteristic feature of the Polystyliphoridae.

In the Nematoplanidae there are some unique peculiarities, shown by EM studies. These are the structure of the epidermal adhesive organs (pit-shaped appearance, large numbers of anchor cells, the nuclei of which may be sunk, etc: Sopott-Ehlers 1979), and a specific organization of the proximal parts of the protonephridial system (broad cytoplasmic processes from the terminal cell and its adjacent canal cell leading between the cilia and the basket-like extensions of these two cells; unpublished observations).

A close relationship between the Nematoplanidae and Polystyliphoridae is also demonstrated by fine structural features. Specific epidermal sense cells, called collar receptor cells, are widespread within free-living platyhelminths (Ehlers 1977). The collar receptors in the Lithophora and Unguiphora are situated within the epidermis and represent the distal part of a sense cell whose nucleus lies in deeper parts of the body. Each collar receptor consistently bears one kinocilium, which lacks a typical rootlet and which is always surrounded by eight specialized microvilli, called stereocilia (Ehlers and Ehlers 1977). At the level of the basal body of the kinocilium runs a fine striped annular rootlet. The collar receptors of free-living Rhabdocoela are similarly differentiated (Ehlers and Ehlers 1977).

The collar receptors of the Unguiphora show a specialization in that in the distal region an electron-dense cuff is developed, which originates at the cell membrane at the level of the septate desmosomes and protrudes deeply into the cytoplasm of the receptor (Sopott-Ehlers 1984). Such a structure does not exist in any other free-living Platyhelminthes.

Another difference concerns the passage of the collar receptors through the epidermis. In the Lithophora the receptor cells always penetrate epidermal cells, a feature also existing in other Neoophora, as in the Rhabdocoela. In the Unguiphora the receptors do not penetrate epidermal cells, but run between epidermal cells to surface, a condition not found in other Neoophora.

4. Proseriata Meixner 1938

In several free-living platyhelminth groups (Macrostomida, Polycladida, 'Typhloplanoida', 'Dalyellioida'), united in the taxon Rhabditophora by Ehlers (this volume), secretion granules with a specific appearance of one to several concentric striated fibrillar layers occur and are called lamellated rhabdites (Smith, Tyler, Thomas, and Rieger 1982). These lamellated rhabdites can also be found within the Seriata namely in the Bothrioplanidae (Reisinger and Kelbetz 1964) and in at least some Tricladida, for example *Procotyla fluviatilis* Leidy 1857 and *Phagocata paravitta* (Reisinger 1923).

Our own EM studies on more than twenty species of Lithophora and Unguiphora have shown that lamellated rhabdites are missing from these groups. This absence of lamellated rhabdites, which are typical for the Macrostomida, Polycladida and Rhabdocoela, is supposedly secondary. Their loss could illustrate a close relationship between the Lithophora and the Unguiphora, which together form the taxon Proseriata.

Further EM studies, especially on the Seriata, which are discussed below, are designed to show whether or not more specific conformities exist between the Lithophora and Unguiphora.

Taxon N.N., including Bothrioplanida and Tricladida (from Steinböck 1925: Alithophora)

Within the Bothrioplanidae von Graff 1904-08, called Bothrioplanida by Steinböck (1925) (syn. Cyclocoela von Graff 1904-08), and the Tricladida Lang 1881, some fundamental differences appear relating to the intestine. The primarily rod-shaped unbranched intestine splits into two branches at the level of the root of the pharynx, i.e. the intestine of the Bothrioplanida and Tricladida always lies lateral to the pharynx tubiformis, whereas it runs dorsal to the pharynx in the Proseriata. While the intestine of the Bothrioplanida is split only at the level of the pharynx and caudally has an unpaired form as in the Proseriata, the Tricladida possess an intestine which is separated into two branches extending to the caudal end. This trifurcate intestine (one cranial, two caudal branches) represents a characteristic attribute of the Tricladida. However, this arrangement is sometimes secondarily changed, when there is either partial fusion of the caudal branches or a differentiation of additional cranial branches (Westblad 1935).

Steinböck (1925), who previously pointed out the close relationship between the Bothrioplanida and Tricladida, and united them under the unsuitable name 'Alithophora' (alithophoral species also exist in the Proseriata, the Unguiphora), mentions the change of the layers of musculature in the pharynx as a further common feature. According to Reisinger, Cichocki, Erlach, and Szyskowitz (1974) some significant similarities exist with respect to embryonic development of the Bothrioplanida and Tricladida. However, until EM studies reveal more specific features common to the Bothrioplanida and Tricladida, it would be inappropriate to suggest a common taxon. At present collar receptors are not known in any representatives of the Bothrioplanida or Tricladida. Assuming this observation will be confirmed by future EM studies, this negative criterion may also be regarded as a unifying feature of the Bothrioplanida and Tricladida. Since extensive EM studies on *Bothrioplana* have not yet been undertaken, possible distinctive characters await confirmation. A wide-ranging reduction of the male copulatory organ, and, resulting from this, parthenogenetic reproduction, represent specific features of the Bothrioplanida by which this group differs from the Tricladida, or at least from the stem species of Tricladida, respectively.

The fact that *Bothrioplana* is not a triclad is also shown by other characteristics. While the position of paired germaria near the pharynx is considered a basic pattern in the Seriata and also exists in *Bothrioplana*, the paired germaria of the Tricladida have a constant cranial position, mostly a short distance behind the brain. This feature is certainly a specific attribute of the Tricladida. Furthermore the Tricladida, so far as is known, show larger numbers of serially arranged excretory pores, while the Proseriata and Bothrioplanida possess only one pair, or by fusion of these two pores an unpaired one. This condition is typical for most Platyhelminthes.

Another remarkable feature of the Tricladida appears during embryonic development. In an early phase of development the embryo forms a muscular apparatus for swallowing the yolk material (Carlé 1935; Le Moigne 1963, 1966; Benazzi and Gremigni 1982). In a later phase of development before the definitive pharynx plicatus is differentiated, the embryonic pharynx is completely reabsorbed. Such a transitory organ neither occurs in other Platyhelminthes nor does it exist in the Proseriata or Bothrioplanida (Reisinger, Cichocki, Erlach, and Szyskowitz 1974).

Characterization of the different taxa

Based on the information given here, and in companion papers (see Ax and Ehlers, this volume), together with data recorded by Karling (1974), it can be deduced that the common stem species of the Seriata must have possessed the following characteristics:

1. The basic features of the Neoophora, such as female gonads with separate germ and yolk cell producing parts and combined with this an ectolecithal egg formation; spermatozoa with two flagella of the trepaxonematid type (9 + '1' type); a rod-shaped unbranched intestine; protonephridia with one pair of nephridial pores and a two-cell cyrto-

cyte; an epidermal duo-gland adhesive system; lamellated rhabdites with concentric striated fibrillar layers; epidermal sense cells with monociliary collar receptors; rhabdomeric photoreceptors with a retinula cell and a surrounding mantle (pigment) cell.

2. The specific features of the Seriata (Fig. 10.1, black square 1) such as a horizontally orientated pharynx tubiformis and follicular differentiation of the gonads, especially of the vitellaria in a serial arrangement.

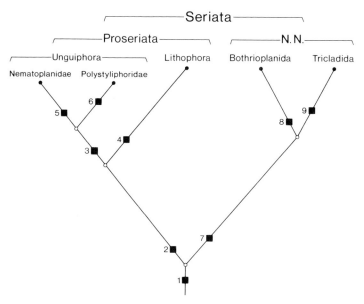

Fig. 10.1. Phylogenetic relationships among the Seriata. For the defining specific characters for each taxon (numbered black squares) see text.

A special feature of the Proseriata (Fig. 10.1, square 2), the lack of lamellated rhabdites, must be noted. The special characters of the Unguiphora (3) are: claw-shaped stylet, more than one pair of germaria (two to six pairs), and collar receptors with an electron-dense cuff. The Lithophora (4) (Monocelididae, Coelogynoporidae, Otoplanidae, Monotoplanidae, Otomesostomatidae) possess a statocyst.

In the Nematoplanidae (5) specific modifications exist in the duo-gland adhesive system and protonephridia, whilst in the Polystyliphoridae (6) there is a multiplication of the male organ.

The taxon N.N. (Bothrioplanida + Tricladida) (7) presumably lacks collar receptors, the intestine is split at the level of the pharynx, and further possible characteristics are mentioned above.

The Bothrioplanida (8) are characterized by parthenogenetic reproduction and a rudimentary male organ. The Tricladida (9) are characterized by the trifurcate intestine, the cranial position of the germaria, serial arrangement of many nephridial pores and differentiation of a transitory embryonic pharynx.

References

Aronova, M. (1974). Electron microscopic observation on the aboral organ of Ctenophora. I. The gravity receptor. *Z. mikrosk. anat. Forsch.* 88, 401-12.

Ax, P. (1956) Monographie der Otoplanidae (Turbellaria). Morphologie und Systematik. *Abh. math.-naturw. Kl. Akad. Wiss. Mainz* 13, 499-796.

—— (1958). Vervielfachung des männlichen Kopulationsapparates bei Turbellarien. *Verhandl. Dt. zool. Ges. Graz* 1957, 227-49.

—— (1963). Relationships and phylogeny of the Turbellaria. In *The lower Metazoa* (ed. E. C. Dougherty) pp. 191-224. University of California Press, Berkeley, Ca.

—— (1966). Das chordoide Gewebe als histologisches Lebensformmerkmal der Sandlückenfauna des Meeres. *Naturw. Rdsch., Braunschweig* 19, 282-9.

Ball, I. R. (1981). The phyletic status of the Paludicola. *Hydrobiologia* 84, 7-12.

—— and Reynoldson, T. B. (1981). British planarians (Platyhelminthes: Tricladida). *Synopses of the British fauna* (eds D. M. Kermack and R. S. K. Barnes) Vol. 19. Cambridge University Press, Cambridge.

Barnes, R. D. (1980). *Invertebrate zoology.* (4th edn). Saunders College, Philadelphia.

Beauchamp, P. de (1961). Classe des Turbellariés. In *Traité de zoologie* (ed. P.-P. Grassé) Vol. 4, pp. 35-212. Masson et Cie, Paris.

Benazzi, M. and Benazzi Lentati, G. (1976). Platyhelminthes. In *Animal cytogenetics* (ed. B. John) Vol. 1. Gebrüder Borntraeger, Berlin.

—— and Gremigni, V. (1982). Developmental biology of triclad turbellarians. In *Developmental biology of freshwater invertebrates* (eds F. W. Harrison and R. C. Cowden) pp. 151-211. Alan R. Liss, New York.

Carlé, R. (1935). Beiträge zur Embryologie der Landplanarien. I. Frühentwicklung, Bau und Funktion des Embryonalpharynx. *Z. Morph. Ökol. Tiere* 29, 527-58.

Crezée, M. (1982). Turbellaria. In *Synopsis and classification of living organisms* (ed. S. P. Parker) Vol. 1, pp. 718-40. McGraw-Hill, New York.

Ehlers, U. (1977). Vergleichende Untersuchungen über Collar-Rezeptoren bei Turbellarien. *Acta zool. fenn.* 154, 137-48.

—— and Ehlers, B. (1977). Monociliary receptors in interstitial Proseriata and Neorhabdocoela (Turbellaria Neoophora). *Zoomorphologie* 86, 197-222.

Horridge, G. A. (1969). Statocysts of medusae and evolution of stereocilia. *Tissue Cell* 1, 341-53.

Karling, T. G. (1974). On the anatomy and affinities of the turbellarian orders. In *Biology of the Turbellaria* (eds N. W. Riser and M. P. Morse) pp. 1-16. McGraw-Hill, New York.

Krisch, B. (1973). Über das Apikalorgan (Statocyste) der Ctenophore *Pleurobrachia pileus*. *Z. Zellforsch. mikrosk. Anat.* 142, 241-63.

Le Moigne, A. (1963). Étude du développement embryonnaire de *Polycelis nigra* (Turbellarié—Triclade). *Bull. Soc. zool. Fr.* 88, 403-22.

—— (1966). Étude du développement embryonnaire et recherches sur les cellules de régénération chez l'embryon de la planaire *Polycelis nigra* (Turbellarié, Triclade). *J. Embryol. exp. Morph.* 15, 39-60.

Meixner, J. (1938). Turbellaria (Strudelwürmer). *Die Tierwelt der Nord- und Ostsee* (ed. G. Grimpe) Vol. 4, 1. Akadem. Verlagsgesellschaft, Leipzig.

Reisinger, E. and Kelbetz, S. (1964). Feinbau und Entladungsmechanismus der Rhabditen. *Z. wiss. Mikrosk.* 65, 472-508.

—— Cichocki, I., Erlach, R., and Szyskowitz, T. (1974). Ontogenetische Studien an Turbellarien: ein Beitrag zur Evolution der Dotterverarbeitung im ektolecithalen Ei, II Teil. *Z. zool. Syst. Evolutionsforsch.* 12, 241-78.

Singla, C. L. (1975). Statocysts of hydromedusae. *Cell Tissue Res.* 158, 391-407.

Smith, J., Tyler, S., Thomas, M. B., and Rieger, R. M. (1982). The morphology of turbellarian rhabdites: phylogenetic implications. *Trans. Am. microsc. Soc.* 101, 209-28.

Sopott, B. (1972). Systematik und Ökologie von Proseriaten (Turbellaria) der deutschen Nordseeküste. *Mikrofauna Meeresboden* 13, 1-72.

Sopott-Ehlers, B. (1979). Ultrastruktur der Haftapparate von *Nematoplana coelogynoporoides* (Turbellaria, Proseriata). *Helgoländer wiss. Meeresunters.* 32, 365-73.

Sopott-Ehlers, B. (1984). Epidermale Collar-Receptoren der Nematoplanidae und Polystyliphoridae (Plathelminthes, Unguiphora). *Zoomorphology* 104, 226-30.

Steinböck, O. (1925). Zur Systematik der Turbellaria metamerata, zugleich ein Beitrag zur Morphologie des Tricladen-Nervensystems. *Zool. Anz.* 64, 165-92.

Vinnikov, Y. A., Aronova, M. Z., Khakeevich, T. A., Tsirulis, T. P., Lavrova, E. A., and Natochin, Y. V. (1981). Structural and chemical features of the invertebrate otoliths. *Z. mikrosk.-anat. Forsch.* 95, 127-40.

Westblad, E. (1935). *Pentacoelum fucoideum* m., ein neuer Typ der Turbellaria metamerata. *Zool. Anz.* 111, 65-82.

11. The position of the Gnathostomulida and Platyhelminthes in the phylogenetic system of the Bilateria

PETER AX

II Zoologisches Institut und Museum der Universität, Göttingen, Federal Republic of Germany

Abstract

The Gnathostomulida and the Platyhelminthes represent adelphotaxa (sister groups) and are united in a taxon with the name Platyhelminthomorpha. The Platyhelminthomorpha form the adelphotaxon of all other bilaterian taxa, which together constitute the monophyletic group Eubilateria.

A phylogenetic system for the Bilateria derived from these empirically testable hypotheses is presented in two equivalent forms; in a diagram of phylogenetic relationships and as a hierarchical systematization with the adelphotaxa indented to the same degree.

Introduction

In nature two sets of living entities exist which have resulted from the processes of speciation and phylogenesis: evolutionary species (in the sense of Simpson 1961; Wiley 1978, 1981; Ax 1984) as closed reproductive communities; groups of evolutionary species as closed-descent communities with a single stem species and including all of its descendants. Accordingly, in a strict phylogenetic system of organisms, we can only accept two distinct forms of taxa:

(i) the taxon species as an equivalent of the evolutionary species; and

(ii) the supraspecific monophyletic taxon (= monophylum) as an equivalent of the closed-descent community of nature.

The principal aim of phylogenetic systematics is to discover the first degree of phylogenetic relationships between evolutionary species and closed-descent communities and then to present this information as a hierarchically arranged systematization. Any two most closely related species form sister species; two corresponding monophyletic groups of species are sister groups; I call them both adelphotaxa (Ax 1984).

These basic statements can be illustrated in a formal diagram of phylogenetic relationships (Fig. 11.1). Two evolutionary species A and B, an evolutionary species A and a closed-descent community B or two closed-descent communities A and B, all represent adelphotaxa if they are the only descendants of the stem species z. As taxa with an identical origin in time, they have to be assigned the same systematic rank. Denoting their adelphotaxon relation formally as level 3 of the hierarchy, we must join A and B on the next level 2 in a new taxon I with a correspondingly higher rank. On this level, taxa I and II form adelphotaxa with identical ranks if they have had a stem species y common only to them earlier in time. Consequently, taxa I and II are united on level 1 of the hierarchy by establishing a monophyletic species group that we call taxon α in this example.

What kind of information entitles us to formulate empirically testable hypotheses about adelphotaxa relationships? We have to demonstrate

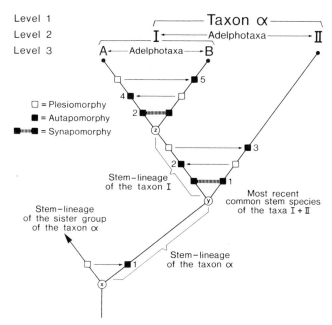

Fig. 11.1. Formal diagram of phylogenetic relationships between hypothetical evolutionary species and closed-descent communities of nature. Further explanation in text.

the common possession of a very distinct form of identity or similarity
in those species or presumed monophyletic species groups suspected to
be adelphotaxa. In our example, I and II can be hypothesized as
adelphotaxa only if we are able to demonstrate at least one identical or
similar character between them which can be interpreted without con-
flict as having arisen only once as an evolutionary novelty (an apomorphic character) in their common stem lineage. The evolutionary
novelties of this stem lineage are the autapomorphies 1 of taxon α. The
common presence of every one of these apomorphic characters con-
stitutes synapomorphies between taxa I and II.

We can only hypothesize taxa A and B as adelphotaxa if we can find
other specific identities or similarities (synapomorphies 2) which may
be interpreted as new characters (autapomorphies 2) of their most
recent common stem species z, i.e. characters which evolved only once
in the stem lineage of taxon I.

Finally, if we assume in Fig. 11.1 that taxa A, B and II do not represent single evolutionary species but monophyletic species groups like
taxa I (A + B) and α (I + II), then it follows that these three taxa must
be justified as monophyla by their own evolutionary novelties. In the
example, these are the autapomorphies 3 for the taxon II, autapomorphies 4 for taxon A and autapomorphies 5 for taxon B.

Hypotheses

By means of these established principles and methods of phylogenetic
systematics (Hennig 1966; Eldredge and Cracraft 1980; Wiley 1981;
Ax 1984) the following testable hypotheses are proposed (Fig. 11.2):

(a) The Gnathostomulida and the Platyhelminthes are both monophyletic species groups.

(b) As taxa with the first degree of phylogenetic relationship the
Gnathostomulida and Platyhelminthes represent adelphotaxa. Consequently they are united on the next higher level of the phylogenetic
system in a new monophylum named Platyhelminthomorpha.

(c) The Platyhelminthomorpha form the adelphotaxon of all other
bilaterian taxa, a monophyletic assemblage for which the name Eubilateria is introduced.

(d) The adelphotaxa Platyhelminthomorpha and Eubilateria together
constitute the monophylum Bilateria within the Metazoa.

Discussion

The arguments underlying the formulation of these hypotheses are
taken from a comprehensive analysis of the mosaic of Platyhelmintho-

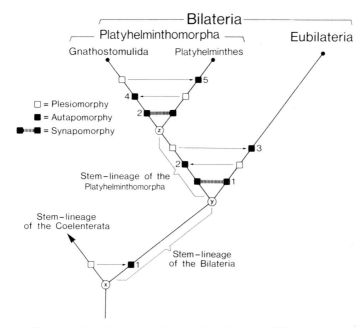

Fig. 11.2. Diagram of phylogenetic relationships within the Bilateria.

morpha characters believed to be present in the most recent common stem species of the adelphotaxa Gnathostomulida and Platyhelminthes (Ax 1984). The following survey refers to those characters which are suggested as having arisen as autapomorphies once in the stem lineage of the Bilateria (1), Platyhelminthomorpha (2), Eubilateria (3), Gnathostomulida (4), and Platyhelminthes (5) (Tables 11.1, 11.2; Fig. 11.2).

1. Autapomorphies of the Bilateria

The following characters evolved in the stem lineage of the closed descent community Bilateria and were present in the most recent common stem species y of the adelphotaxa Platyhelminthomorpha and Eubilateria.

(a) Bilaterally symmetrical body

(b) Anterior brain

Formation of a brain as an accumulation of neurons in connection with the evolution of unidirectional locomotion.

Table 11.1. Interpretation of selected characters of the basic patterns of the taxa Bilateria, Platyhelminthomorpha and Eubilateria. A, Autapomorphies; P, Plesiomorphies.

Bilateria		Platyhelminthomorpha		Eubilateria	
A 1	Bilateral symmetry Brain in the anterior end Subepidermal musculature Primary body cavity with few cells Protonephridia Spiral quartet cleavage	=	Synapomorphies between these two taxa		
P	External fertilization Gonochorism Primitive sperm	A 2	Internal fertilization Hermaphroditism Filiform sperm	P	External fertilisation Gonochorism Primitive sperm
P	Intestine without anus	P	Intestine without anus	A 3	Intestine with anus

(c) Subepidermal musculature

Evolution of subepidermal musculature with an outer layer of loosely arranged circular fibres and an inner layer of longitudinal fibres.

(d) Primary body cavity

Spaces of the weakly developed primary body cavity are filled with few cells (stem cells, ? parenchyma cells).

I reject all speculations about the Platyhelminthes being secondarily small organisms which lost a coelom and other organs through an evolutionary diminution in body size. The main argument against these unjustifiable speculations is derived from correlations that exist between pharynx structure and body size. All members of the platyhelminth taxa Catenulida, Acoela, Nemertodermatida, and Macrostomida which possess a mouth pore in the epidermis or a pharynx simplex have a body size of only a few millimetres, the diameter being only fractions of a millimetre. Larger free-living flatworms with massive parenchyma like the Polycladida, the Tricladida, and some Rhabdocoela always possess a complex apomorphic pharynx compositus, the evolution of which allows an increase in size and volume as a consequence of intensified carnivorous feeding. With the plesiomorphic simple mouth pore or a pharynx simplex, only microscopic body dimensions are attainable. Accordingly, I conclude that the small size of the platyhelminth taxa mentioned and the corresponding microscopic size of the gnathostomulids represent the primary condition within the Platyhelminthomorpha and so must be postulated as being characteristic of the stem species of all bilaterians. Only minute spaces between ectoderm and endoderm are possible with these body dimensions.

(e) Evolution of a nephridial organ

Without a sufficient body cavity there is only room for the action of nephridial cilia which are closed off from the surrounding tissues. A corresponding condition occurs in the terminal cells of protonephridia. Consequently I consider the protonephridia as the primary nephridial organ of the Bilateria and postulate its evolution in the common stem lineage of the Platyhelminthomorpha and Eubilateria. As will be shown later, the most primitive state is retained in the taxon Gnathostomulida.

(f) Developmental mode of the spiral quartet cleavage

The gnathostomulids, most platyhelminth taxa with endolecithal egg production (Catenulida, Macrostomida, Polycladida) and numerous eubilaterian taxa (Nemertea, Entoprocta, Sipuncula, Echiura, Mollusca, Articulata) possess the spiral quartet cleavage as either the only, or at least the primary, mode of development. Therefore, we have to postulate the evolution of the spiral quartet cleavage in the stem lineage of the Bilateria, a pattern which must have been realized in the most recent common stem species of the Platyhelminthomorpha and Eubilateria. It is a logical consequence that all the differing patterns of cleavage within the Platyhelminthes and within the Eubilateria are evolutionary apomorphies.

2. Autapomorphies of the Platyhelminthomorpha

The following characters evolved in the stem lineage of the closed-descent community Platyhelminthomorpha and were present in the most recent common stem species z of the adelphotaxa Gnathostomulida and Platyhelminthes.

(a) Direct sperm transfer and internal fertilization

A free discharge of gametes and external fertilization are plesiomorphic conditions in the reproduction of the Metazoa. They undoubtedly belong to the basic pattern of the Bilateria. We postulate a unique change to direct sperm transfer with internal fertilization in the stem lineage of the Platyhelminthomorpha.

(b) Evolution of hermaphroditism

In comparison with gonochorism, which is widely distributed in the Metazoa, hermaphroditism of the gnathostomulids and the platyhelminths represents an apomorphy. There is no difficulty in postulating that hermaphroditism in these two taxa represents a synapomorphic condition not congruent with the hermaphroditism of other metazoan taxa.

(c) Evolution of a filiform sperm

Correlated with external fertilization is the so-called primitive sperm of the Metazoa (Franzén 1956). It consists of a rounded-conical head, a short middle piece with usually four mitochondria, and a tail consisting of a cilium (9 + 2 pattern of microtubules). Compared with this condition, the sperm of all platyhelminthomorph taxa are greatly modified.

Within the scope of this discussion, we are interested in the most plesiomorphic state of the Platyhelminthomorpha, which is to be found in the gnathostomulid genera *Haplognathia* and *Pterognathia*, and also in *Nemertoderma* and *Meara* of the platyhelminth taxon Nemertodermatida (Sterrer 1974; Hendelberg 1977; Tyler and Rieger 1977). In these taxa we find a thread-like sperm with an elongated head followed by a cylindrical middle piece and a primitive tail with one cilium (9 + 2 pattern). Connected with the change from external to internal fertilization, an evolutionary alteration of the primitive metazoan sperm to a filiform sperm in the stem lineage of the Platyhelminthomorpha is here postulated.

3. Autapomorphy of the Eubilateria

The following character evolved in the stem lineage of the closed-decent community Eubilateria.

One-way intestine with anus

As in the coelenterates, the gnathostomulids and platyhelminths possess a blind-ending intestine without an anal pore. There is no empirical justification for speculations about a secondary loss of the anus in the Platyhelminthomorpha. A blind intestine can be considered plesiomorphic not only for the platyhelminthomorphs but also for the basic bilaterian pattern, maintained in their stem species as a plesiomorphy from the common stem lineage of the Coelenterata and Bilateria.

The evolution of a one-way intestine must be interpreted as an important functional progression in the phylogenesis of the Bilateria. On grounds of parsimony, this step can be hypothesized as a unique evolutionary event in the common stem lineage of all bilaterian taxa with the exception of the Platyhelminthomorpha. Accordingly, I consider an intestine with an anus as the essential autapomorphy for establishing the monophyletic taxon Eubilateria.

4. Autapomorphy of the Gnathostomulida

The following character evolved in the stem lineage of the closed-descent community Gnathostomulida.

Pharynx apparatus with a pair of cuticular jaws and a cuticular basal plate

The minute hard structures of the pharynx and mouth cavity (Fig. 11.3 (c, d)) used for grazing bacteria and other micro-organisms in the interstitial milieu of marine sand represent a convincing apomorphic character complex for the establishment of a monophyletic taxon Gnathostomulida. In comparison with the number of playhelminth autapomorphies, it must be emphasized that the jaws and basal plate together form the only known gnathostomulid apomorphy. As explained below, the gnathostomulids have retained the plesiomorphic conditions in all other divergent characters.

Table 11.2. Interpretation of selected characters of the basic patterns of the taxa Platyhelminthomorpha, Gnathostomulida and Platyhelminthes. A, Autapomorphies; P, Plesiomorphies.

Platyhelminthomorpha		Gnathostomulida		Platyhelminthes
	Simple mouthpore	A 4	Pharynx with jaws and basal plate	P	Simple mouthpore
P	Monociliary epidermis	P	Monociliary epidermis	A 5	Multiciliary epidermis
	Cilia of epidermis cells with accessory centriol		Cilia of epidermis cells with accessory centriol		Cilia of epidermis cells with one basal body, without accessory centriol
	Protonephridia with one cilium		Protonephridia with one cilium		Protonephridia with two cilia
	Terminal cell of protonephridia with 3 rods surrounding the cilium		Terminal cell of protonephridia with 8 rods surrounding the cilium		Terminal cell of protonephridia without rods (microvilli)

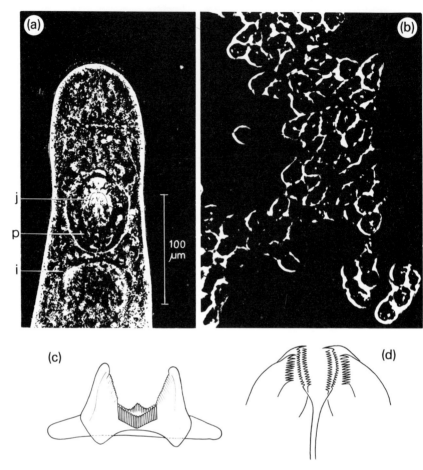

Fig. 11.3. Gnathostomulida. (a) Anterior end of *Gnathostomula paradoxa* Ax with rostral sensorium and pharynx. (b) Monociliary epidermis of *Gnathostomaria lutheri* Ax, detached from the body (phase contrast). (c, d) Basal plate (c) and jaws (d) of *Gnathostomula paradoxa*. Abbreviations: i, intestine; j, jaws; p, pharynx. From Ax 1964a, b, 1965.

5. Autapomorphies of the Platyhelminthes

The following characters evolved in the stem lineage of the closed-descent community Platyhelminthes and were present in the most recent common stem species of the adelphotaxa Catenulida and Euplatyhelminthes (Ehlers 1985, this volume; Ax 1984).

(a) *Multiciliary epidermal cells and* (b) *Cilia without accessory centriole*

The structure of the body wall in the gnathostomulids represents the plesiomorphic condition postulated for the stem species of all bilaterians. The Gnathostomulida are the only bilaterian taxon with an epidermis

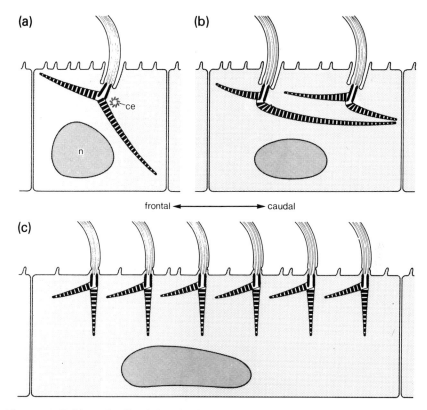

Fig. 11.4. Epidermal cells of the Platyhelminthomorpha. (a) Monociliary cell of the Gnathostomulida with the nucleus and an accessory centriole. (b) Weakly multiciliary cell of the Catenulida. (c) Stronger multiciliary cell of the Rhabditophora. Abbreviations: ce, accessory centriole; n, nucleus. From Ax 1984.

that consists exclusively of monociliary cells (Ax 1964a; Rieger and Mainitz 1977). The cilium of every epidermal cell possesses an accessory centriole besides its basal body (Fig. 11.3(b), Fig. 11.4(a)).

In contrast to this, all taxa of the platyhelminths have a multiciliary epidermis and the cilia do not possess an accessory centriole. Interestingly, the Catenulida with several other plesiomorphic characters develop only a few cilia per epidermal cell (Fig. 11.4(b)) whereas in all other platyhelminth taxa, which we unite as a monophylum with the name Euplatyhelminthes, each cell is equipped with numerous cilia (Fig. 11.4(c)). As a parsimonious interpretation, the evolution of a multiciliary epidermis with reduction of the accessory centrioles is postulated as a single event in the stem lineage of the Platyhelminthes. Non-homologous phenomena occurred in different lines of the Eubilateria.

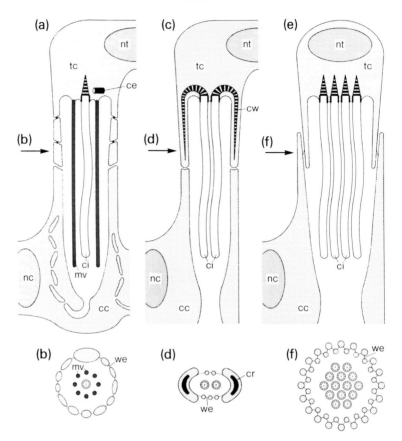

Fig. 11.5. Protonephridia of the Platyhelminthomorpha: terminal cell and neighbouring canal or tubule cell. (a, c and e) Sagittal sections; (b, d and f) Cross sections. (a, b) Gnathostomulida (*Gnathostomula*); (c, d) Catenulida (*Retronectes*). (e, f) Rhabditophora. Abbreviations cc, canal or tubule cell; ce, accessory centriole; ci, cilium; cr, ciliary rootlet; mv, microvilli (rods); nc, nucleus or canal cell; nt, nucleus of terminal cell; tc, terminal cell; we, weir. From Ax 1984.

(c) Protonephridial terminal cell with two cilia and (d) Terminal cell without rods

The structure of the gnathostomulid protonephridium is of remarkable evolutionary interest (Fig. 11.5(a, b)). It is composed of only two cells, the terminal cell and a short tubule cell, and can be regarded as the basic pattern of platyhelminthomorphs and for the most recent common stem species of the Platyhelminthomorpha and Eubilateria. The terminal cell of the gnathostomulid protonephridium, with a single cilium, an accessory centriole and eight rods around the cilium illustrates the evolutionary path from an ectoderm cell to a protonephridium.

This terminal cell is simply a slightly modified monociliary epidermal cell with eight extremely elongated microvilli.

To recognize the minimal evolutionary modifications of this plesiomorphic protonephridium in the stem lineage of the Platyhelminthes, the condition in the taxon Catenulida is of decisive importance (Fig. 11.5(c,d)). The catenulids invariably possess two cilia in the terminal cell and all lack the eight rods around the cilia. Therefore, it is concluded that the doubling of cilia and the reduction of rods (microvilli) are characteristic autapomorphies in the basic construction of the platyhelminth protonephridium.

Further steps in the evolution of the protonephridia within the Platyhelminthes are the formation of cilia flames by multiplication of the cilia and the formation of a weir from two new rows of cytoplasmatic projections, the inner row originating from the terminal cell, the outer one from the adjacent tubule cell (Fig. 11.5(e,f)).

Conclusions

The adelphotaxa relationship between two closed descent communities can be recognized only if the following two conditions are fulfilled. First, a hypothesis that two taxa have had a stem species common only to them must be justified by synapomorphies. In other words, the taxa together must form a monophylum on the next higher level of hierarchy. Secondly, at least one autapomorphy must be demonstrated in either of both taxa for which the other must have retained the corresponding plesiomorphic condition.

In this paper, an attempt has been made to demonstrate that these two conditions have been fulfilled, allowing the pairs Platyhelminthomorpha / Eubilateria and Gnathostomulida / Platyhelminthes to be designated adelphotaxa.

The exigencies of phylogenetic systematics require a strict equivalence between a diagram of phylogenetic relationships and a corresponding written systematization, being interchangeable ways of presenting the same segment of the phylogenetic system. Accordingly, the following is a simple translation of Fig. 11.2 into a hierarchical record with the adelphotaxa indented to the same degree.

Bilateria
Platyhelminthomorpha
Gnathostomulida
Platyhelminthes
Eubilateria

A comparable analysis of the as yet unresolved highest adelphotaxa relations within the Eubilateria is now urgently needed.

References

Ax, P. (1964a). Das Hautgeißelepithel der Gnathostomulida. *Verhandl. Dt. zool. Ges. München* 1963, 452-61.

—— (1964b). Die Kieferapparatur von *Gnathostomaria lutheri* Ax. *Zool. Anz.* 173, 174-81.

—— (1965). Zur Morphologie und Systematik der Gnathostomulida. Untersuchungen an *Gnathostomula paradoxa* Ax. *Z. zool. Syst. Evolutionsforsch.* 3, 259-76.

—— (1984). *Das Phylogenetische System. Systematisierung der lebenden Natur aufgrund ihrer Phylogenese.* G. Fischer, Stuttgart.

Ehlers, U. (1985). *Das Phylogenetische System der Plathelminthes.* Akademie der Wissenschaften und der Literatur, Mainz. G. Fischer, Stuttgart.

Eldredge, N. and Cracraft, J. (1980). *Phylogenetic patterns and the evolutionary process. Methods and theory in comparative biology.* Columbia University Press, New York.

Franzén, Å (1956). On spermiogenesis, morphology of the spermatozoon, and biology of fertilization among invertebrates. *Zool. Bidr. Upps.* 31, 355-482.

Hendelberg, J. (1977). Comparative morphology of turbellarian spermatozoa studied by electron microscopy. *Acta zool. fenn.* 154, 149-62.

Hennig, W. (1966). *Phylogenetic systematics.* University of Illinois Press, Urbana, IL.

Rieger, R. M. and Mainitz, M. (1977). Comparative fine structure study of the body wall in Gnathostomulida and their phylogenetic position between Platyhelminthes and Aschelminthes. *Z. zool. Syst. Evolutionsforsch.* 15, 9-35.

Simpson, G. G. (1961). *Principles of animal taxonomy.* Columbia University Press, New York.

Sterrer, W. (1974). Gnathostomulida. In *Reproduction of marine invertebrates. I. Acoelomate and pseudocoelomate metazoans* (eds A. C. Giese and J. S. Pearse) pp. 345-57. Academic Press, New York.

Tyler, S. and Rieger, R. M. (1977). Ultrastructural evidence for the systematic position of the Nemertodermatida (Turbellaria). *Acta zool. fenn.* 154, 193-207.

Wiley, E. O. (1978). The evolutionary species concept reconsidered. *Syst. Zool.* 27, 17-26.

—— (1981). *Phylogenetics. The theory and practice of phylogenetic systematics.* Wiley, New York.

12. Gnathostomulida: enigmatic as ever

WOLFGANG STERRER

Bermuda Biological Station, Ferry Reach, Bermuda

MARLENE MAINITZ

Universitäts-Hautklinik, Wien, Austria
and

REINHARD M. RIEGER

Department of Biology and Curriculum of Marine Sciences, University of North Carolina, Chapel Hill, North Carolina, USA

Abstract

Since the first summary of gnathostomulid systematics (Sterrer 1972), a number of new taxa, together with data particularly on ultra-structure, have added considerably to our knowledge of this recently discovered phylum (Ax 1956). This presentation reviews major organ systems in the light of new evidence. The systems covered include: the body wall, the ultrastructure of gland cells and the monociliary cell; sense organs, especially the spiral-ciliary organ; the nervous system in relation to the epidermis, basement membrane and intestine; the digestive tract, now known to possess an anus in at least several genera; excretory organs and their probably unique similarity with those of gastrotrichs; the reproductive system, particularly the male apparatus, and its implications for interpreting relationships within the phylum; and sperm ultrastructure. The paper concludes with a critical assessment of our data base, briefly evaluates phylogenetic relationships, and suggests directions for future research.

Introduction

Since the first summary of gnathostomulid systematics (Sterrer 1972) our knowledge of this recently identified animal phylum (Ax 1956) has grown considerably. Twelve additional species, five genera and two families have been described (Farris 1973, 1977; Ehlers and Ehlers 1973; Kristensen and Nørrevang 1977, 1978; Sterrer 1973, 1976; Sterrer and Farris 1975), and much has been learned about their fine structure (Graebner 1969a; Graebner and Adam 1970; Rieger 1976; Rieger and Mainitz 1977; Mainitz 1977, 1979; Kristensen and Nørrevang 1977, 1978; Knauss 1979; Knauss and Rieger 1979; Lammert 1983). A few summary articles have appeared (Sterrer 1974, 1982; Mainitz 1983; Alvestad-Graebner and Adam 1983) or are about to appear (Mainitz, in press). We also know more about their ecology and physiology (Schiemer 1973), an attempt has been made at understanding population structure and speciation (Sterrer 1973, 1977), and the thiobios controversy, fueled by new facts and fancies, rages on (Fenchel and Riedl 1970; Boaden 1977; Reise and Ax 1979; Powell, Crenshaw, and Rieger 1979; Powell and Bright 1981; Reise 1981; Powell, Bright, Woods, and Gittings 1983).

At present there are 17 genera and 54 described species. In spite of this impressive increase in our knowledge of diversity and fine structure, the Gnathostomulida remain much of an enigma. This paper mainly concentrates on data appearing since 1972, especially those containing new facts, and concludes with some thoughts as to possible relationships within and without the phylum.

Epidermis

The monociliated, monolayered epidermis remains one of the most positive distinguishing characters of the phylum within the vermiform Bilateria. It has been investigated in detail by Rieger and Mainitz (1977) and compared with the epidermis of Gastrotricha in an attempt to establish its primitiveness within the Metazoa (Rieger 1976). The basement membrane is a distinct, homogeneous layer separating the epidermis from underlying tissue, and is not continuous with any intercellular matrix (Rieger and Mainitz 1977). The structure of the ciliated epidermal cells is rather uniform within the phylum: the single cilium originates in a ciliary pit bordered by a collar of eight microvilli and is connected to a diplosomal basal body with a rostral and caudal striated rootlet. Significant differences are found in the type and distribution of non-ciliated glandular cells. Sausage-shaped mucous cells occur in Conophoralia and in higher Scleroperalia; in the latter they are arranged in rows alternating with epidermal cells which give the entire epidermis a striped appearance (Sterrer 1972). In addition rhabdoid glands occur in Conophoralia (Rieger and Mainitz 1977). Although the latter group has the ability to adhere to the substratum, none of these glands could be positively identified as adhesive organs on the basis of their ultra-structure (Tyler 1975). A number of other glandular formations identi-fied with the light microscope, such as the 'round epidermal inclusions' of Filospermoidea (Sterrer 1969) and the 'bundles of inclusions' of Onychognathiidae (Sterrer 1972) await more detailed analysis.

Sense organs

Paired sensory bristles, characteristic of Bursovaginoidea, are invariably composed of monociliary cells (Graebner 1968b; Kristensen and Nørrevang 1977). However, they usually sit more deeply in the epider-mis, are innervated and have a somewhat modified and longer funnel than regular epidermal cells. Lammert (1983) reported a concentration of monociliary receptors in *Haplognathia rosea* on the rostrum, around the mouth opening and in the buccal cavity. In the buccal cavity, one cilium that runs parallel to and between the medio-dorsal edge of the jaws has also been identified in *Semaeognathia* and *Labidognathia* (Mainitz, unpublished). There is no information on the fine structure of what has been called ciliary pits (Sterrer 1972) or the ventral ciliary organ (Kristensen and Nørrevang 1977). A new discovery, however, is the spiral-ciliary organ, first found in *Rastrognathia* (Kristensen and Nør-revang 1977) and since confirmed in four other families (Kristensen and Nørrevang 1978). It consists of one or two pairs of cells located in the epidermis on either side of, or anterior to, the mouth. Each cell encloses a spiral cavity which, in cross section, shows 8-10 ciliary pro-files. Kristensen and Nørrevang (1978) suggest that these cross-sections represent many separate cilia all originating from one single cell; the spiral ciliary organ would thus be the first evidence of a multiciliary cell in this phylum. On the other hand, Lammert (1983) has shown that at least in *Haplognathia rosea* all ciliary cross-sections are of only one long, spirally coiled cilium. In general, the similarity between the arrange-ment of sensory bristles in Gnathostomulida-Bursovaginoidea and cer-tain Archiannelida such as *Trilobodrilus* sp. (Rieger and Rieger 1975) is quite striking.

Nervous system

Since Ax's (1956) first observations the gnathostomulid nervous system has been considered as largely or exclusively basiepithelial (intraepi-dermal). Indeed in *Rastrognathia* the nervous system consists of a large brain mass anterior to the mouth, three pairs of longitudinal nerves, and one pair of buccal nerves which link the brain with what has been called a buccal ganglion located at the posterior end of the pharynx (Kristensen and Nørrevang 1977). While the longitudinal nerves are always at the base of the epidermis, hugging the basement membrane, there is no clearly defined basement membrane in the brain region where epithelial, muscle and nerve cells are in intimate contact. Kristensen and Nørrevang (1977) pointed out that the situation is different in *Haplognathia* where (as first shown by Sterrer (1969) and since confirmed in TEM pictures provided by E. B. Knauss) the brain in cross section seems subepithelial, on the inside of the basement membrane (Fig. 12.1). While there is also close contact between nervous and muscle tissue, sometimes with matrix resembling a basement membrane sandwiched in (Fig. 12.1), it is difficult to interpret the *Haplognathia* brain as 'sunken in' since it would then be expected to be enclosed by another basement membrane. The longitudinal nerves originating from the brain in *Haplognathia* traverse the basement membrane and then run the length of the animal in a basiepithelial position. To what extent these differences are connected with the very long rostrum in *Haplognathia* awaits analysis of serial TEM sections.

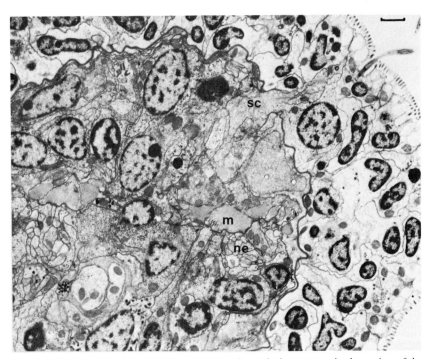

Fig. 12.1. *Haplognathia* cf. *rosacea*. Cross-section through the rostrum in the region of the brain. Note the interruption of the basement membrane by the sensory cell, and the location of muscles with regard to nerves. Abbreviations: m, muscles; n, nerves; sc, sensory cell. Scale bar represents 1 μm. Micrograph courtesy of E. B. Knauss.

Excretory organs

Paired groups of up to three organs with presumably excretory function occur in *Gnathostomula* and *Austrognathia* (Graebner 1968a) and other genera (Kristensen and Nørrevang 1977; Rieger and Mainitz 1977; Knauss 1979). They have been found in three body regions: pharynx, female bursa and male copulatory organ. Embedded in the body wall below the basement membrane each organ consists of three cells: a terminal cell with a single cilium and eight rods, a canal cell and an outlet cell, all of which Graebner (1968b) interpreted as derivatives of the epidermis. Ax (this volume) described the excretory organ as consisting of only two cells (terminal cell and outlet cell), and Lammert (personal communication) reported in *Gnathostomula paradoxa* a second, outer weir in addition to the inner rods in the terminal cell.

Musculature

The musculature is without exception cross-striated. It is found throughout the body between the epidermal basement membrane and the gut epithelium, with circular fibres to the outside of three pairs of longitudinal bundles (Knauss 1979). Locally, muscle concentrations are found in the pharynx and the male copulatory organ; in higher Scleroperalia both are particularly complex (Kristensen and Nørrevang 1977, 1978; Mainitz 1977, 1979) as a result of multiple infoldings of the body wall.

Parenchyma

The virtual absence of a parenchyma, i.e. a mass of unspecialized cells located between the body wall and internal organs, has been postulated on the basis of light microscopy (Sterrer 1969) and confirmed by TEM. The recent observation (Kristensen and Nørrevang 1978) of large translucent cells apparently located between gut and epithelium in *Valvognathia* awaits TEM investigation; they may in fact correspond to what has been described as dorsally vacuolated gut cells in *Tenuignathia* (Sterrer 1976).

Digestive tract

The hard structures of the pharynx have traditionally furnished the greatest number of species characteristics. In addition, some interesting new data have recently been provided by ultrastructural studies. Knauss (1979) has shown in two species of *Haplognathia* the presence of a tissue connection between the dorsal epidermis and the posterior end of the gut which she interpreted as a functional anal pore. Since an identical tissue connection was also found (Mainitz 1977) in the much shorter species *Gnathostomula jenneri*, the anal pore cannot be seen as merely an adaptation to extreme body length as it has for certain Turbellaria (Karling 1966) but may well turn out to be a diagnostic feature of the phylum, as suggested by Knauss (1979).

The buccal cavity is surrounded by a basement membrane, whereas the intestine is devoid of a basement membrane (in *Haplognathia*, see Rieger, this volume) or is provided with a more delicate membrane than that of the epidermis (in *Semaeognathia*, Mainitz, unpublished). Isolated monociliated cells, in addition to those of the buccal cavity, have been identified in the gut epithelium of at least two species (*Haplognathia* cf. *rosacea*, and *Gnathostomula jenneri*): their elongate shape and lack of a striated rootlet suggest a sensory function.

A number of glands associated with the digestive tract have been reported by light microscopy (Sterrer 1972), some of which are also known in finer detail (Graebner 1968b; Kristensen and Nørrevang 1977, 1978; Knauss 1979). They are particularly diverse in the region of the buccal cavity (Fig. 12.2 (a,b)) and pharynx, but require further investigation. The pharyngeal musculature of *Haplognathia rosea* has recently been clarified with TEM by Lammert (1983).

The fine structure of pharynx and mouthparts has been studied by Kristensen and Nørrevang (1977, 1978), Mainitz (1979), Riedl (unpublished), and Knauss (unpublished). While the detailed organization and function of the complex muscular apparatus continues to defy analysis, we can provide a generalized picture.

Jaws and basal plate are cuticular secretions of the ventral pharynx epidermis. They are laid down *in situ* rather than from a growth zone, moulded by epidermal fold templates. In *Haplognathia rosea* the basal plate is produced by one unpaired and two pairs of cells, the jaws by two unpaired and six pairs of cells (Lammert 1983). In Filospermoidea, these folds and the resulting hard parts are comparatively simple and rather two-dimensional; in the more lamellar, cone-shaped jaws of higher Bursovaginoidea the folds (consisting of epithelium, basement membrane and musculature) become increasingly convoluted and bent back against each other. This would explain the curious position of the basement membrane in relation to the musculature, as reported by Kristensen and Nørrevang (1977). A distinctive feature of all jaws studied with TEM is the stabilization of the inner lamella by a vertical sheet of cuticular tubes with an electron-dense core (Fig. 12.2(b); cf. Fig. 14 in Kristensen and Nørrevang 1978). It is composed of 7-11 tubes in *Haplognathia* (Lammert 1983) but numbers about 16 in *Gnathostomula* and 29 in *Labidognathia* (Fig. 12.12(b)). Its function might be to provide jaws with enough rigidity to snap back into a closed position after having been actively opened by musculature. In both structure and position this tubular material is strikingly similar to that described from rotifer jaws (Koehler and Hayes 1969). More recent TEM work, however, establishes the rotifer mastax as a primarily intracellular structure, which would suggest that the similarity in jaw tubes is analogous (Lammert 1983).

The non-lamellar composition of gnathostomulid teeth, together with the large size of the conodont animal (Briggs, Clarkson, and Aldridge 1983), should lay to rest any comparison with conodonts, as first suggested by Ochietti and Cailleux (1969).

Despite its comparative fragility in hypochlorite treatment (Riedl and Rieger 1972) the jugum appears to be a cuticular 'stiff upper lip' similar in structure and origin to the other mouth parts, but less dense, more massive, and traversed by long microvilli (Fig. 12.2(a)).

Reproductive system

1. Male organs

Testes, copulatory organ, and sperm are morphologically the most discontinuous of gnathostomulid structures, much of them as yet insufficiently clarified. Located in the posterior end of the body, testes are paired in some Filospermoidea (*Haplognathia*) and in all Scleroperalia; they seem secondarily fused into one dorsal testis in some Filospermoidea and all Conophoralia (Sterrer 1972). Sperm mature posteriorly, towards the male copulatory organ. The fine structure of male organs is known in *Haplognathia* (Knauss and Rieger 1979) and several Scleroperalia (Mainitz 1977, 1979) but not in Conophoralia. The simplest male organ, as described for two species of *Haplognathia* (Knauss and Rieger 1979), consists of paired lateral testes which grade into sperm ducts. Both are enclosed by epithelial and gland cells and the sperm duct lumen may be lined with microvilli. The sperm ducts may unite before entering the penis, a bulbous structure made up of few large glandular cells some of which form a tissue connection (an interruption of the basement membrane) with the ventral epidermis, the male gonopore. While there are species differences in the structure and distribution of glands and musculature associated with the penis, it is remarkable that there is never a basement membrane associated with any of these structures (Knauss and Rieger 1979).

The scleroperalian copulatory organ, in contrast, is complex and seems derived from an invagination of the body wall (Mainitz 1979).

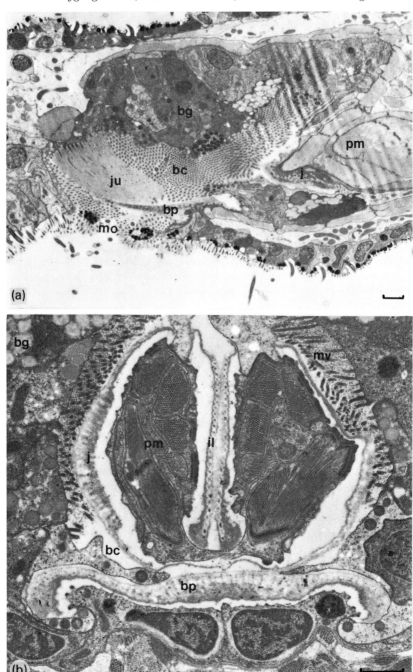

Its main feature is a stylet surrounded by gland cells and muscles. The stylet consists of eight (*Labidognathia*) to ten (Gnathostomulidae) radially arranged parallel rods which are elongated cellular extensions filled with microtubules (inner part of rods) and 'crystalline inclusions' (outer part). Inner and outer rods are separated, at least proximally, by a continuation of the basement membrane which also surrounds, continuous with that of the epidermis, the entire stylet apparatus. In addition, Gnathostomulidae possess a stylet sheath enclosing the rods. This sheath, which seems to originate from the two anterior gland cells, must be regarded as an extracellular secretory product of epithelial cells, and can thus be defined as cuticular (Mainitz 1977), in contrast to some superficially similar, yet entirely intracellular rod-type stylets of an acoel turbellarian studied by the same author. On the basis of light microscopy, it appears that all Scleroperalia follow more or less this general pattern, except *Gnathostomaria* which is characterized by a much elongated, even twisted stylet apparently not composed of rods, and the possession of a conspicuous seminal vesicle (Ax 1956; Sterrer, unpublished). What little is known about the copulatory organ of Conophoralia suggests a simple structure at least superficially similar to that of *Haplognathia*, which prompted Mainitz (1979) to propose that Conophoralia may in fact be a sister group to, rather than derived from, Scleroperalia.

Although much has been learned about sperm structure, homology of sperm components within the phylum has not been completely resolved. There are three, possibly four different types of sperm. The first, or filiform type, is up to 100 μm long and under the light microscope appears to consist of a mostly spiralized head, an elongated, smooth middle piece, and a tail (Sterrer 1969). TEM (Figs 12.3, 12.4) reveals the head to contain nuclear material tipped by what may be an acrosomal vesicle (Fig. 12.4(a, e, f)). The nucleus is a solid spiral posteriorly (Fig. 12.4(b, g)) but separates into a hollow core and a surrounding spiral anteriorly (Fig. 12.4(c)). The middle piece is made up of what are presumably mitochondrial derivatives completely enveloping the single flagellar axoneme (Figs 12.3, 12.4(d)); the latter continues into the tail. Light microscopy suggests there are minor variations of this basic pattern particularly in the spiralization of the head.

Fig. 12.2. (a) *Semaeognathia sterreri*. Median sagittal section through the buccal cavity; anterior to left. Note the large jugum, basal plate, jaws, buccal glands, pharynx musculature, and mouth opening. (b) *Labidognathia longicollis*. Cross-section through jaws and basal plate. Note the massive pharynx musculature inside the cone-shaped jaws, and the row of tubes which forms the vertical inner jaw lamellae. The jaws are surrounded by a buccal cavity with long microvilli. Abbreviations: bc, buccal cavity; bg, buccal glands; bp, basal plate; il, inner jaw lamellae; j, jaws; ju, jugum; mo, mouth opening; mv, microvilli; pm, pharynx musculature. Scale is 1 μm.

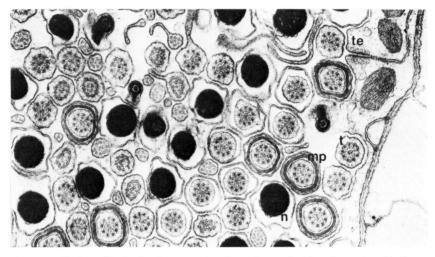

Fig. 12.3. *Haplognathia simplex.* Cross-section through a testis. Note the testis epithelium, and cross sections of sperm in the region of the nucleus, middle piece and tail. Abbreviations: mp, middle piece; n, nucleus; t, tail; te, testis epithelium. Micrograph courtesy of R. M. Kristensen.

Filiform sperm are uniflagellate and all those investigated show an axoneme with a 9 + 2 pattern (Fig. 12.3, 12.4(h)). The filiform sperm is typical of and exclusively found in the order Filospermoidea.

The second, or 'dwarf' type, has been analysed ultrastructurally in three species of *Gnathostomula* (Graebner 1968b, 1969; Riedl 1969; Graebner and Adam 1970). Only 2-3 μm long and rounded, oval or droplet-shaped, it is devoid of any flagellar structures, including spermatogenesis stages. It is further characterized by a basally located nucleus, 1-2 mitochondria, a single dense layer of microtubules immediately under the plasma membrane, and tubular protuberances of the plasma membrane, 0.18-1.0 μm in length ('Füsschen', or micropodia). Other than the shape of the sperm it is the length and arrangement of micropodia that seems to be species-specific. Extrapolating from *Gnathostomula*, all other Scleroperalia may follow this basic 'dwarf' pattern, with the exception of *Gnathostomaria* (see below).

The third, or 'conulus' type, characteristic of the suborder Conophoralia, is presented here for the first time with TEM pictures of *Austrognathia* sp. (Figs 12.5, 12.6). Usually large (9-45 μm long) and in the shape of a laterally compressed cone or mushroom, it consists of hat, cingulum, body, and matrix (Sterrer 1970). Shaped like half a car tyre, the thin, multilayered, electron-dense hat (Fig. 12.6(a), inset) partially surrounds the blunt end of the cone body. The cingulum (girdle) consists of an inner layer of mitochondrial derivatives and an outer labyrinth of what appear to be infoldings of the plasma mem-

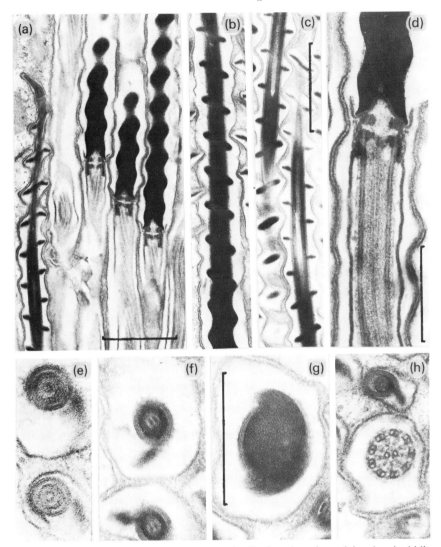

Fig. 12.4. *Haplognathia rosacea*, sperm. (a) Longitudinal sections through head and middle piece; (b,c) Longitudinal sections through the head. Note the lumen. (d) Longitudinal section through head-middle piece; (e-h) Cross-sections through the tip (e,f), head (g) tip (above) and tail (below) (h). Scales are 1 μm in (a-c), 0.5 μm in (d-h).

brane. The cone-shaped body itself is made up of electron-lucid material arranged as more or less concentric lamellar layers around two longitudinal axes. The matrix, as seen above the hat, is fairly electron-dense. Neither the mature conulus nor any of its developmental stages known so far show any recognizable flagellar structures. Short of a complete series in the development of conuli we believe the hat and its

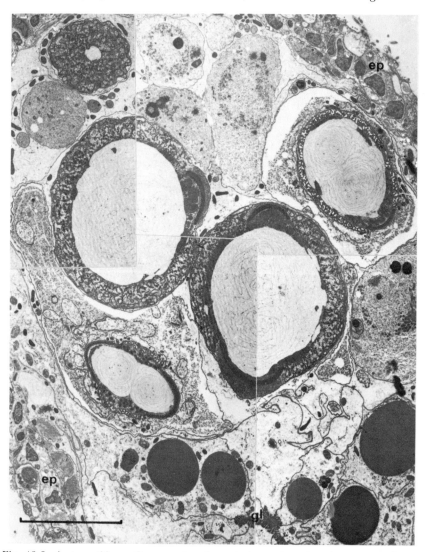

Fig. 12.5. *Austrognathia* sp. Cross-section through the testis and various stages of conulus differentiation. Abbreviations: ep, epidermis; gl, gut lumen. Scale is 10 μm.

surrounding matrix to represent nuclear and/or acrosomal material, and the body to be reserve matter, possibly to be used for the nutrition of the sperm in the interim between copulation and fertilization. The fact that the cone body is closely surrounded by mitochondria and a labyrinth of canals, structures often associated with high metabolic activity and transport, supports its role as a nutrient store.

The observation of 'conulus tetrads' would identify the conulus as a single sperm rather than a spermatophore (Sterrer 1965). The double

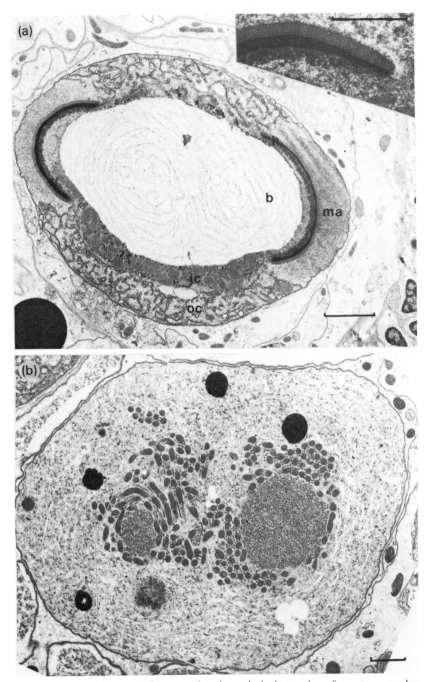

Fig. 12.6. *Austrognathia* sp. (a) Cross-section through the hat region of a mature conulus. Inset shows the layers of the hat. (b) Stage of conuligenesis. Abbreviations: b, body; h, hat; ic, inner layer of cingulum; ma, matrix; oc, outer layer of cingulum. Scales in both are 1 μm.

structure in the cone body as shown here (Fig. 12.6(a)), however, suggests a need to re-examine previously recorded multiple structures such as 'Siamese twin conuli' (Sterrer 1971; Farris 1973), conuli bisected in the hat (Sterrer 1965) or cone area (Kirsteuer 1970), or with the cone area appearing as 2-4 fingers or rods (Sterrer 1970; Ehlers and Ehlers 1973). It is conceivable that each of these 'multiple conuli' represents two or all four spermatozoa from a spermatogenesis tetrad that failed to separate completely, perhaps as the rule rather than the exception. The observation that in Filospermoidea (Knauss and Rieger 1979) spermatid meiosis apparently first proceeds without cytokinesis (resulting in four spermatid nuclei within a single cell) may support this interpretation. The function of such double or quadruple (diploid or quadruploid?) sperm in fertilization remains unknown; that it is transferred to the partner can be concluded from the regular occurrence of conuli (of somewhat altered appearance) in the bursa; at least one of these records is of a Siamese twin conulus (Farris 1973).

A possible fourth sperm type is found only in the genus *Gnathostomaria*. Its ultrastructure is not known. While superficially similar to the round 'dwarf' type, it is larger (8 μm diameter) and its filaments (which may correspond to the dwarf type's micropodia) are much longer (up to 6 μm). Fine structural analysis is particularly desirable as it might provide clues to relationships between the other gnathostomulid sperm types. The filiform type, as the one that conforms most closely to the primitive sperm as postulated by Franzén (1956), certainly deserves detailed comparison with the sperm of other lower Bilateria, particularly Gastrotricha and the turbellarian *Nemertoderma* (Tyler and Rieger 1975, 1977).

2. Female organs

The ovary is the most uniform organ throughout the phylum and is always single, located dorsally in about the mid-body region, with eggs maturing caudally. Eggs are endolecithal; there are no yolk cells or nutritive tissue, nor is there a tunica covering the ovary nor any ducts or structures for oviposition (Mainitz 1983).

Whereas Filospermoidea lack female organs other than the ovary, in Bursovaginoidea there is a bursa system immediately posterior to the ovary. In Scleroperalia the bursa system consists of a bullet-shaped to rounded bursa and an adjoining, round prebursa. The bursa, in Gnathostomulidae, is made up of three groups of layered cells which meet in three longitudinal ridges (cristae); these unite anteriorly to form a nozzle-like 'mouthpiece'. Bursa, cristae, and mouthpiece appear hard due to strongly developed desmosomes (Mainitz 1983). The mouthpiece surrounds a narrow fertilization channel which usually contains elongated sperm. In lower Scleroperalia the bursa is similar but simpler, more rounded and with fewer or no cristae. The bursa is often seen tightly packed with sperm.

The prebursa, derived from female tissue (Mainitz 1983) rather than injected male tissue (Riedl 1971a; Sterrer 1974), seems to receive but not store the sperm from copulation; it also may be involved in digesting male injectory substances and excessive sperm, and store excretory products (Mainitz 1983). It often contains crystal-like bodies (Sterrer 1976) and structures that may be remnants of male stylets (Sterrer 1973).

The bursa system in Conophoralia consists of a simple soft pouch that may be present only temporarily. It usually contains only one or two ('Siamese twin') conuli of characteristic appearance: they are shorter and their cone body is often hollow and covered with small droplets (Sterrer 1970). While the ultrastructure of the bursa conulus is not known, the change in the cone body would again suggest a nutritive role after copulation.

A vagina, located dorsally behind the bursa, seems to be a permanent feature, clearly visible in the light microscope, in Conophoralia. In Scleroperalia a vagina is not permanent (except possibly in *Onychognathia*, Riedl 1971b). In the electron microscope it appears as an interruption of the basement membrane of the dorsal body wall (Mainitz 1983). Whether this tissue connection is the result of either copulation or oviposition, or represents a primary 'organ' remains to be shown in immature specimens (Mainitz 1983).

Conclusion

The new evidence has not altered our view that within the phylum the main evolutionary direction was one of the progressive body and rostrum shortening and increasing complexity of sensory organs, pharyngeal structures and reproductive organs (Sterrer 1972). Adjustments may be necessary over two points. Firstly, the genus *Gnathostomaria*, because of its sperm structure and non-rod male stylet, lies somewhat apart from the rest of Scleroperalia. Secondly, there is doubt concerning the relationship of Conophoralia as a whole. With regard to the body wall (Rieger and Mainitz 1977), and particularly the male copulatory organ (Mainitz 1979), Conophoralia are sufficiently different from higher Scleroperalia as to suggest a derivation closer to the base, possibly near *Gnathostomaria*. Such an adjustment would allow a redefinition of Scleroperalia which, as it is presently constituted, is a paraphyletic group with Conophoralia.

Regarding the position of Gnathostomulida within lower Bilateria we welcome Ax's hypothesis (this volume) as an invitation to uncover new data that may shed further light on the phylum's origins. There are peculiarities that need examining, such as the frequency of tissue connections instead of permanent openings, e.g. anus, male pore in Filospermoidea, vagina in Scleroperalia. Further, there remain striking discrepancies with Platyhelminthes, such as the indication of an anal pore, the bilateral symmetry of the pharynx and the cross-striation of all musculature, features widespread among Aschelminthes. There are also similarities with taxa other than Platyhelminthes that need further investigation: the monociliary epithelium and the structure of proto-nephridia in Gastrotricha, or the tubular reinforcement of the inner jaw lamella in rotifers (Koehler and Hayes 1969). Rather than having found their permanent resting place within lower invertebrates Gnatho-stomulida continue therefore, in our opinion, to remain an enigmatic taxon, 'eine rätselhafte Wurmgruppe', that may well keep us puzzled for years to come.

Acknowledgements

This update would not have come about without generous help from G. Rieger, R. Riedl, R. M. Kristensen, E. B. Knauss, V. Lammert, E. E. Ruppert, and E. N. Powell, most of whom provided unpublished TEM pictures. We also acknowledge NSF Grant DEB 8119652 to R. M. Rieger and financial support by project No. 3115 of the Öster-reichischer Fonds zur Förd d. wiss. Forschung to Dr R. Riedl. The chapter is contribution No. 975 from the Bermuda Biological Station for Research, Inc.

References

Alvestad-Graebner, I. and Adam, H. (1983). Gnathostomulida. In *Reproductive biology of invertebrates* (eds K. G. and R. G. Adiyodi) Vol. 2, pp. 171-80. John Wiley, New York.

Ax, P. (1965). Die Gnathostomulida, eine rätselhafte Wurmgruppe aus dem Meeressand. *Abh. math.-naturw. Kl. Akad. Wiss. Mainz* 8, 1-32.

Boaden, P. J. S. (1977). Thiobiotic facts and fancies (aspects of the distribution and evolution of anaerobic meiofauna). *Mikrofauna Meeresbod.* 61, 45-63.

Briggs, D. E. G., Clarkson, E. N. K. and Aldridge, R. J. (1983). The conodont animal. *Lethaia* 16, 1-14.

Ehlers, B. and Ehlers, U. (1973). Interstitielle Fauna von Galapagos II. Gnathostomulida. *Mikrofauna Meeresbod.* 22, 175-99.

Farris, R. (1973). On *Austrognathia strunki* nov. spec. from the Florida Keys (Gnathostomulida). *Int. Rev. ges. Hydrobiol.* 58, 463-72.

—— (1977). Three new species of Gnathostomulida from the West Atlantic. *Int. Rev. ges. Hydrobiol.* 62, 765-96.

Fenchel, T. M. and Riedl, R. J. (1970). The sulphide system: a new biotic community underneath the oxidised layer of marine sand bottoms. *Mar. Biol.* 7, 255-63.

Franzén, Å. (1956). On spermiogenesis, morphology of the spermatozoon, and biology of fertilization among invertebrates. *Zool. Bidr. Upps.* 31, 355-482.

Graebner, I. (1968a). Erste Befunde über die Feinstruktur der Exkretionszellen der Gnathostomulidae (*Gnathostomula paradoxa*, (Ax, 1956) und *Austrognathia riedli*, (Sterrer, 1965)). *Mikroskopie* 23, 277-92.

—— (1968b). Ergebnisse einer elektronenmikroskopischen Untersuchung von Gnathostomuliden. *Verhandl. Dt. zool. Ges. Innsbruck* 1968, 580-99.

—— (1969a). Ergebnisse einer licht—und elektronenmikroskopischen Untersuchung der Spermiogenese einiger *Gnathostomula*—Arten. *Zool. Anz.*, Suppl. 33, Verh. Zool. Ges. 1969, 371-82.

—— (1969b). Vergleichende elektronenmikroskopische Untersuchung der Spermienmorphologie und Spermiogenese einiger *Gnathostomula*-Arten: *Gnathostomula paradoxa* (Ax, 1956), *Gnathostomula axi* (Kirsteuer, 1964), *Gnathostomula jenneri* (Riedl, 1969). *Mikroskopie* 24, 131-60.

—— and Adam, H. (1970). Electron microscopical study of spermatogenesis and sperm morphology in gnathostomulids. In *Comparative spermatology* (ed. B. Baccetti) pp. 375-82. Academic Press, New York.

Karling, T. G. (1966). On the defecation apparatus in the genus *Archimonocelis* (Turbellaria, Monocelididae). *Sarsia* 24, 37-44.

Kirsteuer, E. (1970). On some species of meiobenthic worms of the class Gnathostomulida from Barbados, West Indies. *Am. Mus. Novit.* 2432, 1-10.

Knauss, E. B. (1979). Indication of an anal pore in Gnathostomulida. *Zool. Scr.* 8, 181-6.

—— and Rieger, R. M. (1979). Fine structure of the male reproductive system in two species of *Haplognathia* (Gnathostomulida, Filospermoidea). *Zoomorphologie* 94, 33-48.

Koehler, J. K. and Hayes, T. L. (1969). The rotifer jaw: a scanning and transmission electron microscope study. I. The trophi of *Philodina acuticornis odiosa. J. Ultrastruct. Res.* 27, 402-18.

Kristensen, R. M. and Nørrevang, A. (1977). On the fine structure of *Rastrognathia macrostoma* gen. et sp. n. placed in Rastrognathiidae fam. n. (Gnathostomulida). *Zool. Scr.* 6, 27-41.

—— —— (1978). On the fine structure of *Valvognathia pogonostoma* gen. et sp. n. (Gnathostomulida, Onychognathiidae) with special reference to the jaw apparatus. *Zool. Scr.* 7, 179-86.

Lammert, V. (1983). Ultrastruktur des Pharynx von *Haplognathia rosea (Sterrer, 1968) (Gnathostomulida). Diplomarbeit, Universität Göttingen.*

Mainitz, M. (1977). The fine structure of the stylet apparatus in Gnathostomulida Scleroperalia and its relationship to turbellarian stylets. *Acta zool. fenn.* 154, 163-74.

—— (1979). The fine structure of gnathostomulid reproductive organs. I. New characters in the male copulatory organ of Scleroperalia. *Zoomorphologie* 92, 241-72.

—— (1983). Gnathostomulida. In *Reproductive biology of invertebrates* (eds K. G. Adiyodi and R. G. Adiyodi) Vol. 1, pp. 169-80. John Wiley, New York.

—— (in press). Gnathostomulida. In *Reproductive biology of invertebrates* (eds K. G. Adiyodi and R. G. Adiyodi), Vol. 4, John Wiley, New York.

Ochietti, S. and Cailleux, A. (1969). Comparaison des Condodontes et des machoires de Gnathostomulides. *C. r. hebd. Séanc. Acad. Sci. Paris* 268, 2664-6.

Powell, E. N. and Bright, T. J. (1981). A thiobios does exist—gnathostomulid domination of the canyon community at the East Flower Garden brine seep. *Int. Rev. ges. Hydrobiol.* 66, 675-83.

—— —— Woods, A., and Gittings, S. (1983). Meiofauna and the thiobios in the East Flower Garden brine seep. *Mar. Biol.* 73, 269-83.

Powell, E. N., Crenshaw, M. A., and Rieger, R. M. (1979). Adaptations to sulfide in the meiofauna of the sulfide system. I. ^{35}S—sulfide accumulation and the presence of a sulfide detoxification system. *J. exp. mar. Biol. Ecol.* 37, 57,-76.

Reise, K. (1981). Gnathostomulida abundant alongside polychaete burrows. *Mar. Ecol. Prog. Ser.* 6, 329-33.

—— and Ax, P. (1979). A meiofaunal 'thiobios' limited to the anaerobic sulfide system of marine sand does not exist. *Mar. Biol.* 54, 225-37.

Riedl, R. (1969). Gnathostomulida from America. First record of the new phylum from North America. *Science, N.Y.* 163, 445-52.

—— (1971a). On the genus *Gnathostomula* (Gnathostomulida). *Int. Rev. ges. Hydrobiol.* 56, 385-496.

—— (1971b). On *Onychognathia*, a new genus of Gnathostomulida from the tropical and subtropical West Atlantic. *Int. Rev. ges. Hydrobiol.* 56, 201-14.

—— and Rieger, R. M. (1972). New characters observed on isolated jaws and basal plates of the family Gnathostomulidae (Gnathostomulida). *Z. Morph. Tiere* 72, 131-72.

Rieger, R. M. (1976). Monociliated epidermal cells in Gastrotricha: Significance for concepts of early metazoan evolution. *Z. zool. Syst. Evolutionsforsch.* 14, 198-226.

—— and Mainitz, M. (1977). Comparative fine structure study of the body wall in Gnathostomulida and their phylogenetic position between Platyhelminthes and Aschelminthes. *Z. zool. Syst. Evolutionsforsch.* 15, 9-35.

—— and Rieger, G. E. (1975). Fine structure of the pharyngeal bulb in *Trilobodrilus* and its phylogenetic significance within Archiannelida. *Tissue Cell* 7, 267-79.

Schiemer, F. (1973). Respiration rates of two species of gnathostomulids. *Oecologia, Berlin* 13, 403-6.

Sterrer, W. (1965). *Gnathostomula axi* Kirsteuer and *Austrognathia* (ein weiteres Gnathostomuliden-Genus) aus der Nordadria. *Z. Morph. Ökol. Tiere* 55, 783-95.

—— (1969). Beiträge zur Kenntnis der Gnathostomulida. I. Anatomie und Morphologie des Genus *Pterognathia* Sterrer. *Ark. Zool.* 22, 1-125.

—— (1970). On some species of *Austrognatharia, Pterognathia* and *Haplognathia* nov. gen. from the North Carolina coast (Gnathostomulida). *Int. Rev. ges. Hydrobiol.* 55, 371-85.

—— (1971). On the biology of Gnathostomulida. *Vie Milieu*, Suppl. 22, 493-508.

—— (1972). Systematics and evolution within the Gnathostomulida. *Syst. Zool.* 21, 151-73.

—— (1973). On *Nanognathia,* a new gnathostomulid genus from the east coast of the United States. *Int. Rev. ges. Hydrobiol.* 58, 105-15.

—— (1974). Gnathostomulida. In *Reproduction of marine invertebrates* (eds A. C. Giese and J. S. Pearse) Vol. 1, pp. 335-57. Academic Press, New York.

—— (1976). *Tenuignathia rikerae* nov. gen. nov. spec., a new gnathostomulid from the West Atlantic. *Int. Rev. ges. Hydrobiol.* 61, 249-59.

—— (1977). Jaw length as a tool for population analysis in Gnathostomulida. *Mikrofauna Meeresbod.* 61, 247-56.

—— (1982). Gnathostomulida. In *Synopsis and classification of living organisms* (ed. S. P. Parker) Vol. 1, pp. 847-51. McGraw-Hill, New York.

—— and Farris, R. (1975). *Problognathia minima* nov. gen. nov. spec., representative of a new family of Gnathostomulida, Problognathiidae n. fam. from Bermuda. *Trans. Am. microsc. Soc.* 94, 357-67.

Tyler, S. (1975). Comparative ultrastructure of adhesive systems in the Turbellaria and other interstitial animals. Ph.D. Dissertation, University of North Carolina.

—— (1977). Ultrastructural evidence for the systematic position of the Nemertodermatida (Turbellaria). *Acta zool. fenn.* 154, 193-207.

—— and Rieger, R. M. (1975). Uniflagellate spermatozoa in *Nemertoderma* (Turbellaria) and their phylogenetic significance. *Science, N.Y.* 188, 730-2.

13. *Annulonemertes* gen. nov., a new segmented hoplonemertean

GUNNAR BERG

Department of Zoology, University of Göteborg, Sweden

Abstract

A new monostiliferous hoplonemertean, *Annulonemertes minusculus* gen. et sp. nov., is described. The species is characterized by annulations of the body surface which correspond to segmented divisions of some of the internal organs, particularly the intestine. The phylogenetic implications of *Annulonemertes* gen. nov. are discussed.

Introduction

Nemerteans are generally regarded as unsegmented, acoelomate worms. However, Friedrich (1949) described one specimen as a new species of *Arenonemertes*, *A. minutus*, which among other characters possessed a few constrictions in the posterior body regions. Friedrich was unable to offer any insight into the nature of these constrictions. The aim of the present investigation is to present information on a new species of aberrant nemertean with a constant number of annuli along its body, and to discuss its phylogenetic position within the Nemertea.

Material and methods

The material examined was collected from Balsfjord, west of Eriksjord, near the Marine Biological Station at Tromsö, northern Norway. After *in vivo* studies the worms were anaesthetized in MS-222 and fixed in Bouin's fluid. Paraffin sections, 6-7 μm thick, were stained with Heidenhain's azan.

Annulonemertes gen. nov.

Diagnosis

Interstitial marine monostiliferous hoplonemerteans with body externally divided into segments by distinct constrictions; eyes absent; rhynchocoel short, restricted to first two body segments, with wall containing interwoven longitudinal and circular muscle fibres; proboscis slender, with normal monostiliferous construction; body wall musculature without diagonal layer; dorsoventral muscles absent; frontal organ present; cephalic glands well developed; cerebral sensory organs absent; nervous system with neither neurochords nor accessory lateral nerve; foregut without distinct pyloric tube, intestine with short anterior caecum which lacks forwardly directed diverticula, main intestinal canal divided into segments which communicate with each other via funnel-shaped apertures; excretory system rudimentary; sexes separate.

Annulonemertes minusculus sp. nov.

1. Type specimens

Type specimens deposited with the Museum of Natural History, Göteborg, Sweden. Holotype, mature female, complete series of longitudinal sections, Registration Number Nemert. 70; paratype, mature female, complete series of transverse sections, Registration Number Nemert. 71.

2. Type locality

Tromsö, Balsfjord, west of Eriksjord, northern Norway, 80-90 m depth in coarse sand and shells.

3. Additional material

Two further specimens, retained in the author's collection, were obtained from Tromsö, Åsgård, 45 m depth in shell gravel, sand and clay.

4. External features

The overall external appearance of the nemerteans is shown in Fig. 13.1(a). In all the specimens examined there are seven distinct

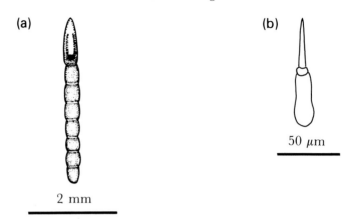

(a)

(b)

50 μm

2 mm

Fig. 13.1. *Annulonemertes minusculus* gen. et sp. nov. (a) Dorsal view of complete individual, drawn from life. (b) Central stylet and basis of proboscis armature.

transverse constrictions or annuli, such that behind the head the body is divided into seven segments. The nemerteans are between 3-4 mm long. Eyes are absent. The internal organs are visible through the body wall, and give the worms an orange-yellow colour. At higher magnifications white pigment spots can be distinguished; these are scattered all over the body surface but are particularly numerous near the intersegmental constrictions.

5. Body wall and musculature

The body wall shows the characteristic hoplonemertean construction and sequence of layers (Figs 13.3, 13.4). Neither a diagonal muscle layer nor dorsoventral muscle strands are present.

6. Digestive system

The anterior portion of the alimentary canal possesses a typical hoplonemertean arrangement (Fig. 13.2) except that there is no distinct pyloric tube. Posteriorly the stomach narrows to form a short duct which opens into the dorsal wall of the intestine. An extremely short anterior caecum, which does not possess diverticula, extends below and to one side of the stomach.

The remaining part of the alimentary canal differs strikingly from the usual hoplonemertean form. It has no lateral diverticula. Instead, the intestine is divided into a number of barrel-shaped segments or sections (Figs 13.2-13.4), each communicating with its neighbour through a funnel-shaped opening. This subdivision of the intestine exactly follows the external annulation. At the borders between adjacent segments there is an internal connective tissue membrane traversing the

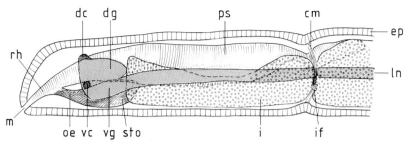

Fig. 13.2. *Annulonemertes minusculus* gen. et sp. nov. Schematic reconstruction of the most anterior body region in lateral aspect. Abbreviations: cm, connective tissue membrane; dc, dorsal commissure; dg, dorsal ganglion; ep, epithelium; i, intestine; if, funnel-shaped opening between intestinal compartments; ln, longitudinal nerve; m, mouth; oe, oesophagus; ps, proboscis sheath (rhynchocoel); rh, rhynchodaeum; sto, stomach, vc, ventral commissure; vg, ventral ganglion.

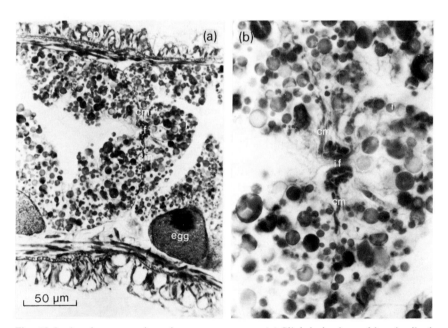

Fig. 13.3. *Annulonemertes minusculus* gen. et sp. nov. (a) Slightly horizontal longitudinal section of a part of the intestinal region. (b) Longitudinal section of a funnel-shaped opening between adjacent intestinal regions at a higher magnification (about 2.5 × larger than (a)). Abbreviations as in Fig. 13.2.

body (Figs 13.2, 13.3). Around the intestinal funnels, which penetrate these membranes, are many circular muscle fibres which apparently regulate the size of the funnel apertures.

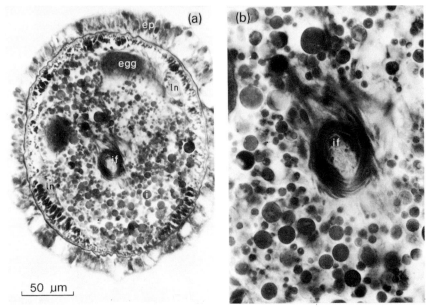

Fig. 13.4. *Annulonemertes minusculus* gen. et sp. nov. (a) Cross-section of the intestinal region through the boundary between two parts of the intestine. (b) Cross-section of the funnel-shaped opening in higher magnification (about 2.5× larger than (a)). Abbreviations as in Fig. 13.2.

7. Rhynchocoel and proboscis

The rhynchocoel reaches for about one-third of the body length, i.e. to the posterior part of the first segment behind the cephalic region. The musculature of the rhynchocoel wall is extremely thin and delicate in the head 'segment' but becomes somewhat better developed behind the first connective tissue segmental membrane. Throughout the rhynchocoel length its musculature consists of a single layer of interwoven longitudinal and circular muscle fibres.

The proboscis possesses a normal monostiliferous construction. There are ten proboscis nerves in all the specimens examined. The central stylet and basis are illustrated in Fig. 13.1(b). *In vivo* studies of the stylet apparatus give central stylet:basis length ratios of between 0.93 and 1.14 (mean value = 1.04), and basis breadth:length ratios of 0.36-0.43 (mean 0.40). The stylet bulb contains two accessory stylet pouches, each housing two to four accessory stylets.

8. Circulatory and excretory systems

In the present material the various parts of these systems are difficult to distinguish. All that can be identified are two lateral longitudinal blood vessels running below the main lateral nerve cords, and portions of a rudimentary excretory system situated on either side of the stomach just behind the brain.

9. Nervous system and sense organs

The nervous system shows no special characteristics (Fig. 13.2). The lateral nerve cords have only a single fibre core (i.e. there is no accessory lateral nerve) and neither they nor the brain lobes contain neurochords or neurochord cells.

The frontal organ, consisting of a single ciliated pit, opens at the tip of the head. The cephalic glands, which discharge through this organ, are well developed and reach the posterior regions of the brain lobes.

No evidence of eyes or cerebral sensory organs can be distinguished.

10. Reproductive system

The sexes are separate. The gonads are located dorsolaterally above the lateral nerve cords. In the females each ovary contains only a single large ripe egg. There are two or three gonads in each body segment. Gonoducts have not been discerned.

Systematic discussion

Annulonemertes gen. nov. can be distinguished from all other known hoplonemertean genera by its external annulation and the morphology of its intestine. The only other species possessing some superficial resemblance to the present forms is *Arenonemertes minutus* Friedrich (1949). *A. minutus* was collected from a sandy bottom at 12 m depth in the south-western Baltic Sea; unfortunately, however, the description is based on only a single specimen and there is no information about its internal anatomy. Figure 1 in Friedrich's (1949) article depicts a worm with three 'segments' and a total of six gonads. His drawings also show the form and size of the cerebral ganglia, and a pair of cerebral sensory organs is described in the text. To judge from these characters *Arenonemertes minutus* is clearly separate from *Annulonemertes minusculus* gen. et sp. nov. Friedrich was of the opinion that the two constrictions in the posterior part of *Arenonemertes minutus* were difficult to explain in a functional context. He also considered that the small number of paired gonads and their restriction to the posterior region of the body provided sufficient grounds for *Arenonemertes minutus* to be placed in a new genus. However, Friedrich (1949) believed that until additional morphological evidence became available the generic designation of his species should only be regarded as provisional. In my opinion, however, *Arenonemertes minutus* cannot be included in the genus *Arenonemertes*. Clearly the species show at least an external resemblance to the new genus *Annulonemertes* in that the constrictions in the posterior part of *Arenonemertes minutus* recall those of *Annulonemertes* gen. nov. For as long as nothing is known about the internal morphology of the former taxon, however, the possibility of including it in *Annulonemertes* gen. nov. remains equivocal.

At present the hoplonemertean suborder Monostilifera contains seven families (Gibson 1982). The morphological characters of *Annulonemertes minusculus* gen. et sp. nov., however, make it impossible to include the species within any of these families. The interlaced muscle layers of the rhynchocoel wall provide a point of comparison with the Cratenemertidae (cf. Berg 1972), but other characters are closer to those of the Emplectonematidae or Tetrastemmatidae. In a future paper, where additional *Annulonemertes* gen. nov. material will be described, the problem of the familial relationships will be discussed in more detail.

Phylogenetic discussion

The structure of *Annulonemertes* gen. nov. is reminiscent of metameric segmentation. What is the significance of this? According to Clark (1964) metameric segmentation is not easily defined, but it may be noted that one of its essential features is that the body-wall musculature and the coelom are involved in the serial repetition of parts, although the coelom of many segmented animals may have lost its compartmentalized organization. The nervous system, excretory organs, gonads, and other organs, may all show a comparable seriation, but unless the musculature is also segmentally arranged, or shows signs of having been so at an earlier state of phylogeny, the animal is regarded as pseudometameric. The fact that it is the segmentation of the musculature which is so essential in the definition of metamerism immediately suggests that the locomotory theories of the origin of metameric segmentation should be given special consideration. A flexible, muscular body wall enclosing a hydrostatic skeleton is typical for vermiform animals. Hydroids use the fluid in their coelenteric cavities, whilst in turbellarians and nemerteans the parenchyma serves as a fluid skeleton. According to Clark (1964), this fluid skeleton is of limited importance in nemerteans because the muscular activities used in locomotion either make virtually no demands upon the hydrostatic skeleton, as in pedal locomotion, or they are rarely and relatively ineffectively performed. Peristaltic creeping does demand the use of such a skeleton and the nemerteans which employ this method of locomotion have a parenchyma which is largely gelatinous and therefore approximates a true fluid skeleton. Peristalsis is perfected in coelomate animals.

In *Annulonemertes* gen. nov. a coelomic cavity is lacking, but the voluminous compartments of the intestine possibly have an analogous function. Furthermore, the body-wall musculature is reduced at each internal connective tissue membrane, and the latter appear to be functionally equivalent to a septum. Given that closure of the intestinal funnel junctions seems likely, each segment can be presumed to function in a manner analogous to that of an annelid segment.

It remains, therefore, to establish whether the annulation in *Annulonemertes* gen. nov. is a plesiomorphic or an apomorphic character within the Nemertea. No unequivocal answer can yet be given, but some clues to phylogenetic relationships may be found by comparing the spermatozoa of nemerteans with those of other groups. Whitfield (1972) gives a description of the spermatozoon of the hoplonemertean *Emplectonema neesii*. He compares the spermatozoa of *Emplectonema* with platyhelminth spermatozoan types, and states that the spermatozoa of nemerteans have an acrosome, a single 9 + 2 flagellum, and no cytoplasmic microtubules. It is, therefore, fundamentally different in design from the spermatozoa of all non-acoel platyhelminths, and the only feature in common with some acoel spermatozoa is the presence of a 9 + 2 flagellar tube arrangement. These divergences in the spermatozoan ultrastructure of nemerteans and platyhelminths suggest that if the two groups had a common metazoan ancestor it existed before any of the present-day groups of platyhelminths developed from a basic stem or stem groups, and before their spermatozoa became adapted for internal fertilization. On the basis of light microscopy Franzén (1967) also states that the nemerteans have retained a much more primitive spermatozoan organization than the Platyhelminthes, and he remarks upon the extreme improbability of the specialized spermatozoa of the flatworms having given rise to the relatively unspecialized (or 'primitive') types of higher groups. Whitfield (1972) closes his account with a question: Is it at all possible that the stem group of the nemerteans was closer to the ancestor (or ancestors) of the higher protostomes than was the stem group of the platyhelminths?

Generally similar results to those of Whitfield (1972) were also obtained by Afzelius (1971) in an investigation of the spermatozoa of the bdellonemertean *Malacobdella grossa*.

The phylogenetic significance of *Annulonemertes* gen. nov. can be interpreted in at least three different ways (Fig. 13.5):

1. The segmentation of *Annulonemertes* gen. nov. is an apomorphic character in the nemerteans and is evolutionarily a dead end.

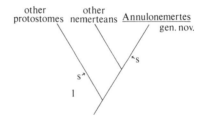

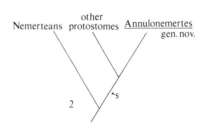

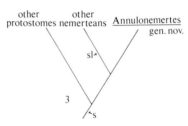

Fig. 13.5. Three different ways to interpret the phylogenetic significance of *Annulo-nemertes* gen. nov. The arrows indicate stages in each interpretation at which segmentation occurred; sl in '3' shows where the secondary loss of segmentation took place, leading to the typical nemertean type of body organization.

2. The segmentation is an apomorphic character in the nemerteans, and is close to the ancestors of metameric protostomes.

3. The segmentation is a plesiomorphic character in the nemerteans. *Annulonemertes* gen. nov. is close to the ancestors of metameric protostomes, but nearly all other nemerteans have lost this segmentation.

Notwithstanding differences in spermatozoan structure, the literature indicates that there is an overwhelming volume of evidence linking nemerteans with flatworms (cf. e.g. Hyman 1951; Gibson 1972). Thus the most plausible and probably most well-founded decision with regard to these three possibilities is likely to be the first one, that is to say that *Annulonemertes* gen. nov. is nothing but a dead end within the Nemertea and is unrelated to the evolution of metameric protostomes.

However, if future investigations, such as ultrastructural or biochemical, provide additional support for a nemertean-annelid relationship, *Annulonemertes* gen. nov. will again warrant attention.

Acknowledgements

I wish to express my most sincere thanks to the director and staff of the Marine Biological Station at Tromsö for their kind help and advice during my stay there, to Dr Christer Erséus for valuable discussion and criticism, and to Mrs Bozena Lewkowics, Mrs Barbro Löfnertz and Mrs Aino Wahlström for invaluable technical assistance. This study has been supported by grants from the Swedish Natural Science Research Council.

References

Afzelius, B. (1971). The spermatozoon of the nemertine *Malacobdella grossa. J. Submicrosc. Cytol.* 3, 181-92.

Berg, G. (1972). Studies on *Nipponnemertes* Friedrich, 1968. I Redescription of *Nipponnemertes pulcher* (Johnston, 1837) with special reference to intraspecific variation of the taxonomical characters. *Zool. Scr.* 1, 211-25.

Clark, R. B. (1964). *Dynamics of metazoan evolution: the origin of the coelom and segments.* Clarendon Press, Oxford.

Franzén, Å. (1967). Remarks on spermiogenesis and morphology of the spermatozoon among the lower Metazoa. *Ark. Zool.* 19, 335-42.

Friedrich, H. (1949). Uber zwei bemerkenswerte neue Nemertinen der Sandfauna. *Kieler Meeresforsch.* 6, 68-72.

Gibson, R. (1972). *Nemerteans.* Hutchinson, London.

—— (1982). Nemertea. In *Synopsis and classification of living organisms* (ed. S. P. Parker) Vol. 1, pp.823-46. McGraw-Hill, New York.

Hyman, L. H. (1951). *The invertebrates.* Vol. 2. *Platyhelminthes and Rhynchocoela.* McGraw-Hill, New York.

Whitfield, P. J. (1972). The ultrastructure of the spermatozoon of the hoplonemertine *Emplectonema neesii. Z. Zellforsch. mikrosk. Anat.* 128, 303-16.

14. Phylogenetic aspects of pseudocoelomate evolution

S. LORENZEN

Zoologisches Institut, Christian-Albrechts-Universität, Kiel, Federal Republic of Germany

Abstract

There is evidence that pseudocoelomates have evolved from relatively large ancestors with body sizes measured in centimetres rather than millimetres, which reproduced by external fertilization (eggs and sperm shed into the water) and had a fluid-filled body cavity. Holophyly (= monophyly *sensu* Hennig; see Lorenzen 1983) for a putative taxon Pseudocoelomata as well as for the Priapulida, Nematomorpha, and Rotatoria, cannot be established. Holophyly for the Kinorhyncha, Gastrotricha, Nematoda, and Acanthocephala is, however, well founded. It is argued that there are probably two main lines of evolution within the pseudocoelomates, one line giving rise to the Priapulida, Kinorhyncha, Gastrotricha, Nematoda, and Nematomorpha, the other to the Rotatoria and Acanthocephala. New evidence is presented that the Acanthocephala are most closely related to the Rotatoria, especially to the Bdelloidea.

Introduction

The following established taxa are regarded as pseudocoelomates: Priapulida, Kinorhyncha, Gastrotricha, Nematoda, Nematomorpha, Rotatoria, and Acanthocephala. A recently discovered new group, the Loricifera, is also a member of this assemblage (Kristensen 1983).

Twenty years ago Remane (1963a) stated: 'The relationship of the so-called Pseudocoelomata (or Pseudocoelia) is one of the most difficult problems in systematic zoology.' This statement is still true, and discussion is continuing on whether they comprise a holophyletic group, how they are related to the other invertebrates, and how they are related to one another.

Over the last two decades phylogenetic problems concerning the pseudocoelomates have been discussed by Remane (1963a), Rieger and Rieger (1977), Clark (1979), Coomans (1981) and others. The following contribution is based on phylogenetic systematics. In it, Hennig's term 'synapomorphy' is replaced by a new term 'holapomorphy', because the former term indicates only the necessary, the latter the necessary *and sufficient* conditions for regarding a taxon as holophyletic (Lorenzen 1983).

Phylogenetic systematics has only rarely been applied to pseudocoelomate groups. As to gross systematics, this has only been performed on free-living nematodes (Lorenzen 1981) and by Ruppert (1982) on the evolution of the gastrotrich pharynx.

In the following discussion, discrete points (or characters) of phylogenetic importance are numbered 1-36; these numbers correspond in part to the tentative phylogenetic scheme of pseudocoelomate relationships illustrated in Fig. 14.1

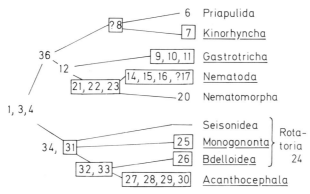

Fig. 14.1. Phylogenetic system for the pseudocoelomates. Numbers refer to the points in the text: boxed numbers are considered to be holapomorphies, the others non-holapomorphies. The holophyly of those taxa underlined is considered to be established.

The origin of the pseudocoelomates

To understand the origin of the pseudocoelomates, the following is considered crucial:

Point 1: Shedding eggs and sperm into the water is regarded as plesiomorphous within Metazoa compared with copulation (Franzén 1956; Jägersten 1972; Olive, this volume) because:

(i) sperm shed into the water are regarded as more primitive than those transferred by copulation;

(ii) many evolutionary shifts from shedding gametes into water to copulation are known, but movement in the opposite direction appears to be highly unusual. It is only in the sessile gastropod *Serpulorbis* (Vermitidae, Prosobranchia) that copulation has been replaced by shedding highly specialized spermatophores into the water (Scheuwimmer 1979), a clearly derived condition.

The following conclusions (Points 2–4) are drawn from Point 1:

Point 2: The thiobios hypothesis (Boaden 1975, 1977) is rejected because it postulates an evolutionary change from copulation to shedding gametes into water.

Point 3: Animals in which both sexes shed their gametes into the water have a relatively large body size (measured in centimetres) and a spacious fluid-filled body cavity (or cavities) in which gametes are stored or into which the gonads extend. Minute animals (measured in millimetres) display copulation. Hence, a relatively large body size is regarded as plesiomorphous within the Metazoa.

Point 4: Eutely, i.e. constancy in number and arrangement of cells or nuclei in certain organs or in entire animals, is only known from minute species or their descendants, all of which display copulation. Therefore, a complete lack of eutely is regarded as plesiomorphous within the Metazoa. Eutely is always correlated with very low numbers of cells or nuclei.

Among the pseudocoelomates, only the larger species of the Priapulida shed their gametes into the water and have primitively a rather large body size. All others display copulation and a minute body size (although parasites may be larger). From Points 1, 2 and 4 it is concluded that pseudocoelomates have evolved from rather large ancestors shedding their gametes into water and provided with a spacious fluid-filled body cavity (or cavities). According to the enterocoel theory, these cavities are coelomic (Remane 1963b).

Typical pseudocoelomate features are the lack of asexual reproduction, little or no ability to regenerate, presence of an anus (reduced in some cases), generally minute body size, tendency towards eutely, and the presence of a body cavity interpreted as a modified coelom. This cavity contains abundant fluid in species provided with eversible organs but less fluid in small forms without eversible organs.

Point 5: None of the just listed pseudocoelomate characters may be interpreted as holapomorphic for pseudocoelomates, because it is uncertain to what extent they are due to homology or analogy. A putative taxon Pseudocoelomata cannot therefore be established as being holophyletic.

The phylogenetic status of individual
pseudocoelomate taxa

Priapulida

This group comprises only about 12 recent species which nevertheless display a relatively wide diversity of form. This suggests that the living Priapulida comprise only the remnant of a once more successful group. This view accords with the fossil evidence from the mid-Cambrian (Conway Morris 1979), presence of external fertilization in larger species and presence of a primarily large body size.

Point 6: Although the diagnostic characters of the Priapulida are clear-cut, there seem to be none which could be interpreted as holapomorphic. This is because the basic position of priapulids within the pseudocoelomates makes it difficult to decide whether their characters are primarily or secondarily absent in other groups of the Pseudocoelomata.

Kinorhyncha

I interpret the following character as holapomorphic:

Point 7: The adult body is primarily divided into 13 somites, the first being the introvert. This character is unique within the invertebrates, and it is unlikely that it is secondarily absent in other groups. Not only the cuticle, but also the underlying epidermis, the longitudinal nerves and the muscular system are segmented. Because of these features, the kinorhynchs have repeatedly been interpreted as an off-shoot of the annelid or even arthropod line of evolution (Remane 1936). These views are not acceptable because:

(i) an introvert is completely unknown within the Articulata, although in many polychaetes the pharynx works like an introvert;

(ii) the segmentation of the kinorhynchs is a functional prerequisite for the effective operation of the introvert and, hence, cannot be derived from the segmentation of Articulata.

Point 8: As in priapulids, the scalids of the introvert are arranged quincunxially and, hence, may be homologous with those of Priapulida. In kinorhynchs, the scalids act as sensory and locomotory organs (Moritz and Storch 1972).

Gastrotricha

Based on a study of the gastrotrich cuticle, Rieger and Rieger (1977) presented evidence that the gastrotrichs possess two unique features:

Point 9: The outer part of the cuticle consists of several layers of unit membrane-like structures.

Point 10: All external cilia, including the locomotory ones, are completely enclosed by layers of the outer part of the cuticle.

Both characters are interpreted as holapomorphies as it seems unlikely that these characters have been lost in other groups.

Gastrotrichs also display another unique feature which is interpreted as holapomorphic:

Point 11: Eggs are released by rupture of the body wall (Teuchert 1968).

Point 12: Remane (1936), Teuchert (1968, 1977), Ruppert (1982) and others have concluded from similarities in the peripheral and central nervous system, the pattern of innervation of body muscles, the pattern of early development, and the structure of the pharynx that nematodes are the closest extant group to the gastrotrichs. Since nematodes also seem to be closely related to nematomorphs the three groups together may represent a holophyletic group. However, this view cannot be substantiated by a reliable holapomorphy.

Point 13: Rieger and Rieger (1977) and Ruppert (1982) have tentatively concluded from similarities in the outer layer of the cuticle and the structure and design of the pharynx, that Paucitubulatina (Chaetonida, Gastrotricha) and Nematoda may be sister groups. This implies that nematodes are specialized gastrotrichs, that separation of the sexes in nematodes evolved from hermaphroditism and that nematodes release their eggs only secondarily through a structural opening. At present, these proposals seem unlikely.

Nematoda

Nematodes are the most successful pseudocoelomate group, due at least to the following two features, which enable them to exploit many varied ecological niches:

(i) Cuticle, epidermis and longitudinal muscles form a highly integrated system that produces a very efficient locomotion (Clark 1964; Crofton 1966).

(ii) There are many different designs of cuticle structure, sometimes occurring even in the different postembryonic stages of a single species. Furthermore, the cuticle is the site of considerable biochemical activity (Bird and Bird 1969).

Reliable holapomorphies of the nematodes are:

Point 14: There are three separate rings of sensillae present at the anterior end, the first two consisting of six and the third of four sensillae.

Point 15: If there are two ovaries, they are opposed to each other and open through a common vulva situated near the middle of the body.

Point 16: During postembryonic development there are always four moults.

It is unlikely that these characters have been lost in other pseudo-coelomate groups.

Three important discoveries have recently been made in nematodes:

Point 17: Riemann (1972) described a new nematode *Kinonchulus* from the Amazon which apparently is able to use its anterior end as an introvert (Fig. 14.2), although observations on living specimens are wanting. Riemann concluded that this could favour the hypothesis of a closer relationship between nematodes and nematomorphs because in the latter the larvae are provided with an eversible proboscis (Fig. 14.3). *Kinonchulus* displays other characters regarded as plesiomorphic within nematodes, such as a weak cuticle and a large distance between second and third circle of anterior sensillae (Riemann 1977; Lorenzen 1981), so that an introvert-like anterior end could be interpreted as plesiomorphic for nematodes. It should be mentioned that Jäger-skiölds's (1909; in Hyman 1951) drawing of the anterior end of the zooparasitic nematode *Hystrichis* also suggests an introvert-like function of it, but observations on living specimens appear to be wanting.

Point 18: Thorough studies of the nerve system of the nematodes *Rhabditis* and *Ascaris* have revealed that contrary to earlier opinion the dorsal nerve cord does not originate from the nerve ring but from

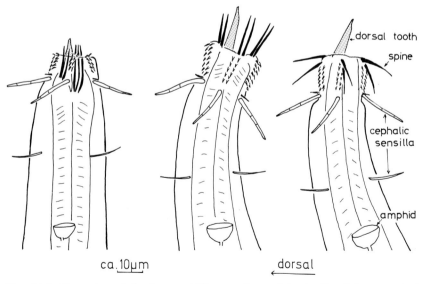

Fig. 14.2. Operation of the introvert-like anterior end in the free-living nematode *Kinochulus*. After Riemann (1972).

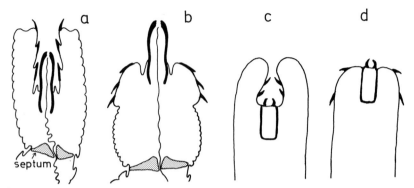

Fig. 14.3. Operation of the introvert-like anterior end in: (a, b) *Chordodes* (Gordioidea); (c, d) *Nectonema* (Nectonematoidea). (a, b) after Inoue (1958); (c, d) after Huus (1932).

several points on the ventral nerve cord (White, Southgate, Thomson, and Brenner 1976; Johnson and Stretton 1980). If this unique character is present in all nematodes, it could also be interpreted as holapomorphic for nematodes.

Point 19: In *Eudorylaimus* (Dorylaimida), cilia in the gut and circular muscles embedded in the epidermis have been discovered (Zmoray and Guttekova 1979; Guttekova and Zmoray 1980). Previously, both characters were unknown in nematodes, and reconfirmation would be helpful in establishing the assumption that nematodes evolved from ciliated ancestors provided with circular muscles. Since this assumption seems likely, both characters would be plesiomorphic for nematodes.

Nematomorpha

Although this group comprises exclusively highly specialized parasites of arthropods, there is no reliable holapomorphy for it. The reasons for this paradoxical situation are that the Nematomorpha share striking similarities with the Mermithoidea (Nematoda):

(i) the juveniles live parasitically in the body cavity of arthropods, whereas the larvae and the adults are free-living in aquatic habitats;

(ii) the intestinal duct is reduced in all stages, and during the parasitic phase nourishment is absorbed through the body wall;

(iii) the inner wall of the pharynx is a slender, cuticular tube extending into the body.

Point 20: If the listed similarities are homologous, either the nematomorphs are highly specialized nematodes or the nematodes have evolved from nematomorph-like ancestors with a reduced gut. The latter possibility is unlikely because in nematodes a functional gut is regarded as plesiomorphic. The former possibility cannot yet be accepted either because there is no evidence that nematomorphs secondarily lack the holapomorphic nematode characters. They are, however, present in Mermithoidea. Both mermithid and nematomorph anatomy are as yet rather poorly known.

The following characters, unique within the pseudocoelomates, are interpreted as holapomorphies of the Nematoda plus Nematomorpha:

Point 21: The body wall contains exclusively longitudinal muscles (but see Point 19).

Point 22: Movements are performed by flexures in the dorsoventral plane.

Point 23: The body wall is covered by a flexible cuticle. The cuticle of adult nematomorphs (Eakin and Brandenburger 1974) is especially similar to that of adult mermithids (Batson 1979).

There may possibly be another homology based on some similarity of the spicular apparatus of nematodes and the copulatory spines of nematomorphs (Kiryanova 1958).

The common ancestors of nematodes and nematomorphs may have had an introvert-like anterior end, because such a character is found in the larvae of nematomorphs and in the nematode genus *Kinonchulus* (Point 17).

Rotatoria

After nematodes, rotifers are the next most successful pseudocoelomate group. Their success is based on their ability to reproduce faster than other invertebrates if conditions are optimal and to form resting stages when environmental conditions deteriorate. The resting stages ensure survival and passive distribution to other localities (Birky and Gilbert 1971). These features are particularly useful in habitats subject to rapidly changing conditions, i.e. fresh water, soil, and coastal waters, where rotifers are abundant.

Point 24: The diagnostic characters of rotifers are very clear-cut: presence of a wheel organ, a retrocerebral organ, a mastax, and a foot provided with adhesive glands. However, no character may be interpreted as holapomorphic for rotifers because Acanthocephala may be considered as highly specialized rotifers which only secondarily lack the important rotifer characters.

Point 25: A holapomorphy of the Monogononta is the presence of only one gonad in both sexes.

Point 26: A holapomorphy of the Bdelloidea is the complete lack of males.

Acanthocephala

This is a group of highly specialized parasites. The adults live in the alimentary tract of vertebrates, the larvae in the body cavity of arthropods.

The holophyly of the Acanthocephala is based at least on the following four holapomorphies:

Point 27: The inner part of the syncytial epidermis has an extensive lacunar system.

Point 28: Females have a uterus bell, i.e. a pumping and sorting organ which passes fully embryonated eggs into the uterus and the remaining eggs back into the body cavity.

Point 29: The proboscis has spines which originate from the basement membrane of the epidermis.

Point 30: During ontogenetic development there is a special juvenile stage, the acanthella.

These characters are not known from other pseudocoelomates; it is unlikely that the latter have lost them during evolution. Spines originating from the basement membrane are also known from certain representatives of the Turbellaria (Rieger and Doe 1975; Doe 1976).

In Acanthocephala there is a problem in distinguishing the dorsal and ventral sides. Traditionally, the ventral side is regarded as the one towards which the everted proboscis is inclined. Hence, the brain would be located ventrally and the gonads dorsally to the lateral plane of the body, in opposition to their occurrence in other pseudocoelomates. Therefore, the traditional determination of dorsoventrality is rejected and the reverse adopted; von Haffner (1950) has already drawn attention to this problem.

Until now Acanthocephala have been regarded as an isolated group, as near relatives of Priapulida or as near relatives of the rotifers (for references see Conway Morris and Crompton 1982). The following evidence suggests that rotifers, especially Bdelloidea, are in fact the closest living relatives of the Acanthocephala:

Point 31: In both Rotatoria and Acanthocephala, the outer part of the syncytial epidermis is condensed into a cuticule-like structure perforated by many radial tubules originating from the outer plasmalemma (Welsch and Storch 1973). This unique character is interpreted as a holapomorphy of the Rotatoria plus Acanthocephala.

Point 32: Lemnisci, a pair of projections of the neck epidermis into the body cavity, are well known from Acanthocephala. It has generally been overlooked that they are also present in certain, if not all, bdelloid rotifers (Fig. 14.4). Almost a century ago Zelinka (1886, 1888) described them from *Mniobia* and *Zelinkiella* (Zelinka placed the species studied in *Callidina* and *Discopus*) but neither he nor other authors called them lemnisci, and they failed to conclude from this observation a close relationship between rotifers and acanthocephalans. In Bdelloidea, the lemnisci originate from the epidermis beneath the wheel organ. In Monogononta, the epidermis of the wheel organ bears at most swellings.

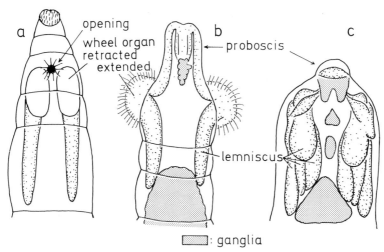

Fig. 14.4. Proboscis and lemnisci in bdelloid rotifers: (a,b) *Mniobia*; (c) *Zelinkiella*. In living specimens, either the wheel organ or the proboscis is extended; the extension of both organs, as in (b), is only found in freshly killed specimens. (a,b) after Zelinka (1886); (c) after Zelinka (1888).

Point 33: A proboscis which works like an introvert, comparable to that of Acanthocephala, is common to all Bdelloidea. It is located dorsally, in front of the pharynx and the wheel organ. When moving over the substratum, it serves as an adhesive and a sense organ (Remane 1932). So far as a proboscis is present in Monogononta it does not operate like an introvert.

Since in both the Bdelloidea and Acanthocephala the basic structure and relative position of the lemnisci and proboscis are the same, the presence of both organs is interpreted as holapomorphic for the Bdelloidea plus Acanthocephala.

Point 34: In most males of the Monogononta, the intestine is reduced to a cord to which the testis is attached. According to von Haffner (1950), the ligament cord of the Acanthocephala is also interpreted as a reduced intestine because the gonads are attached to it. Males are unknown from Bdelloidea.

Point 35: Acanthocephala display eutely in the nervous system and in the epidermis. This feature strongly suggests that the Acanthocephala originated from minute ancestors. Acanthocephala are larger than rotifers so that eutely of the epidermis poses problems because of the small number of available nuclei. These problems have been solved in two different ways:

(i) in smaller species the nuclei become large (up to 2 mm) and irregularly shaped;

(ii) in larger species the nuclei are fragmented amitotically, an extremely unusual feature among metazoans, supporting the conclusions given above.

The arguments presented here favour the view that the groups Acanthocephala, Bdelloidea plus Acanthocephala, and Rotatoria plus Acanthocephala are each holophyletic and that Acanthocephala are specialized rotifers. I prefer to maintain provisionally the classes Rotatoria and Acanthocephala because discussion of the problem is not yet finished. In parenthesis, it should be noted that endoparsitic rotifers are known to live in the guts of invertebrates (Budde 1925).

Although von Haffner (1950) has also argued for a close relationship between rotifers and acanthocephalans, two of his arguments cannot be accepted:

(i) his interpretation of the proboscis receptacle as a modified pharynx, because in rotifers the proboscis is always clearly separated from the pharynx;

(ii) his interpretation of the tip of the proboscis as representing the anterior region of the rotifer wheel organ, because in rotifers the proboscis is always distinctly separated from the wheel organ.

Recently, Conway Morris and Crompton (1982) have argued for a close relationship between Priapulida and Acanthocephala but I am unable to accept the main thesis of their argument (the overall appearance).

Conclusion

Point 36: An introvert occurs in Priapulida, Kinorhyncha, larvae of Nematomorpha, and possibly in some nematodes. From this it may be tentatively concluded that these four taxa are quite closely related to each other. Since the Gastrotricha are regarded as near-relatives of the nematodes, a further tentative conclusion is that gastrotrichs may have evolved from ancestors provided with an introvert. The proboscis of the Rotatoria and Acanthocephala cannot be regarded as an introvert because it lacks a mouth opening.

No holapomorphy could be established for the pseudocoelomates as a whole, nor for the Priapulida, Nematomorpha and Rotatoria. Therefore, the latter three taxa may prove suitable for elucidating the relationships of the pseudocoelomates with remaining invertebrates.

References

Batson, B. S. (1979). Body wall of juvenile and adult *Gastromermis boophthorae* (Nematoda: Mermithidae): ultrastructure and nutritional role. *Int. J. Parasit.* 9, 495-503.

Bird, A. F. and Bird, J. (1969). Skeletal structures and integument of Acanthocephala and Nematoda. In *Chemical zoology* (eds M. Florkin and B. Scheer) Vol. 3, pp. 253-88. Academic Press, New York.

Birky, C. W. and Gilbert, J. J. (1971). Parthenogenesis in rotifers: the control of sexual and asexual reproduction. *Am. Zool.* 11, 245-66.

Boaden, P. J. (1975). Anaerobiosis, meiofauna and early metazoan evolution. *Zool. Scr.* 4, 21-4.

—— (1977). Thiobiotic facts and fancies (aspects of the distribution and evolution of anaerobic meiofauna). In *The meiofauna species in time and space* (eds W. Sterrer and P. Ax), *Mikrofauna Meeresbod.* 61, 45-63.

Budde, E. (1925). Die parasitischen Rädertiere mit besonderer Berücksichtigung der in der Umgegend von Minden i. W. beobachteten Arten. *Z. Morph. Ökol. Tiere.* 3, 706-84.

Clark, R. B. (1964). *Dynamics in metazoan evolution. The origin of the coelom and segments.* Clarendon Press, Oxford.

—— (1979). Radiation of the Metazoa. In *The origin of major invertebrate groups* (ed. M. R. House), Systematics Association Special Vol. 12, pp. 55-102. Academic Press, London.

Conway Morris, S. (1979). The Burgess Shale (Middle Cambrian) fauna. *Ann. Rev. Ecol. Syst.* 10, 327-49.

—— and Crompton, D. W. (1982). The origins and evolution of the Acanthocephala. *Biol. Rev.* 57, 85-115.

Coomans, A. (1981). Aspects of the phylogeny of nematodes. In *Origine dei grandi phyla dei metazoi*, Convegno Internazionale, Atti dei Convegni Lincei 49, pp. 161-74.

Crofton, H. D. (1966). *Nematodes.* Hutchinson, London.

Doe, D. A. (1976). The proboscis hooks in Karkinorhynchidae and Gnathorhynchidae (Turbellaria, Kalyptorhynchia) as basement membrane intracellular specializations. *Zool. Scr.* 5, 105-15.

Eakin, R. M. and Brandenburger, J. L. (1974). Ultrastructural features of a gordian worm (Nematomorpha). *J. Ultrastruct. Res.* 46, 351-74.

Franzén, Å. (1956). On spermiogenesis, morphology of the spermatozoon and biology of fertilization among invertebrates. *Zool. Bidr. Upps.* 31, 355-480.

Guttekova, A. and Zmoray, I. (1980). Ultrastructure of the intestine of the free-living nematode *Eudorylaimus* sp. and *Dorylaimus* sp. *Biológia, Bratislava* 35, 89-96.

Haffner, K. von (1950). Organisation und systematische Stellung der Acanthocephalen. *Zool. Anz.*, Suppl. 145 (Klatt-Festschrift), 243-74.

Huus, J. (1932). Über die Begattung bei *Nectonema munidae* Br. und über den Fund der Larve von dieser Art. *Zool. Anz.* 97, 33-7.

Hyman, L. H. (1951). *The invertebrates.* Vol. 3. *Acanthocephala, Aschelminthes and Entoprocta. The pseudocoelomate Bilateria.* McGraw-Hill, New York.

Inoue, I. (1958). Studies on the life history of *Chordodes japonensis*, a species of Gordiacea. I. The development and structure of the larva. *Jap. J. Zool.* 12, 203-18.

Jägersten, G. (1972). *Evolution of the metazoan life cycle.* Academic Press, London.

Johnson, C. D. and Stretton, A. O. (1980). Neural control of locomotion in

Ascaris: anatomy, electrophysiology, and biochemistry. In *Nematodes as biological models* (ed. B. M. Zuckerman) Vol. 1, pp. 159-95. Academic Press, London.

Kiryanova, E. S. (1958). On the structure of the copulative organs of males of the freshwater hairworms (Nematomorpha, Gordioidea). *Zool. Zh.* 37, 359-72 (in Russian).

Kristensen, R. M. (1983). Loricifera, a new phylum with Aschelminthes characters from the meiobenthos. *Z. zool. Syst. Evolutionsforsch.* 21, 163-180.

Lorenzen, S. (1981). Entwurf eines phylogenetischen Systems der freilebenden Nematoden. *Veröff. Inst. Meeresforsch. Bremerh.*, Suppl. 7, 1-472.

—— (1983). Phylogenetic systematics: problems, achievements and its application to the Nematoda. In *Concepts in nematode systematics* (eds A. R. Stone, H. M. Platt and L. F. Khalil) pp. 11-23. Academic Press, London.

Moritz, K. and Storch, V. (1972). Über den ultrastrukturellen Bau der Skaliden von *Trachydemus giganteus* (Kinorhyncha). *Mar. Biol.* 16, 81-9.

Remane, A. (1932). Rotatoria. In *H. G. Bronn's Klassen und Ordnungen des Tierreichs*, Vol. 4. Akademische Verlagsgesellschaft, Leipzig.

—— (1936). Gastrotricha und Kinorhyncha. In *H. G. Bronn's Klassen und Ordnungen des Tierreichs*, Vol. 4. Akademische Verlagsgesellschaft, Leipzig.

—— (1963a). The systematic position and phylogeny of the Pseudocoelomata. In *The lower invertebrates* (ed. E. C. Dougherty) pp. 247-55. University of California Press, Berkeley, Ca.

—— (1963b). The enterocoelic origin of the coelom. In *The lower invertebrates* (ed. E. C. Dougherty) pp. 78-90. University of California Press, Berkeley, Ca.

Rieger, G. E. and Rieger, R. M. (1977). Comparative fine structure study of the gastrotrich cuticle and aspects of cuticle evolution within the Aschelminthes. *Z. zool. Syst. Evolutionsforsch.* 15, 81-124.

Rieger, R. M. and Doe, D. A. (1975). The proboscis armature of Turbellaria-Kalyptorhynchia, a derivative of the basement lamina? *Zool. Scr.* 4, 25-32.

Riemann, F. (1972). *Kinonchulus sattleri* n. g. n. sp. (Enoplida, Tripyloidea), an aberrant free-living nematode from the lower Amazonas. *Veröff. Inst. Meeresforsch. Bremerh.* 13, 317-26.

—— (1977). Causal aspects of nematode evolution: relations between structure, function, habitat and evolution. In *The meiofauna species in time and space* (eds W. Sterrer and P. Ax), *Mikrofauna Meeresbod.* 61, 217-30.

Ruppert, E. E. (1982). Comparative ultrastructure of the gastrotrich pharynx and the evolution of myoepithelial foreguts in Aschelminthes. *Zoomorphologie* 99, 181-220.

Scheuwimmer, A. (1979). Sperm transfer in the sessile gastropod *Serpulorbis* (Prosobranchia: Vermetidae). *Mar. Ecol. Prog. Ser.* 1, 65-70.

Teuchert, G. (1968). Zur Fortpflanzung und Entwicklung der Macrodasyoidea (Gastrotricha). *Z. Morph. Tiere* 63, 343-418.

—— (1977). The ultrastructure of the marine gastrotrich *Turbanella cornuta* Remane (Macrodasyoidea) and its functional and phylogenetic importance. *Zoomorphologie* 88, 189-246.

Welsch, U. and Storch, V. (1973). *Einführung in die Cytologie und Histologie der Tiere*. G. Fischer, Stuttgart.

White, J. G., Southgate, E., Thomson, J. N. and Brenner, S. (1976). The structure of the ventral nerve cord of *Caenorhabditis elegans*. *Phil. Trans. R. Soc.* B275, 327-48.

Zelinka, C. (1886). Studien über Räderthiere. I. Über die Symbiose und Anatomie von Rotatorien aus dem Genus *Callidina*. *Z. wiss. Zool.* 44, 396-506.

—— (1888). Studien über Räderthiere. II. Der Raumparasitismus und die Anatomie von *Discopus synaptus* n. g. nov. sp. *Z. wiss. Zool.* 47, 353-458.

Zmoray, I. and Guttekova, A. (1979). Notes on the muscular ultrastructure of some free-living nematodes from the genus *Eudorylaimus*. *Biológia, Bratislava* 34, 593-605.

15. The relationships of rotifers

PIERRE CLÉMENT

Laboratoire d'Histologie et Biologie Tissulaire,
Université Lyon 1, Villeurbanne, France

Abstract

In terms of their ultrastructure and behaviour rotifers seem to be a very primitive group, presumably related to the ancestors of the Platyhelminthes. They have no direct relationships with other pseudocoelomate groups, except the Acanthocephala. Aschelminthes represent an artificial and heterogeneous group.

The ultrastructure of the visual cells of rotifers is used as the basis for an original hypothesis on the evolution of visual cells in invertebrates. The classical 'tree of evolution' is perhaps better regarded as a more complex 'bush'.

Problems about origins

1. The rotifers: a success and a cul-de-sac

The abundance of rotifers in different biotopes, mostly in freshwater, represents an unequivocal success in terms of animal evolution. The keys to this success are parthenogenesis, followed by exponential rates of reproduction when conditions are favourable, and survival strategies when conditions become unfavourable, such as diapausing resting eggs of Monogononta and dessication-resistant adults of Bdelloidea.

However, this success is also a cul-de-sac. The eutely of rotifers is perfect, with no cellular division from birth to death. The organs of rotifers are fixed, specialized, and unable to regenerate. There is, of course, some morphological and behavioural variability, but this is always set against a background of inflexibility.

The problem of the relationships of rotifers is thus the problem of their origin.

2. Concerning methodology for phylogeny

The criteria generally used to distinguish a homology from an analogy remain arbitrary since we do not know the precise mechanisms operating in the evolution of zoological species and higher groups. We must argue with logical but perhaps arbitrary rules (the criteria of Remane (1954) or Hennig (1966)). We also can try to formulate some new rules. Nevertheless, phylogenetic hypotheses remain in part at least speculative, reflecting the ideology of their champions (Clément, in press). Some biologists believe in a perfect biological order, either established since the sixth day of the creation, or towards which we are inexorably moving. Others insist on the creative function of a quasi-anarchist disorder, which is correlated with some elementary rules and results in an abundance of living forms, including the lower Metazoa.

The classical disagreements on the origins of the different groups of the lower Metazoa (Dougherty 1963) were mostly argued on the basis of morphological data. New information on ultrastructure, behaviour, population genetics, biochemistry, etc. must now also be incorporated into the arguments. Accordingly I shall present facts and hypotheses which come for the most part from recent ultrastructural and some behavioural investigations on rotifers.

3. The classical disagreement about the origins of rotifers

From the first observations of rotifers by Leeuwenhoek to recent treatises, a series of different relationships have been proposed, including Infusoria, Polypa, Crustacea, Annelida, Mollusca, and Turbellaria (see Hyman (1951) and de Beauchamp (1965)). Remane (1963) suggested that the pseudocoel of rotifers represents a regressed coelom; the rotifers would thus be derived from coelomates (Dinophilidae or trochophore larvae). In later works Remane, Storch, and Welsch (1972, 1976) maintained this hypothesis, which was also favoured by Koste (1978). On the other hand, de Beauchamp (1907, 1909, 1965) favoured a close relationship between rotifers and Turbellaria. Having taken a prudent position in this debate (Ruttner-Kolisko 1963), Ruttner-Kolisko later wrote (in 1974) 'there is no fact indicating a case of reduction from highly developed, coelomate worms: no rudimentary coelom or mesoderm'. I shall try to support this position by providing a short summary of ultrastructural data for each organ and will develop in more detail one example taken from the visual cells. In attempting this I aim (*a*) to establish recurrent features within the taxon Rotifera and (*b*) to compare the primitive characteristics of rotifers with those of other groups.

The epidermis

1. Rotifers

The peripheral skeleton is always an intracytoplasmic lamina located within the syncytial integument. It can be thick and rigid (lorica, toes, integumentary expansions) or soft and supple (articulations). The structural pattern of the intracellular skeletal lamina differs from one species to another, and four categories are known (Clément 1969b, 1980; Fig. 15.3). The diversity of this structure, which always has the same function, is a useful and hitherto overlooked indicator of rotiferan phylogeny.

Other features of the epidermis, such as the different types of ciliary apparatus, presence or absence of the foot, pedal glands, and the retrocerebral apparatus are more well known (de Beauchamp 1907, 1909; Remane 1929-33). Electron microscopy has shown that the retrocerebral organ is composed only of mucous glands (Clément 1977a,b), and that some ultrastructural features of the anterior cilia are original, i.e. singlets or doublets at the base of the cilia, a dense structure near the central tubules in the apical region of the cilia (Clément 1977a,b; Clément, Amsellem, Cornillac, Luciani, and Ricci 1980c).

2. Comparison with other groups

An intracytoplasmic peripheral skeleton seems to exist only in Acanthocephala. In other Aschelminthes, Annelida, Mollusca and Arthropoda, the external skeleton is an extracellular cuticle. The origin of an intracytoplasmic skeleton probably lies in the apical terminal web of many epithelial cells. The ciliary integuments of Platyhelminthes, as well as those of Cnidaria and Ctenophora, possess such a terminal web. The comparative ultrastructure of the epidermis (cilia, ciliary roots of the terminal web) in these different groups may provide new arguments to support a direct relationship between their epidermis.

Pseudocoel and collagen

1. Rotifers

EM has confirmed the absence of a thin membrane surrounding the pseudocoel. Thus, as the pseudocoel is not a coelomic cavity, is it a mesenchymatic or a conjunctive space? The presence of moving amoeboid cells in the pseudocoel has not yet been confirmed by EM.

The structures described by Nachtwey (1925) and Remane (1929-33) may be expansions of muscular or epithelial cells. Moreover, there are neither fibrils nor microfilaments of collagen in the pseudocoel. In the absence of a biochemical assay for hydroxyproline, we cannot assume that a soluble collagen is absent. Some cells are separated from the pseudocoel by a very thin basal lamina (Fig. 15.2(b)).

2. Comparison with other groups

There are many cells, associated with fibrous collagenous structures, in the pseudocoel of Kinorhyncha, Priapulida, and Nematomorpha (Hyman 1951). These cells are scarce or absent in Nematoda (the pseudocoelocytes are fixed and are neither phagocytic nor amoeboid) and Gastrotricha. Fibrils of collagen, which are very widespread in animals, are absent in Ctenophora and in parasitic Platyhelminthes, but collagen is present as a network of microfilaments and precipitates *in vitro* in the shape of distinct fibrils (Franc, Franc, and Garrone 1976). Is it the same in Rotifera?

A special collagen is present in the cuticle of nematodes, nematomorphs, and annelids. But its presence in rotifers, where the cuticle is often absent, very thin, or gelatinous is improbable.

The digestive tract

The different categories of mastax, adapted to particular feeding styles, are well known since the work of de Beauchamp (1909), who also described the digestive tract. EM has recently provided some new information about the digestive tract (Clément *et al.* 1980a,b,c). A previously unknown supple structure, the buccal velum, which functions like a fish-trap, is located between the buccal and the pharyngeal parts in a bdelloid (*Philodina*) as well as in monogononts which have a malleous mastax (*Brachionus*). In these two cases, there are also unique pharyngeal cilia which contain dense and cross-striated material, located outside the axoneme in *Brachionus* and inside it in *Philodina* (Clément *et al.* 1980c).

The pharyngeal epithelial cells are thus very distinctive. Another characteristic feature is the hard structures of the mastax, the trophi. They are often built by the fusion of different materials; some extracellular cuticular material (rami, Koehler and Hayes 1969a,b; Clément 1977a,b) and some intracellular electron-opaque material (Clément 1977a,b) (Fig. 15.4), but the origin of the most important parts, the hardest portions of the trophi, is still unclear. Koehler and Hayes (1969a) suggest that they are extracellular in *Philodina*, but do not discuss this point in *Asplanchna* (1969b). Clément (1977a,b) observed that the fulcrum and manubria are inside a cell in *Trichocerca* (Fig. 15.4), and that in this species the hardest parts of the unci and rami are closely juxtaposed to, or are inside, a lining cell. As in *Asplanchna*, the rami of *Trichocerca* and *Notommata* are hollow structures, with living cytoplasm and a nucleus inside them (Clément 1977a). In *Notommata*, the manubria are also hollow structures containing cytoplasm and the nucleus. In *Trichocerca* the manubria contain only cytoplasm, the nucleus being located outside the hard structure (Fig. 15.4). Are these hardest parts of the trophi intracellular differentiations, at least at their basis, in adult rotifers? Some work supports this interpretation (Clément 1977a, unpublished observations) (Fig. 15.4), but other research suggests that they may be extracellular (Koehler and Hayes 1969b; Clément *et al.* unpublished observations).

A consistent difference between bdelloids and monogononts is the stomach. The bdelloids have very specialized stomachic structures (Mattern and Daniel 1966b; Clément and Amsellem, unpublished) which are not present in monogononts (Clément 1967, 1977a). All the rotifers possess a cloaca, except *Asplanchna* which has no intestine.

Excretory system

1. Rotifers

Huxley (1853) placed rotifers in Vermes because he identified protonephridia with flame-cells. These organs are good phylogenetic indicators. In the flame-cells of rotifers, the nucleus is not apical, and the apical cap bears the cilia of the flame (Figs 15.1, 15.3). The filtering wall is constructed of two concentric grids; outside are the columns which support the filtering membrane, and inside are the skeletal pillars, which contain dense material. The latter is cross-striated only in the flame-cells of bdelloids (Fig. 15.3) (Mattern and Daniel 1966a; Schramm 1978; Clément 1980). In the most primitive flame-cells, columns and pillars coalesce (Figs 15.1, 15.2) (Clément 1980; Clément and Fournier 1981). The protonephridial duct plays an active role in osmoregulation (Braun, Kümmel, and Mangos 1966; Clément 1969a; Clément and Fournier 1981).

2. Comparison with other groups

The flame-cell in platyhelminths is very similar to that of rotifers. Its nucleus is apical, and the filtering wall is little different, so it is possible to imagine a phyletic connection (Fig. 15.1) (Clément and Fournier

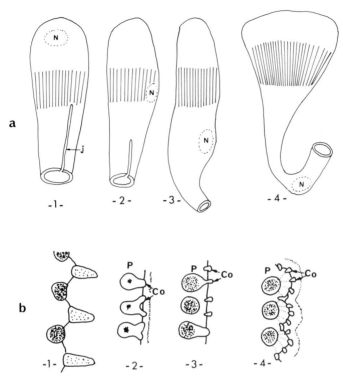

Fig. 15.1. Flame cells. 1, Platyhelminthes. 2-4, Rotifers: 2, *Philodina* (Bdelloidea); 3, *Trichocerca* (Monogononta); 4, *Notommata, Asplancha, Brachionus* (Monogononta). (a) Whole cells, (b) TS of the filtering wall. Abbreviations: Co, columns; j, cellular junction; N, nucleus; P, pilars.

1981). In contrast, the protonephridia of Priapulida and Gastrotricha bear typical solenocytes, paired or grouped (Kümmel and Brandenburg 1961; Brandenburg 1962; Teuchert 1973). Flame-cells do not exist in Nematoda and Nematomorpha. They are pluricellular in Kinorhyncha and Nemertea. The Ctenophora possess pluricellular ciliary rosettes whose function is almost identical to that of a flame-cell (Franc 1972).

Muscles

EM has demonstrated that the circular muscles of rotifers are not anucleate, but uninucleate (Clément 1977a,b). There is a great diversity of muscle cells in rotifers, e.g. more or less striated, slow or fast, etc. (Amsellem and Clément 1977; Clément 1977a,b), but in each case the myofilaments seem to be tubular in cross-section, as do those of crustaceans (Figs 15.3, 15.4) (Atwood 1972; Pringle 1972).

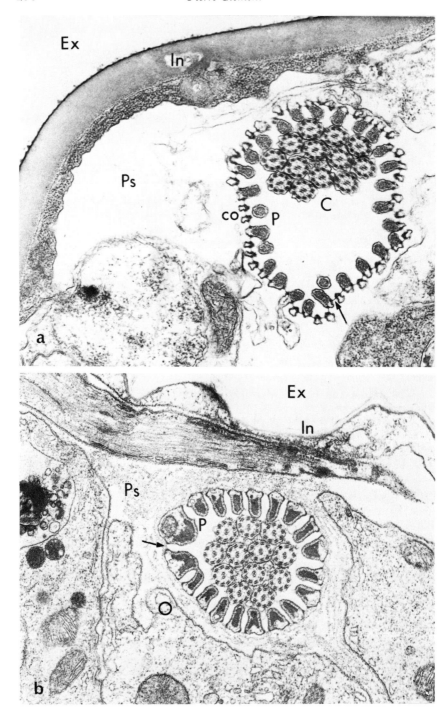

In Annelida and Mollusca, the thick myofilaments are composed of paramyosin. Their diameter is larger, and they are cross-striated. The same applies in Platyhelminthes (Reger 1976; Kryvi 1973; Fournier, personal communication) and nematomorpha (Eakin and Brandenburger 1974; Lanzavecchia, Valvassori, Eguileor, and Lanzavecchia 1979).

Nervous system

The nervous system is centralized and glia-free in rotifers (Clément 1977a, b, 1980). The number of neurons is very few, with 200-250 per animal (150-200 in the brain). Epithelial cells, and sometimes muscular cells, surround some nerves or ganglia, and a significant proportion of the brain. Such an absence of glia is known in primitive Metazoa such as sea anemones, jellyfish, siphonophores, ctenophores (Horridge 1968), but these Metazoa possess nerve nets, with or without ganglia. In Nematoda, there are sheath structures around each anterior sensory neurite. Moreover, they are epithelial, as in rotifers (Ward, Thomson, White, and Brenner 1975; Ware, Clark, Crossland and Russell 1975; Vergin and Endo 1976). But Ware *et al.* also reported glial cells in the nerve ring of *Caenorhabditis elegans*.

Oogenesis, parthenogenesis, and heterogenetic reproductive cycle

The ultrastructure of the syncytial ovary and vitellarium has been described in some monogononts (Bentfeld 1971a, b; Clément 1977a, b) and a bdelloid (Clément 1980; Amsellem and Ricci 1982). The structure is almost identical, the vitellus is synthesized in the syncytial vitellarium and then migrates in an oocyte by the way of a cytoplasmic bridge; the oocyte then synthesizes a shell, and the egg is laid. We need more information in order to understand the difference between oogenesis in bdelloids and in monogononts. In the latter, parthenogenesis produces mictic or amictic females; the former if they are not fertilized will lay haploid males, the latter diploid females.

Fig. 15.2. Pseudocoel and flame cells. TEM transverse section of the filtering wall of a flame cell. (a) *Trichocerca rattus*, × 30 000. (b) *Philodina roseola*, × 30 000. The membrane of the filtering wall (arrowed) is situated between the columns; pillars and columns are often bound together. A thin basal lamina is present in the pseudocoel of *P. roseola* around the flame cell. Abbreviations: C, cilia of the flame cell; co, columns; Ex, external medium; In, integument; P, pillars; Ps, pseudocoel.

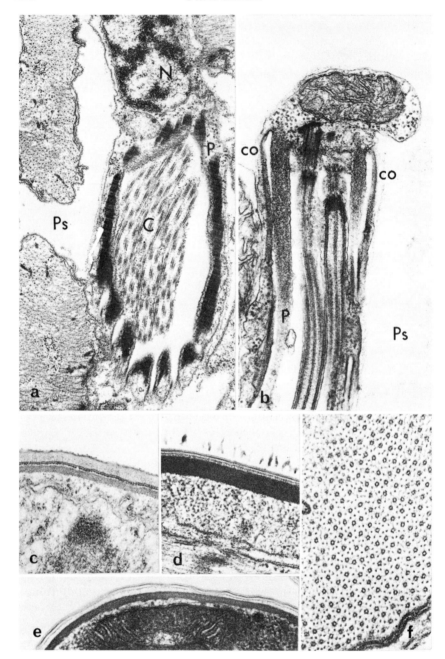

It is probable that the simultaneous production of mictic and amictic females is primitive in monogononts. In the more evolved species, production of mictic females appears only after an appropriate stimulus which signals the probability of a forthcoming deterioration of the environment. Such stimuli include a long photoperiod in *Notommata copeus*, vitamin E in *Asplanchna*, population density in some *Brachionus* (see Pourriot and Clément 1981; Clément, Luciani, and Pourriot 1981).

Seison is a primitive rotifer and lacks parthenogenesis. It is generally considered that parthenogenesis is a secondary acquisition in the evolution of a group. With the onset of parthenogenesis new evolutionary constraints arise in rotifers. The total absence of males in bdelloids probably explains their great homogeneity.

Some examples of parthenogenesis occur in Turbellaria, but the main mode of reproduction in most species is asexual, with sometimes bisexual reproduction.

A great diversity of eyes and ocelli

There are about ten different sensory organs in *Trichocerca rattus*, each one of which has been described by EM (Clément 1977a,b). The number of sensory organs is greater in *Asplanchna* (Wurdak, Clément, and Amsellem 1983). Currently we have more information on the considerable ultrastructural diversity of sensory structures in rotifers than on their sensitivity and function (Clément, Wurdak, and Amsellem 1983). Here I shall discuss only one such example, the photoreceptors.

1. The phyletic line of the cerebral eyes of monogonont rotifers

The four cerebral eyes as seen by EM (Fig. 15.5) are very different in their organization, as well as in their presumed sensory photoreceptor

---

Fig. 15.3. TEM sections of flame cell, integument, and muscle. (a) Oblique section of the flame cell in *Philodina roseola* (Bdelliodea). The pillars are cross-striated. The nucleus is located at the level of the base of the pillars; × 24 000. (b) Axial section of the flame cell in *Rhinoglena frontalis* (Monogononta). The insertion of the columns and pillars is shown. The columns are cytoplasmic cylindrical structures, in continuity with the cytoplasm of the cap and base of the flame cell. Pillars are cylindrical structures, inserted like the cilia, on the cap of the flame cell, × 30 000. (c) Integument in young *Brachionus calyciflorus*, × 40 000. (d) Integument in adult *Philodina roseola*, × 40 000. (e) Integument in young *Philodina roseola*, × 54 000. N.B. The internal layer of the intracytoplasmic skeleton is thin in monogononts (c) and young bdelloids (e), but thick in adult bdelloids (d). (f) Transverse section of the principal muscle in the mastax of *Trichocerca rattus.*. The myosin-like myofilaments are tubular, the actin-like filaments are classical, × 100 000. Note their ratio, and the bridges between these filaments.

Abbreviations: C, cilia; co, columns; N, nucleus; P, pillars; Ps, pseudocoel.

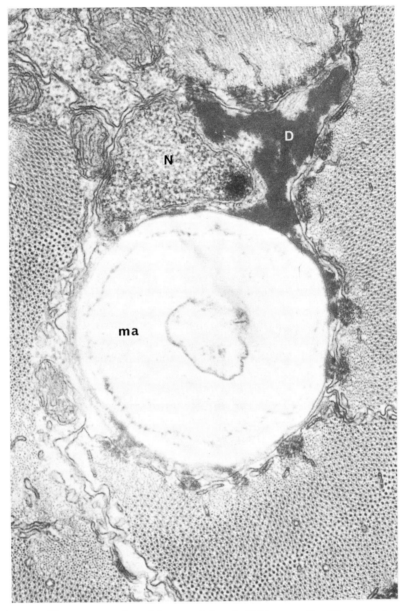

Fig. 15.4. Transverse section of the manubrium in *Trichocerca rattus*. The hardest structure of the manubrium is cylindrical; some cytoplasm is present in the middle. This
structure is located inside a cell, the nucleus of which is visible. This cell contains
electron-dense material on which are inserted some muscles of the mastax. This dense
material is in close contact with the hardest part of the manubrium; it is uncertain
whether this hardest material is extra- or intracellular × 47 000. Abbreviations: D,
electron-dense material; ma, manubrium; N, nucleus.

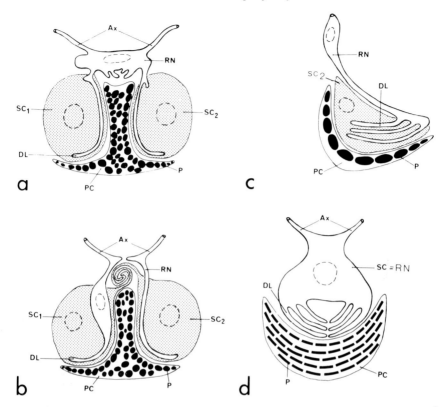

Fig. 15.5. Diagram of the cerebral eye in four species of monogonont rotifers. (a) *Brachionus calyciflorus*; (b) *B. plicatilis*; (c) *Trichocerca rattus*; (d) *Asplancha brightwelli*. An epithelial cell contains the pigment cup composed of pigment granules or platelets. The primitive sensory neurons are stippled. The relay neuron whose axons lead to the neuropile of the brain is shown in white. It bears dendritic lamellae, lodged in the sensory neuron in *Brachionus* and *Trichocerca*, or piled up in the extracellular space in *Asplanchna*; in this latter case, the relay neuron is also the sensory one. Abbreviations: Ax, axon; DL, dendritic lamellae; P, pigment granules; PC, pigment cup; RN, relay neuron; SC, sensory neuron.

organelles. Nevertheless, it is possible to imagine a phyletic line involving these four eyes (Fig. 15.5). The structure is first paired (*Brachionus*) and becomes unpaired by the loss of half of the eye (*Trichocerca*) or by the loss of the two sensory cells (*Asplanchna*), where only the median unpaired cell remains; the relay neurone becomes sensory here. The photosensitive structures seem to be formed primitively by the membranes of the endoplasmic reticulum (*Brachionus calyciflorus*, and partly in *B. plicatilis*) (Fig. 15.6); they are then in association with the lamellae of the relay neurone. These lamellae are lodged in the cytoplasm of the sensory neurone(s) in *Brachionus* and *Trichocerca*. They are numerous

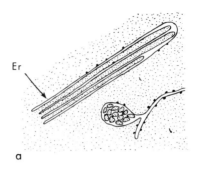

a

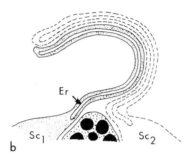

b

Fig. 15.6. Diagram of structures involving the endoplasmic reticulum, present in the sensory neurons of the cerebral eye of *Brachionus*. (a) *B. calyciflorus*. (b) *B. plicatilis*. Abbreviations: Er, endoplasmic reticulum; Sc, sensory neuron.

in *Trichocerca* and *Asplanchna*; they occur in an extracellular space in *Asplanchna*.

The eye of *Asplanchna* is thus highly evolved, and it is not correct to consider it as the prototype eye of other rotifers as Eakin (1963, 1965, 1982) has done.

2. The ampullae cilia with electron-dense material
(Figs 15.7-15.9)

Each of the paired cerebral eyes of *Philodina roseola* (bdelloid) contains two ampullae cilia, whose interiors are electron-dense. They are lodged just within the inside part of an epithelial, red pigmented cell, at the periphery of the brain. (Clément 1980; Fig. 15.8). Almost identical cilia are found in the anterior ocelli of *Trichocerca rattus* (Clément 1977a,b, 1980) and *Asplanchna brightwelli* (Clément *et al.* 1983, Clément and Wurdak 1984) (Fig. 15.9). In these 'ocelli' there is no accessory-screen pigment. The same cilia, however, have not been found in *Brachionus* or *Rhinoglena*. Nevertheless, I do not think that they are independent acquisitions in the three genera where they exist. This is an original phyletic line of photoreceptors, until now unrecognized in

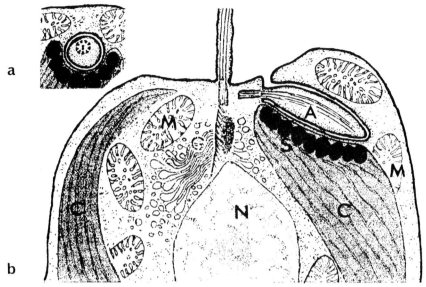

Fig. 15.7. Anterior portion of the phytoflagellate *Chromulina psammobia*. (a) T.S. of the ampulla-shaped cilium. It contains an electron-dense matrix around the axoneme. It is lodged in an invagination of the plasma membrane. (b) Axial section. The ampulla-shaped cilium is apposed to the accessory pigment granules which make up the stigma. Abbreviations: A, ampulla-shaped cilium; C, chloroplast; M, mitochondrion; N, nucleus; S, stigma. (Modified from Fauré-Frémiet and Rouiller 1957).

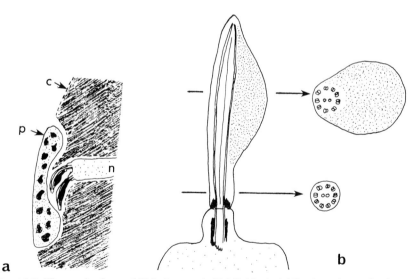

Fig. 15.8. The cerebral eye of *Philodina roseola* (Bdelloidea). (a) Section of a cerebral eye. (b) Detail of an ampulla-shaped cilium (axial and transverse sections). Abbreviations: C, brain; n, sensory neurite bearing two ampulla-shaped cilia; P, pigmented epithelial cell.

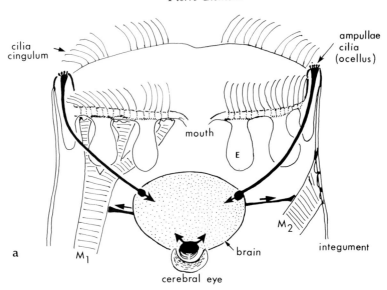

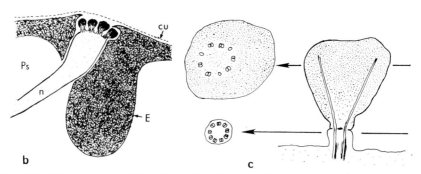

Fig. 15.9. The anterior ocelli of *Asplanchna brightwelli* (Monogononta). (a) Diagram of the anterior part of an animal. (b) Diagram of an anterior ocellus. (c) Detail of an ampulla-shaped cilium (axial and transverse section). Abbreviations: cu, thin anterior cuticle; E, epithelial-supporting cell; M_1, a longitudinal muscle; M_2, a circular muscle; n, sensory neurite bearing the ampulla-shaped cilia; Ps, pseudocoel.

Metazoa, but previously known in phytoflagellates (Fig. 15.7) (Fauré-Frémiet and Rouiller 1957).

3. A phaosome (Fig. 15.10)

A highly specialized sensory structure has been found in *Philodina roseola* (Clément 1980; Clément *et al.* 1983, unpublished; Fig. 15.10). Flattened cilia and ciliary expansions are piled up in the spherical cavity of the sensory nerve ending. By analogy with Annelida (see Verger-

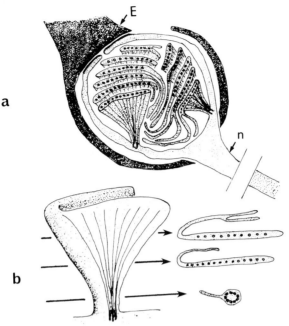

Fig. 15.10. Diagram of the phaosome of *Philodina roseola*. (a) The sensory nerve ending, linked to an epithelial cell by desmosomes, contains a closed spherical cavity in which flattened cilia are piled up. (b) Detail of a flattened cilia. Abbreviations: E, epithelial cell; n, sensory nerve ending.

Bocquet 1984) we call this structure a 'phaosome'. We have no direct proof that it functions as a photoreceptor, but the same structure is common in the eyes of Hirudinea and Oligochaeta; it has also been described in pogonophores (Nørrevang 1974) and platyhelminths (see Fournier, in press). In platyhelminths and rotifers the phaosome contains numerous cilia. These cilia are at first unmodified (Ehlers and Ehlers 1977); they are then flattened and bear lamellar membranes. In Annelida the cilia regress, and the membranes do not come from the cilia. In pogonophores, there are neither cilia nor ciliary roots. I consider the phaosome as a phyletic line of photoreceptors (Fig. 15.12).

4. The anterior ocelli of *Rhinoglena frontalis*

The ocelli were previously described as epithelial differentiations (Stossberg 1932; Salvini-Plawen and Mayr 1977). They are in fact rhabdomeric nervous structures borne on the end of a sensory neurite, and juxtaposed to a red pigmented epithelial cup (Fig. 15.11) (Clément 1980; Clément *et al.* 1983). The pericaryon of the sensory neurone is located at the periphery of the brain.

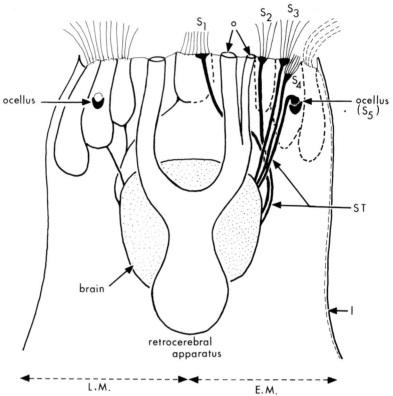

Fig. 15.11. Contrast in resolution of rotifer structures under light microscope *in vivo* (left) (after Stossberg 1932), and on the right with observations carried out with the aid of an E.M. Abbreviations: I, integument; O, aperture of the canals of the retrocerebral apparatus; S_1-S_4, tips of the sensory neurons; S_5, a sensory nerve ending, only the accessory pigment cup is epithelial; ST, sensory nerve fibres.

5. The pigmented cup of ocelli and eyes

In *Philodina roseola*, the pigment granules are not ordered in the pigmented epithelial cell. The same holds for the eyes of *Brachionus*, where nevertheless the pigmented granules often coalesce under the cytoplasmic membrane at the level of the two cups. In the eye of *Trichocerca* and in the ocelli of *Rhinoglena* there is only one layer of coalesced large pigment granules, as in the stigma of phytoflagellates. In the eye of *Asplanchna* the pigments are located in platelets, arranged in concentric layers which function as a multilayered mirror (reflected tapetum), as Land (1979) described in *Pecten*. The observed distance between the concentric layers supports this function in *Asplanchna* (Cornillac 1982).

This kind of mirror is common amongst Metazoa. Fournier and Combes (1978) described it in the eye of a monogenean. It is frequent in Annelida and Mollusca, and is also present in the nauplius eye of some copepods (Dudley 1969).

I believe that, as for the photoreceptor cells, the frequency of this specialized structure in different zoological groups argues in favour of the existence and transmission of a genotype which induces it.

6. Comparison with other groups (Fig. 15.12)

The phylogenetic significance of the photoreceptor structures in Metazoa is a subject of considerable disagreement. Eakin (1963, 1965, 1982) defends a diphyletic theory, Vanfleteren (1982) and Vanfleteren and Coomans (1975) a monophyletic theory, while Salvini-Plawen and Mayr (1977) and Salvini-Plawen (1982) support an 'aphyletic' theory.

I propose to reconcile these alternatives with a 'polyphyletic' theory, which is partly illustrated in Fig. 15.12; there are different phyletic lines of photoreceptors. Three of them are present in rotifers; one is primitive in Metazoa (ampullae cilia) and derives from phytoflagellates. The other two are also present in other Metazoa, with small variations (phaosomes), or with more substantial differences amongst cerebral eyes which diversify into a great variety of original forms in rotifers.

Nevertheless, Fig. 15.12 is not complete as the phyletic lines of accessory pigment (stigma, tapetum etc.) are not indicated. It would be necessary also to include biochemical and physiological data about the different visual pigments, and the depolarization or hyperpolarization of the visual cell after illumination. The data in this field are presently too restricted to be used for a phylogeny. The multiplication of the visual cells and their organization in a visual organ also provide very important information, which explain the visual performance of each type of eye or ocellus (Clément, in press).

In Fig. 15.12 the rotifers are depicted as an isolated and very primitive group. Their closest relationships seem to be with platyhelminths, which in several respects are more evolved than rotifers.

Conclusions

Comparing the embryogenesis of different pseudocoelomates Joffe (1979) distinguished three groups:
   (i)   Priapulida and Nematomorpha;
   (ii) Nematoda and Gastrotricha;
   (iii) Rotifera and Acanthocephala, not far from Turbellaria.
I agree with Joffe because, (*a*) Rotifera and Acanthocephala have the same skeletal integument, which can be derived from Turbellaria; (*b*) Rotifera have several points in common with Platyhelminthes: ciliated integument with mucous gland, protonephridia with flame-cells,

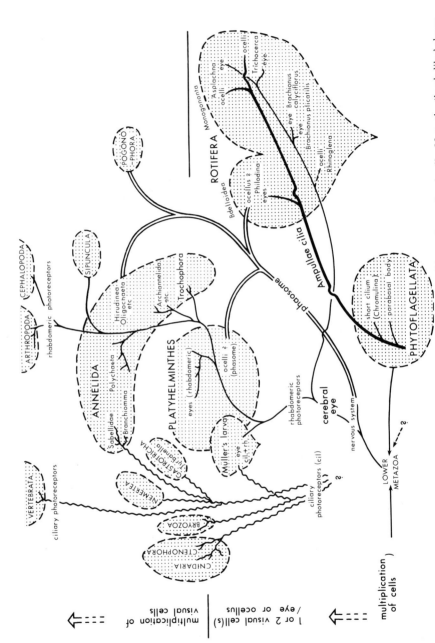

Fig. 15.12. The evolutionary relationships of lower Metazoa as based on evidence from visual cells. Note the 'bush-like' shape, in contrast to earlier 'tree-like' reconstructions.

phaosomes, cerebral eye with pigmented cup, etc; (*c*) other groups of pseudocoelomates differ more from Rotifera than the latter differ from Platyhelminthes.

Nevertheless, the Rotifera differ from Platyhelminthes in several aspects:

(i)   possession of a pseudocoel, without mobile cells, fibrous or filamentous collagen or a nervous network;

(ii) eutely and the absence of regeneration and asexual reproduction;

(iii) a glia-free nervous system, and the absence of learning and imprinting in behaviour;

(iv) muscles without paramyosin;

(v) the classical presence of a cloaca;

(vi) an intracellular peripheral skeleton, and partly intracellular trophi.

So it seems that there was a very early separation between Platyhelminthes and Rotifera. After this separation, Rotifera became an evolutionary cul-de-sac, with the success that we know. On the other hand, the evolutionary potential of Platyhelminthes remains important and this may explain the superior behavioural patterns of some Platyhelminthes.

Finally, this work on rotifers (Fig. 15.12) provides some new rules for the game of phylogeny. The classical image of the tree of the evolution is not, in my opinion, helpful. The picture of a bush with more complex interactions between the branches, seems to be nearer to the reality than I imagine for evolution. To understand better the nature of the 'branches' of this irregular 'bush' of evolution we need more information:

(*a*) on the different origins of the modifications of the genome; is sexual reproduction the only rule?

(*b*) On the genetics of each embryogenesis, and the role of autoconstruction, and of the direct influence of the environment, sometimes transmissible during several generations (extra-chromosomal inheritance).

Rotifers are simple, diverse and isogenic in each clone. They are a good model to use for constructing a new image of the old tree of evolution.

References

Amsellem, J. and Clément, P. (1977). Correlations between ultrastructural features and contraction rates in rotiferan muscles. I. Preliminary observations on longitudinal retractor muscles in *Trichocerca rattus*. *Cell Tissue Res.* 181, 81-90.

—— and Ricci, C. (1982). Fine structure of the female genital apparatus of *Philodina* (Rotifera, Bdelloidea). *Zoomorphologie* 100, 89-105.

Atwood, H. D. (1972). Crustacean muscles. In *The structure and function of muscle* (2nd edn.) (ed. C. H. Bourne) Vol. 1, pp. 422-90. Academic Press, New York.

Beauchamp, P. de (1907). Morphologie et variatións de l'appareil rotateur des Rotifères. *Archs Zool. exp. gén.* 6, 1-29.

—— (1909). Recherches sur les Rotifères: les formations tégumentaires et l'appareil digestif. *Archs Zool. exp. gén.* 10, 1-410.

—— (1965). Classe des Rotifères. In *Traité de Zoologie.* Vol. 4. *Anatomie, Systématique, Biologie* (ed. P.-P. Grassé) pp. 1225-379. Masson et Cie, Paris.

Bentfeld, M. E. (1971a). Studies of oogenesis in the rotifer *Asplanchna*. I. Fine structure of the female reproductive system. *Z. Zellforsch. mikrosk. Anat.* 115, 165-83.

—— (1971b). Studies of oogenesis in the rotifer *Asplanchna*. II. Oocyte growth and development. *Z. Zellforsch. mikrosk. Anat.* 115, 184-95.

Brandenburg, J. (1962). Elektronenmikroskopische Untersuchung des Terminalapparatus von *Chaetonotus* sp. (Gastrotrichen) als ersten Beispiels einer Cyrtocyte bei Askelminthen. *Z. Zellforsch. mikrosk. Anat.* 57, 136-44.

Braun, G., Kümmel, G. and Mangos, J. A. (1966). Studies on the ultrastructure and function of a primitive excretory organ, the protonephridium on the rotifer *Asplanchna priodonta*. *Pflügers Arch. ges. physiol.* 289, 141-54.

Clément, P. (1967). Ultrastructure du système osmorégulateur d'un Rotifère, *Notommata copeus*. Conclusions physiologiques et phylogénétiques. Thèse Doctorat Spécialité, Université Lyon.

—— (1969a). Ultrastructures d'un Rotifère *Notommata copeus*. II—Le tube protonéphridien. *Z. Zellforsch. mikrosk. Anat.* 94, 103-17.

—— (1969b). Premières observations sur l'ultrastructure comparée des téguments de Rotifères. *Vie et Milieu* A20, 461-82.

—— (1977a). Introduction à la photobiologie des Rotifères dont le cycle reproducteur est contrôlé par la photopériode. Approches ultrastructurale et expérimentale. Thèse Doctorat Etat, Université Lyon.

—— (1977b). Ultrastructural research on rotifers. *Arch. Hydrobiologia* 8, 270-97.

—— (1980). Phylogenetic relationships of rotifers, as derived from photoreceptor morphology and other ultrastructural analyses. *Hydrobiologia* 73, 93-117.

—— (in press). Organisation biologique et evolution. In *Temps et devenir; en hommage à I. Prigogine.* Colloque de Cerisy, June 1983.

—— and Fournier, A. (1981). Un appareil excréteur primitif: les protonéphridies (Platyhelminthes et Némathelminthes) *Bull. Soc. zool. Fr.* 106, 55-67.

—— and Wurdak, E. (1984). Photoreceptors and photoreceptions in rotifers. In *Photoreception and vision in invertebrates* pp. 241-88. Plenum Press, New York.

—— Luciani, A. and Pourriot, R. (1981). Influences exogènes sur le cycle reproducteur des Rotifères. *Bull. Soc. zool. Fr.* 106, 255-62.

—— Wurdak, E., and Amsellem, J. (1983). Behavior and ultrastructure of sensory organs in rotifers. *Hydrobiologia* 104, 89-130.

—— Amsellem, J., Cornillac, A., Luciani, A. and Ricci, C. (1980a). An ultrastructural approach to feeding behaviour in *Philodina roseola* and *Brachionus calyciflorus* (Rotifers). I. The buccal velum. *Hydrobiologia* 73, 127-31.

—— —— —— —— —— (1980b). An ultrastructural approach to feeding behaviour in *Philodina roseola* and *Brachionus calyciflorus* (Rotifers). II. The oesophagus. *Hydrobiologia* 73, 133-6.

—— —— —— —— —— (1980c). An ultrastructural approach to feeding behaviour in *Philodina roseola* and *Brachionus calyciflorus* (Rotifers). III. Cilia and muscles. Conclusions. *Hydrobiologia* 73, 137-41.

Cornillac, A. M. (1982). Yeux cérébraux et réponses motrices à la lumière chez *Brachionus calyciflorus* et *Asplanchna brightwelli* (Rotifères). Thèse Doctorat Spécialité, Université Lyon.

Dougherty, E. (ed.) (1963). *The lower Metazoa. Comparative biology and phylogeny.* University of California Press, Berkeley, Ca.

Dudley, P. L. (1969). The fine structure and development of the nauplius eye of the copepod *Duropygus seclusus* Illg. *La Cellule* 68, 6-42.

Eakin, R. M. (1963). Lines of evolution of photoreceptors. In *General physiology of cell specialization* (eds D. Mazia and A. Tyler) pp. 393-425. McGraw-Hill, New York.

—— (1965). Evolution of photoreceptors. *Cold Spring Harb Symp. quant. Biol.* 30, 363-70.

—— (1982). Continuity and diversity in photoreceptors. In *Visual cells in evolution* (ed. J. A. Westfall) pp. 91-105. Raven Press, New York.

—— and Brandenburger, J. (1974). Ultrastructural features of a Gordian worm (Nematomorpha). *J. Ultrastruct. Res.* 46, 331-74.

Ehlers, B. and Ehlers, U. (1977). Ultrastruktur pericerebraler cilienaggregate bei *Dicoelandropora atriopapillata* Ax und *Notocaryoplanella glandulosa* Ax (Turbellaria, Proseriata). *Zoomorphologie* 88, 163-74.

Fauré-Frémiet, E. and Rouiller, C. (1957). Le flagelle interne d'une chrysomonadale: *Chromulina psammobia. C. r. hebd. Séanc. Acad. Sci., Paris* 244, 2655-7.

Fournier, A. (1984). Photoreceptors and photosensitivity in Platyhelminthes. In *Photoreception and vision in invertebrates.* pp. 217-40. Plenum Press, New York.

—— and Combes, F. (1978). Structure of photoreceptors of *Polystoma intergerrimum* (Platyhelminths, Monogena). *Zoomorphologie* 91, 147-55.

Franc, J. M. (1972). Activités des rosettes cilieés et leurs supports ultrastructuraux chez les Cténaires. *Z. Zellforsch. mikrosk Anat.* 130, 527-44.

Franc, S., Franc, J. M. and Garrone, R. (1976). Fine structure and cellular origin of collagenous matrices in primitive animals: Porifera, Cnidaria and Ctenophora. *Front. Matrix. Biol.* 3, 143-56.

Hennig, W. (1966). *Phylogenetic systematics.* University of Illinois Press, Urbana, Il.

Horridge, C. A. (1968). *Interneurons. Their origins, action, specificity, growth and plasticity.* W. H. Freeman, London.

Huxley, T. H. (1853). *Lacinularia socialis.* A contribution to the anatomy and physiology of the rotifers. *Trans. microsc. Soc. Lond.* 1, 1-9.

Hyman, L. H. (1951). *The invertebrates,* Vol. 3. McGraw-Hill, New York.

Joffe, B. I. (1979). The comparative embryological analysis of the development of Nemathelminthes. *Dokl. Akad. Nauk SSSR* 84, 39-62 (in Russian).

Koehler, J. K. and Hayes, T. L. (1969a). The rotifer jaw: a scanning and transmission electron microscope study. The trophi of *Philodina acuticornis adiosa. J. Ultrastruct. Res.* 27, 402-18.

—— —— (1969b). The rotifer jaw: a scanning and transmission electron microscope study. The trophi of *Asplanchna sieboldi. J. Ultrastruct. Res.* 27, 419-34.

Koste, W. (1978). *Rotatoria.* Borntraeger, Berlin.

Kryvi, H. (1973). Ultrastructural studies of the sucker cells in *Hemiurus communis* (Trematoda). *Norw. J. Zool.* 21, 273-80.

Kümmel, G. and Brandenburg, J. (1961). Die Rensengeisselzellen (Cyrtocyten). *Z. Naturf.* B16, 692-7.

Land, M. F. (1979). Des animaux dotés d'yeux à miroir. *Pour la Science* 16, 28-37.

Lanzavecchia, G., Valvassori, R., Eguileor, M. de, and Lanzavecchia, P. (1979). Three dimensional reconstruction of the contractile system of the Nematophore muscle fiber. *J. Ultrastruct. Res.* 66, 201-23.

Mattern, C. F. T. and Daniel, W.A. (1966a). The flame-cell of Rotifers. Electron microscope observations of supporting rootlets structures. *J. Cell Biol.* 29, 547-51.

—— —— (1966b). The stomach cell of Rotifer. Electron microscope observations of the terminal web. *J. Cell Biol.* 29, 547-51.

Nachtwey, R. (1925). Untersuchungen über die Keimbahn organogenese und Anatomie von *Asplanchna priodonta* gosse. *Z. wiss. Zool.* 126, 239-492.

Nørrevang, A. (1974). Photoreceptors of the phaosome (hirudinean) type in pogonophore. *Zool. Anz.* 193, 297-304.

Pourriot, R. and Clément, P. (1981). Action de facteurs externes sur la reproduction et le cycle reproducteur des rotifères. *Acta Oecologica, Oecologia Gen.* 2, 135-51.

Pringle, J. W.S. (1972). Arthropod muscles. In *The structure and function of muscle* (2nd edn) (ed. G. H. Bourne) Vol. 1, pp. 491-542. Academic Press, New York.

Reger, J. F. (1976). Studies on the fine structure of cercarial tail muscle of *Schistosoma* sp. (Trematoda). *J. Ultrastruct. 'Res.* 57, 77-86.

Remane, A. (1929-1933). Rotatoria. In *Dr H. G. Bronn's Klassen und Ordnungen des Tierreichs.* pp. 1-448. Akademische Verlagsgesellschaft.

—— (1954). Morphologie als Homologienforschung. *Verhandl. Dt. zool. Ges.: Zool. Anz.* Suppl. 18, 159-83.

—— (1963). The systematic position and phylogeny of the pseudocoelomates. In *The lower Metazoa* (ed. E. C. Dougherty) pp. 247-55. University of California Press, Berkeley, Ca.

—— Storch, V. and Welsch, U. (1972). *Kurzes Lehrbuch der Zoologie.* G. Fischer, Stuttgart.

—— —— —— (1976). *Systematische Zoologie.* G. Fisher, Stuttgart.

Ruttner-Kolisko, A. (1963). The interrelationships of the Rotatoria. In *The lower Metazoa* (ed. E. C. Dougherty) pp. 263-72. University of California Press, Berkeley, Ca.

—— (1974). Plankton Rotifers. Biology and taxonomy. *Binnengewässer* 26 (1, suppl.), 99-234.

Salvini-Plawen, L. von (1982). On the polyphyletic origin of photoreceptors. In *Visual cells in evolution* (ed. J. A. Westfall) pp. 137-54. Raven Press, New York.

—— and Mayr, E. (1977). On the evolution of photoreceptors and eyes. *Evolut. Biol.* 10, 207-63!.

Schramm, U. (1978). On the excretory system of the Rotifer *Habrotrocha rosa* Donner. *Cell Tissue Res.* 189, 515-24.

Stossberg, K. (1932). Zur Morphologie der Rädertiergattungen *Euchlanis, Brachinonus* und *Rhinoglena. Z. wiss. Zool.* 142, 313-424.

Teuchert, G. (1973). Die Feinstrucktur des Protonephridialsystems von *Turbanella cornuta* Remane, einem marinen Gastrotrisch der Ordnung Macrodasyoidea. *Z. Zellforsch. mikrosk. Anat.* 136, 277-89.

Vanfleteren, J. R. (1982). A monophyletic line of evolution? Ciliary induced photoreceptor membranes. In *Visual cells in evolution* (ed. J. A. Westfall) pp. 107-36. Raven Press, New York.

—— and Coomans, A. (1975). Photoreceptor evolution and phylogeny. *Z. zool. Syst. Evolutionsforsch.* 14, 157-69.

Verger-Bocquet, M. (1984). Photoréception et vision chez les Annélides. In *Photoreception and vision in invertebrates.* Plenum Press, New York.

Vergin, W. P. and Endo, B. Y. (1976). Ultrastructure of a neurosecretory organ in a root-knot nematode. *J. Ultrastruct. Res.* 56, 258-76.

Ward, S., Thomson, N., White, J. G. and Brenner, S. (1975). Electron microscopical reconstruction of the anterior sensory anatomy of the nematode *Caenorhabditis elegans. J. comp. Neurol.* 160, 313-37.

Ware, R. W., Clark, D., Crossland, K. and Russell, R. L. (1975). The nerve ring of the nematode *Caenorhabditis elegans:* sensory input and motor output. *J. comp. Neurol.* 162, 71-110.

Wurdak, E., Clément, P. and Amsellem, J. (1983). Sensory receptors involved in the feeding behaviour of the rotifer *Asplanchna brightwelli. Hydrobiologia* 104, 203-12.

16. Why is a gastrotrich?

P. J. S. BOADEN

The Queen's University of Belfast,
Marine Biology Station,
Portaferry, Co. Down, Northern Ireland

Abstract

Recent work on mainly marine gastrotrichs is reviewed and an attempt is made to find a reason for their existence. Recent research has centred on comparative morphology and ultrastructure. The gastrotrichs are believed to be archaecoelomate. Most species possess duo-glandular adhesive organs and many are monociliated. The integument has an outer lamellar layer and a basal layer which is penetrated by some microvilli. Cuticular sculpture evolved within the group and many of these integumentary features appear related to the instability of marine sand as a habitat. Marine gastrotrichs are characteristically gregarious and tend to occur in small patches. Some new information concerning this is presented for species from a Northern Ireland beach. Very little information concerning physiology and biochemistry is available. Anaerobiosis is probable in some species but the group is predominantly aerobic. Oxygen consumption is reduced via a gregarious response as population density increases. A Precambrian origin of Gastrotricha from amongst the earliest Protostomia is likely, the closest existing relatives being the Gnathostomulida (plesiomorphic to Gastrotricha) and the Nematoda (apomorphic). The macrodasyid gastrotrichs are *lebens-ort* types for marine sand; the success of the order Chaetonotida is linked to later exploitation of freshwater.

Introduction

The title of this paper is derived from the old English riddle 'Why is a duck?'. The traditional answer is, 'Because one of its legs is both the same'. Hopefully the answer in this paper is less surrealistic since the question is intended to introduce discussion of why the Gastrotricha came into existence and why they have persisted.

Definition

It may be best to begin by answering the apparently simpler question '*What* is a gastrotrich?'. At present the Gastrotricha may be regarded as a phylum or subphylum within the Aschelminthes. Since there does not seem to be as much evolutionary distance between Gastrotricha and Nematoda as, for example, between fish and mammals, the lesser ranking is preferred by the author.

The Gastrotricha may be defined as non-segmented spiralian Metazoa, 75-500 μm in length, with a through-gut consisting of a pharynx (triangular in cross section) and hindgut, with at least some external ciliation, with adhesive tubules, and a tegumentary lamellar layer extending around the cilia. The tegument may be sculptured. A circulatory system is absent. The nervous system has a pair of frontal ganglia joined by a pharyngeal commissure and two ventro-lateral nerve tracts. Protonephridial excretory organs are probably always present. Marine species are generally hermaphrodite but partheno-genesis occurs in many freshwater forms. The marine forms are pre-dominantly benthic but freshwater species are benthic, phytal or pelagic. At present about 450 species are known, about half from fresh-water (Balsamo 1983), but there are many as yet undescribed species.

Systematics

It is generally accepted that the Gastrotricha are comprised of two orders, the marine Macrodasyida and the mainly freshwater Chaeto-notida.

The Macrodasyida usually possess numerous adhesive tubules (Fig. 16.1), one branch of the pharynx lumen is mid-dorsal and there is usually a pair of pharyngeal pores which lead laterally from the pharynx to the exterior. Six families are recognized (d'Hondt 1971; Hummon 1974a).

The Chaetonotida lack pharyngeal pores and the pharynx lumen has a mid-ventral branch. There are two sub-orders. The Multitubulatina have small lateral adhesive tubules together with several tubules set on two posterior pedicles; there is one family, the Neodasyidae. The Paucitubulatina have only a single or rarely a double, pair of posterior tubules; six families are recognized (d'Hondt 1971).

Recent studies

Although many new species have been described, the systematic outline given in the classical work by Remane (1936) has been little altered.
However, there have been various advances in knowledge of the group
through ultrastructural, ecological, behavioural and other studies.

1. Ultrastructure

Brandenburg (1962) published the first ultrastructural information
(concerning protonephridia) about gastrotrichs. Boaden (1968) published pictures of the tegument and adhesive tubules. Teuchert (1973)
also studied protonephridia but later (e.g. Teuchert 1977; Teuchert
and Lappe 1980) studied other aspects of fine structure. The latter two
papers demonstrate that there is a coelomic body cavity, although the
function of a hydrostatic skeleton is probably served by the paired
Y-organ (a series of lateral vacuolar cells) in most species.

The paper on gastrotrich fine structure by Rieger, Ruppert, Rieger,
and Schoepfer-Sterrer (1974) marked the advent of an important
series, through which run various themes. The common occurrence of
monociliated epidermal cells with a structure similar to those in the
Gnathostomulida is a plesiomorphic feature and indicates a very early
evolutionary origin, as cogently argued by Rieger (1976). Tyler and
Rieger (1980) have studied the structure of adhesive organs in gastrotrichs. They are mostly of duo-glandular construction with one cell
producing adhesive and the other a releasant; in some species a cilium
(presumed to be sensory) is associated with the duo-glandular tubule
(Figs 16.3, 16.4). It is thought that the simpler structure of tubules in
the multitubulatid *Neodasys* may be a more primitive condition and thus
that the duo-glandular type in macrodasyids and paucitubulatids may
not be strictly homologous (Tyler, Melanson, and Rieger 1980). However *Neodasys* is regarded as primitive in various other characteristics
(Remane 1936), including a monociliated epidermis (Rieger 1976) and
thought should be given to separating it from the otherwise apomorphic
Chaetonotida.

The basic structure of gastrotrich cuticle, first observed by Erwin (in
Boaden 1968) in *Turbanella* as consisting of an outer lamellar layer and
an inner basal layer (Fig. 16.5) has been confirmed in many species
(Rieger and Rieger 1977). The lamellar layer extends around the cilia
and, in forms with a smooth cuticle, may consist of from three to
twenty-five bilayered lamellae. In some species the bilayering is not
apparent. The basal layer often has an outer striated zone.

In species bearing spines, scales or hooks the lamellar layer is often
reduced, sometimes to only one bilayer. The various 'sculpturings'

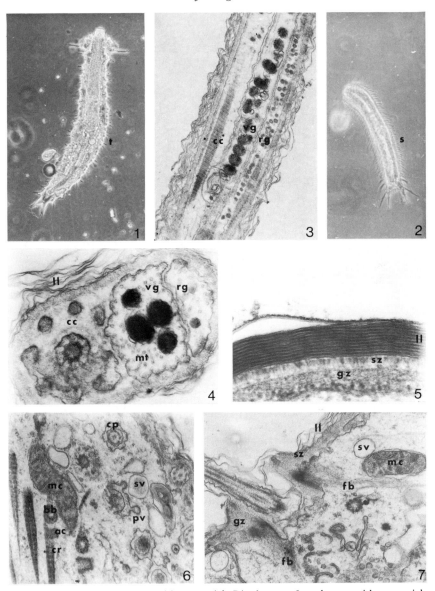

Figs. 16.1-16.7. 1, A macrodasyid gastrotrich *Dinodasys* sp. 2, a chaetonotid gastrotrich *Chaetonotus* sp.; both low-power phase contrast. 3-7 transmission electron micrographs: 3, longitudinal and 4, transverse sections of an adhesive tubule of *Turbanella hyalina* Schultze; 5, tegument of *Thiodasys sterreri* Boaden; 6, oblique-horizontal and 7, vertical sections of an epidermal cell of *T. hyalina*. Abbreviations: ac, accessory centriole; bb, basal body; cc, ciliary cell; cp, ciliary pit; cr, ciliary rootlet; fb, fibrils; gz, granular zone; ll, lamellar layer; mc, mitochondrion; mt, microtubules; pv, pinocytotic vesicle; rg, releasing gland; s, spines; sv, surface vesicle; sz, striated zone; t, tubules; vg, viscid gland.

mentioned are extensions of the basal layer. Such sculpturing seems to have evolved on at least three different occasions within the gastrotrichs' evolutionary history. The scales and plates of the genera *Chordodasys* and possibly *Xenodasys* surround epidermal evaginations (Rieger *et al.* 1974) and are regarded as a fourth line of cuticular differentiation (Rieger and Rieger 1977).

In the pharynx of all species so far studied the lamellar layer is well developed but the bilayers may be differently packed than elsewhere. The striated zone is often missing but the inner basal layer may be underlain by electron-dense spheres. The ultrastructure of the pharyngeal cuticle in the Paucitubulatina is strong evidence (Rieger and Rieger 1977) in support of the close relationship between the Gastrotricha and Nematoda (Remane 1936).

Rieger and Rieger (1977) also argue that the lamellar layer of the cuticle could function in regulating ion exchange rather than serving as an adaptation against abrasion, as suggested by Boaden (1968). Although it is possible that dilution could occur in interlamellar spaces and thus help even out environmental ionic fluctuations, I still think it likely that evolution of the gastrotrich cuticle was a response to the sand habitat and, furthermore, that this trend was continued by production of sculpturing which in marine gastrotrichs is often found in forms from sediments prone to disturbance (Boaden 1968). It is possible that the lamellar layer might have an absorptive nutritional role but it is hard to see how this would function. However, judging from the number of vesicles, mitochondria, and microtubules, the epidermal layer is very active (Figs 16.6, 16.7). The vesicles at the base of the cilia (Fig. 16.6) show the typical clathrin structure associated with protein uptake (DeRobertis and DeRobertis 1980). Formation of the layers is presumably fairly costly in energetic terms, hence their function must be of some importance.

The epidermis is known to be cellular in the Macrodasyida and in *Neodasys* and at least partially so in one species of the Paucitubulatina (Rieger 1976). Thus the generally assumed syncytial nature of the gastrotrich epidermis (Remane 1936) has not stood the test of electron microscopy. The epidermal cells may have various interdigitations or protrusions; in particular microvilli are common on the ventral ciliated cells. The microvilli extend through the basal cuticular layer but rarely further. Microvilli are often associated with the base of the cilia; pinocytotic vesicles also occur in this region (Rieger 1976; Rieger and Rieger 1977).

2. Reproduction

The question 'Why is a gastrotrich?', apart from implying that a gastrotrich is an identifiable entity, also implies its present existence.

Gastrotrichs in order to exist must have survived and reproduced since their evolutionary origin. They have not been inconvenienced by our lack of understanding of their reproductive organs. This problem was highlighted by Schoepfer-Sterrer (1974) who demonstrated that the organ previously called the penis in *Macrodasys* was a bursa and, similarly, that the so-called bursa was probably part of the male system. Hummon (1974b) reviews previous knowledge of reproduction and development in Gastrotricha.

Ruppert and Shaw (1977) showed that at least in the Macrodasyida the 'male' copulatory organ (better termed the caudal organ) was not connected to the testes. Later Ruppert (1978a) demonstrated conclusively that in *Macrodasys* sperm are transferred externally to the caudal organ from the copulant's own male pore and are then transferred from this organ to the co-copulant's frontal organ. The frontal organ consists of a bursa (i.e. seminal receptacle) and a spermatheca. In the Thaumastodermatidae Ruppert (1978b) showed that a direct testis to caudal organ connection has been acquired and, in some cases, that the caudal and frontal organs are contiguous. He pointed out some structural similarities between the macrodasyid *Lepidodasys* and the chaetonotid *Neodasys*. A regrouping of the higher taxa in Gastrotricha seems imminent. The suggestion that *Neodasys* should not be so closely allied to the 'true' chaetonotids is reiterated.

The latter group (i.e. Paucitubulatina) are clearly apomorphic to other Gastrotricha, as evidenced by their much lower diversity, their occurrence in freshwater and various other features (Remane 1936). Many freshwater chaetonotids are known to reproduce parthenogenetically but recently sperm have been found in laboratory and field populations although it has yet to be demonstrated that they are functional (Weiss and Levy 1979; Kisielewska and Kisielewska 1984; Hummon in press).

3. Gregariousness and patch size

Ruppert (1978a) records gregariousness of *Macrodasys* in culture dishes as part of mating behaviour and mentions similar clumping in *Dolichodasys* and *Turbanella*. Nixon (1976) reported on the patchy field distribution of *Tetranchyroderma bunti* (Thane-Fenchel)—a factor probably linked to gregariousness. Boaden and Erwin (1971) briefly reported a positive gregarious response in *Turbanella hyalina* Schultze. Hummon (1972, 1975b) discussed aggregation in various marine species. The probability of successful reproduction is obviously enhanced by behaviour which helps ensure encounters in the three-dimensional maze of sand interstices.

Recent work on a beach in Northern Ireland (for site details see Boaden and Elhag 1984), with a grid of 64 samples within 40 × 40 cm quadrats, has indicated field patch size patterning of various scales. These may be calculated in several ways, e.g. plotting mean square (variance) against sample block size (Kershaw 1973). In *Turbanella varians* Maguire patches of less than 25 cm², of 100-200 cm², and of about 800 cm² are apparent. In *Neodasys chaetonotoides* Remane there are patches of <50 cm² and 200-400 cm². Such patches could be due to a number of abiotic and biotic factors. Gray and Johnson (1970) showed that patchiness in *Turbanella cornuta* Remane could be linked to the quality and quantity of bacterial sand-flora.

However, patches are also produced under laboratory conditions in which the response is clearly due to gregarious behaviour. For example, the position of individual animals within populations in dishes of sterile seawater can be recorded with reference to a grid pattern. The tendency to aggregate can then be determined visually and confirmed statistically. Experiments undertaken during the same period as the previous field work, using dishes of 8 × 8 cm containing 50 ml of seawater and with populations of between 20-30 *T. varians*, indicated patch sizes of 2-7 cm². With 100-120 individuals, patch sizes were 14-21 cm², although in these cases occasionally two smaller patches of 5-8 cm² were formed rather than one larger patch. Similarly, 20-80 *N. chaetonotoides* form patches of 10-16 cm². Patches form rapidly, usually within 20-40 min, but the population may disperse and reaggregate once or twice within 140 min. The patches are maintained by an increase in turning rate rather than reduced speed of active individuals within the patch, although a few individuals may remain stationary.

The gregarious response is linked with respiration rate. Thus the respiration rate of individual *N. chaetonotoides* is lower when determined from batches of 40 or more animals than when determined from 30 or less in the respiration chamber (Boaden and Elhag 1984). When individual respiration rates are determined from batches of 15 *Neodasys* in 6 ml of oxygenated sterile seawater at 12°C the rate is considerably higher than that similarly determined using seawater previously inhabited by *Neodasys* (6 ml taken from 10 ml which had contained 30 individuals overnight was used after re-oxygenation). The values obtained per individual were 18.6 and 11.7 nl min⁻¹

4. Interspecific relationships

Little information is available regarding trophic relationships of Gastrotricha. Although they possess epidermal microvilli it is not known whether they show active epidermal absorption. Gastrotrichs are generally regarded as feeding on bacteria, protozoans and diatoms; it is not known whether they are subject to selective predation though they form part of the diet of some meiofaunal predators and of macrofauna (d'Hondt 1971; Hummon 1975c). The latter paper also examined evidence for competition between gastrotrich species. Hogue (1978) demonstrated sub-tidal gastrotrich patch size of 4-10 cm across and showed that within-sediment vertical distribution helped partition available habitat between potentially competitive species. Boaden and Erwin (1971) demonstrated chemical antagonism between a gastrotrich and archiannelid species. Gray and Johnson (1970) illustrated a relationship with bacteria.

5. Physiology and biochemistry

Even less information is available on the physiology and biochemistry of gastrotrichs. Some data regarding tolerances, for example of various salinities (Remane 1936), are derived from field observations and these are supplemented by a little laboratory work which includes salinity and temperature effects (Boaden and Erwin 1971; d'Hondt 1971; Hummon 1972, 1975a,b). Hummon (1975a) studied respiration and osmoregulation in a *Turbanella* species. Maguire and Boaden (1975) demonstrated that *Thiodasys sterreri* Boaden has a capacity for anaerobic respiration, fixing CO_2 for use in a reversed Krebs cycle sequence. Such work and field observations suggest that several gastrotrich species are true anaerobes (Fenchel and Riedl 1970; Boaden 1974; Powell, Bright, Woods, and Gittings, in press). Powell, Crenshaw, and Rieger (1980) have demonstrated a sulphide detoxification system in one such gastrotrich.

Discussion

Unfortunately, our present understanding of relationships within the Gastrotricha leaves much to be desired. It is clear that some major taxonomic revision is required although Hummon (1974a) has made some progress towards this. Since knowledge of the systematics of the group is incomplete it is difficult to construct evolutionary relationships and discern trends. This hampers the search for the least-derived forms and gastrotrich origins. However some pointers are available. The Paucitubulatina are clearly apomorphic and we may assume this is also true of their main habitat. Ruppert (1977) has suggested that the evolution of this sub-order may be linked to the adoption of flotation as a dispersal mechanism and this linked to size, the occurrence of spines (Fig. 16.2), and parthenogenesis enabled colonization of freshwater. The search for plesiomorphic forms must therefore concentrate on the Macrodasyida and Neodasyidae.

It is reasonably certain that paired gonads, smooth cuticle, and monociliated epidermis are plesiomorphic. This indicates that the Thaumastodermatidae and Lepidasyidae, which mostly lack these features, are apomorphic. The Turbanellidae may be regarded as apomorphic in the possession of specialized groups of adhesive tubules and the Dactylopodolidae also appear specialized in possession of a rather definite head and their bifid hind end; they are also unusual in lacking anterior tubules. We are thus left with the Neodasyidae, Macrodasyidae, and Planodasyidae as the possibly most plesiomorphic families. The distinction between the latter two taxa is unclear, especially in view of the genera (or genus) *Thiodasys* and *Megadasys* (Boaden 1974), forms which, like the other particularly elongate genus *Dolichodasys*, display several apparently primitive features.

It is tempting to propose that *Neodasys*, with its mixture of macrodasyid and paucitubulatid characters (Remane 1936), its monociliated smooth epidermis, and its possibly primitive monoglandular tubules, is closest to the ancestral condition. If *Neodasys* has retained a constellation of plesiomorphic characters, then the pharyngeal arrangement of a Y-shaped lumen and a lack of pores may be primitive although this is different from most macrodasyids.

The evolution of spermatozoan structure within the Gastrotricha needs further consideration. I have previously argued (Boaden 1977) that internal fertilization, which is plesiomorphic in gastrotrichs, rather than external fertilization by Franzén's (1956) 'primitive sperm type', may have been the original mode of reproduction in the Metazoa. If this is so then a trend from the apparently less complex sperm of *Dolichodasys* and *Neodasys* towards the spirally wound sperm of many macrodasyids could be demonstrated. Support for this idea may be found in the non-flagellate nature of sperm in gnathostomulids as well as some gastrotrichs and turbellarians, i.e. in three of the most 'primitive' metazoan taxa.

It is clear that the Gastrotricha are *lebens-ort* types (Riedl 1963) for marine sand. The Gnathostomulida, which appear similarly plesiomorphic and the closest related group (other than the apomorphic Nematoda) to the Gastrotricha within the Metazoa, occur predominantly in anaerobic to micro-oxic fine sand. The Turbellaria are also at their most diverse in the marine sand habitat although it has been argued that they may have arisen in mud (Riedl 1963).

Inspection of the distribution of mono- and multi-ciliated genera combining data in d'Hondt (1971) and Rieger (1976) shows apparently little difference in occurrence in relation to granulometry. A scoring scheme of one per fine, two per medium and three per coarse sand record gives non-significantly different means of 1.90 and 1.98 for the multiciliated and monociliated condition respectively. However, the impression gained is that the coarse sand occurrence of monociliated forms tends to be in deeper water, whereas multiciliated forms occur regularly in coarse sand both littorally and sublittorally.

It is therefore suggested that the original habitat of Gastrotricha was fine to medium sand with moderate to low detritus content. Further, it is suggested that the Gastrotricha evolved in response to food shortage in the Precambrian era (see habitat evolution in Boaden 1977). Their pharynx and through-gut were necessary to concentrate and efficiently process the rather more sparse food available in their habitat in contrast to the detritus-rich fine sand habitat of their sister group the Gnathostomulida. The latter were able to graze effectively prokaryote food from particle surfaces. Both groups probably retained some ancestral ability to absorb dissolved organic matter epidermally. The group which gave rise to the gnathostomulids probably also contained turbellariomorph forms specializing in protein-rich food obtained by predation or scavenging.

The Nematoda are closely related to the Gastrotricha (Remane 1936; Rieger 1976). It is suggested that the main impetus for their evolution from the Gastrotricha or gastrotrich precursor was response to habitat instability. The Gastrotricha overcame this difficulty through evolution of adhesive tubules, which also retained their advantage in the sand habitat, and of sculpturing devices, whereas the nematodes specialized in overall strengthening of the cuticle. The latter process however largely denied the Nematoda planktonic habitats which to some extent become available to the Paucitubulatina, at least in freshwater. Perhaps the latter success was partly due to lack of competition from planktonic cnidarians.

The answer to 'Why is a gastrotrich?' could obviously be because its ancestors survived at least until reproducing. The same could be said of many other taxa, including ducks. A better answer more allied to the gastrotrichs' success and our ignorance of reality would be 'One of its tubules is both the same'.

Acknowledgements

My thanks to former G. and M. Williams Research Fund bursary students for assistance with field work and to D. G. Erwin and C. Maguire (who both left meiofauna for greater things) for use of unpublished TEM photographs.

References

Balsamo, M. (1983). *Gastrotrichi (Gastrotricha)*. *Guide per il riconoscimento dell specie animali delle acque interne italiane, 20*. Consiglio Nazionale delle Richerche, Verona.

Boaden, P. J. S. (1968). Water movement—a dominant factor in interstitial ecology. *Sarsia* 34, 126-36.

—— (1974). Three new thiobiotic Gastrotricha. *Cah. biol. mar.* 15, 367-78.

—— (1977). Thiobiotic facts and fancies (Aspects of the distribution and evolution of anaerobic meiofauna). *Mikrofauna Meeresbod.* 61, 45-63.

—— and Elhag, E. A. G. (1984). Meiobenthos and the oxygen budget of an intertidal sand beach. *Hydrobiologia*, 118, 39-47.

Boaden, P. J. S. and Erwin, D. G. (1971). *Turbanella hyalina* versus *Protodriloides symbioticus*: a study of interstitial ecology. *Vie Mileu* Suppl. 22, 479-92.

Brandenburg, J. (1962). Elektronmikroskopische Untersuchung des Terminalapparates von *Chaetonotus* sp. (Gastrotricha) als erstes Beispiel einer Cyrtocyte bei Aschelminthen. *Z. Zellforsch. mikrosk. Anat.* 57, 136-44.

DeRobertis, E. D. P. and DeRobertis, E. M. F. (1980). *Cell and molecular biology*. Holt Saunders, Philadelphia.

Fenchel, T. and Riedl, R. J. (1970). The sulfide system; a new biotic community underneath the oxidized layer of marine sandy bottoms. *Mar. Biol.* 7, 225-68.

Franzén, Å. (1956). On spermiogenesis, morphology of the spermatozoon, and biology of fertilization among invertebrates. *Zool. Bidr. Upps.* 31, 355-480.

Gray, J. S. and Johnson, R. M. (1970). The bacteria of a sandy beach as an ecological factor affecting the inertidal gastrotrich *Turbanella hyalina*. *J. exp. mar. Biol. Ecol.* 4, 119-33.

Hogue, E. W. (1978). Spatial and temporal dynamics of a subtidal estuarine gastrotrich assemblage. *Mar. Biol.* 49, 211-22.

d'Hondt, J.-L. (1971). Gastrotricha. *Oceanogr. mar. Biol. Ann. Rev.* 9, 141-92.

Hummon, M. R. (in press). Ultrastructure of sperm development in postparthenogenic *Lepidodermella squammata*, a freshwater gastrotrich. *Hydrobiologia*.

Hummon, W. D. (1972). Dispersion of Gastrotricha in a marine beach of the San Juan Archipelago, Washington. *Mar. Biol.* 16, 349-55.

—— (1974a). Some taxonomic revisions and nomenclatural notes concerning marine and brackish-water Gastrotricha. *Trans. Am. microsc. Soc.* 93, 194-205.

—— (1974b). Gastrotricha. In *Reproduction of marine invertebrates* (eds A. C. Giese and J. S. Pearse) Vol. 1, pp. 485-504. Academic Press, New York.

—— (1975a). Respiratory and osmoregulatory physiology of a meiobenthic marine Gastrotrich, *Turbanella ocellata* Hummon 1974. *Cah. biol. mar.* 16, 255-68.

—— (1975b). Habitat suitability and the ideal free distribution of Gastrotricha in a cyclic environment. In *Proceedings of the Ninth European Marine*

Biology Symposium (ed. H. Barnes) pp. 495-525. Aberdeen University Press, Aberdeen.

—— (1975c). Seasonal changes in secondary production, faunal similarity and biological accommodation, related to stability among the Gastrotricha of two semi-enclosed Scottish beaches. In *Proceedings of the Tenth European Marine Biology Symposium* (eds G. Persoone and E. Jaspers) Vol. 2, pp. 309-36. Universa Press, Wetteren.

Kershaw, K. A. (1973). *Quantitative and dynamic plant ecology* (2nd edn). Edward Arnold, London.

Kisielewska, G. and Kisielewska, J. (1984). On sexual reproduction of freshwater gastrotrichs of the family Chaetonotidae. *Hydrobiologia.*

Maguire, C. and Boaden, P. J. S. (1975). Energy and evolution in the thiobios: the extrapolation from the marine gastrotrich *Thiodasys sterreri*. *Cah. biol. mar.* 16, 635-46.

Nixon, D. E. (1976). Dynamics of spatial pattern for the gastrotrich *Tetranchyroderma bunti* in the surface sand of high energy beaches. *Int. Rev. ges. Hydrobiol.* 66, 211-48.

Powell, E. N., Bright, T. J., Woods, A. and Gittings, S. (in press). The East Flower Garden brine seep: implications for benthic community structure. *Hydrobiologia.*

—— Crenshaw, M. A. and Rieger, R. M. (1980). Adaptations to sulfide in sulfide-system meiofauna. Endproducts of sulfide detoxification in three turbellarians and a gastrotrich. *Mar. Ecol. Prog. Ser.* 2, 169-77.

Remane, A. (1936). Gastrotricha. In *H. G. Bronn's Klassen und Ordnungen des Tierreichs* Vol. 4, Pt. II, B.1, pp. 1-242.

Riedl, R. (1963). Probleme und Methoden der Erforschung des litoralen Benthos. *Zool. Anz.* suppl. 26, 505-67.

Rieger, G. E. and Rieger, R. M. (1977). Comparative fine structural analysis of the gastrotrich cuticle and aspects of cuticle evolution within aschelminthes. *Z. zool. Syst. Evol. forsch.* 15, 81-123.

Rieger, R. M. (1976). Monociliated epidermal cells in Gastrotricha: significance for concepts of early metazoan evolution. *Z. zool. Syst. Evolutionsforsch.* 14, 198-226.

—— Ruppert, E. E., Rieger, G. E. and Schoepfer-Sterrer, C. (1974). On the fine structure of gastrotrichs with description of *Chordodasys antennatus* sp. n. *Zool. Scr.* 3, 219-37.

Ruppert, E. E. (1977). Zoogeography and speciation in marine Gastrotricha. *Mikrofauna Meeresbod.* 61, 231-51.

—— (1978a). The reproductive system of gastrotrichs II. Insemination in *Macrodasys*: a unique mode of sperm transfer in Metazoa. *Zoomorphologie* 89, 207-28.

—— (1978b). The reproductive system of gastrotrichs. III. Genital organs of Thaumastodermatinae subfam. n. and Diplodasyinae subfam. n. with discussion of reproduction in Macrodasyida. *Zool. Scr.* 7, 93-114.

—— and Shaw, K. (1977). The reproductive system of Gastrotrichs. I. Introduction with morphological data for two new *Dolichodasys* species. *Zool. Scr.* 6, 185-95.

Schoepfer-Sterrer, C. (1974). Five new species of *Urodasys* and remarks on the terminology of the genital organs in Macrodasyidae (Gastrotricha). *Cah. biol. mar.* 15, 229-54.

Teuchert, G. (1973). Die Feinstruktur des Protonephridial systems von *Turbanella cornuta* Remane, einem marinen Gastrotrichen der Ordnung Macrodasyoidea. *Z. Zellforsch. mikrosk. Anat.* 136, 277-89.

—— (1977). Leibeshohlenverhaltnisse von dem marinen Gastrotrich *Turbanella cornuta* Remane (Ordnung Macrodasyoidea) und eine phylogenetische Bewertung. *Zool. Jb. Abt. Anat.* 97, 586-96.

—— and Lappe, A. (1980). Zum sognannten 'Pseudocoel' der Nemathelminthes—Ein Vergleich der Leibeshöhlen von mehreren Gastrotrichen. *Zool. Jb. Abt. Anat.* 103, 424-38.

Tyler, S., Melanson, L. A. and Rieger, R. M. (1980). Adhesive organs of the Gastrotricha II. The organs of *Neodasys. Zoomorphologie* 95, 17-26.

—— and Rieger, G. E. (1980). Adhesive organs of the Gastrotricha I. Duogland organs. *Zoomorphologie* 95, 1-15.

Weiss, M. J. and Levy, D. P. (1979). Sperm in parthenogenetic freshwater Gastrotricha. *Science N. Y.* 205, 302-3.

17. Affinities and intraphyletic relationships of the Priapulida

JACOB VAN DER LAND

Rijksmuseum van Natuurlijke Historie, Leiden, Netherlands

and

ARNE NØRREVANG
Institute of Cell Biology and Anatomy, Universitetsparken, Copenhagen, Denmark

Abstract

Several authors consider that the Priapulida belong to the Coelomata, believing these worms to have a coelomic epithelium. Others place them in the Pseudocoelomata or Aschelminthes because of the non-cellular character or supposed absence of coelomic boundaries. Features which might provide clues to their phylogenetic position are discussed, notably the structure of integument, mesenteries, coelomic lining, muscles, and embryological development. Ultrastructural studies have provided new information, but there are still no convincing arguments in favour of a close relationship with any other phylum. The Priapulida are probably an ancient phylum, of which no close relatives remain.

Fossil finds show that a considerable radiation had already occurred by the Cambrian, but they provide little information on the level of the relationships of the phylum. Most of the larger living species have been known for some time and because of their uniform morphology are placed in a single family. Several meiobenthic species have been discovered recently. They show many specializations and are placed in two families. The extent of evolutionary radiation differs between adults and larvae in that the latter tend to be more conservative.

Introduction

The Priapulida comprise a small phylum of benthic marine worms. At present fifteen living and six fossil species are known. They exhibit a considerable diversity (Table 17.1; Figs 17.2-17.10), but their relation to one another is beyond doubt. There is sufficient circumstantial evidence to include six Cambrian fossil species in the phylum.

Priapulids have been known since the 18th century and for more than a century they have been studied intensively. All aspects of their relationships have been discussed at length. Nevertheless it is worthwhile reinvestigating their relationships, because a number of new developments have taken place. Several new species have been described in recent years, ultrastructural studies have provided new information, and fossil forms have been studied in detail. Despite this there is still a divergence of opinion about the affinities of the group.

Earlier workers included the priapulids in other groups. On account of their radial symmetry they were sometimes considered to be echinoderms, but generally they were placed in the group Gephyrea, together with Sipuncula and Echiura. The history of research has been summarized by Lang (1953, 1963) and van der Land (1970). In more recent times, discussion has concentrated on the possibility of a relationship with groups of worms either belonging to the aschelminths or pseudocoelomates, or to the eucoelomates. The wide divergence of opinion can be demonstrated by simply mentioning some groups that have been considered to be related to the Priapulida: Coelomata (Shapeero 1961), Nematomorpha (Malakhov 1980), Kinorhyncha (Lang 1963; Malakhov 1980), Rotifera (Por and Bromley 1974), Acanthocephala (Conway Morris 1977), and Sipuncula (Candia Carnevali and Ferraguti 1979).

General structure

The priapulid body (Fig. 17.1) is unsegmented. The anterior is an introvert, which can be retracted into the posterior part of the body and acts as a locomotory and feeding organ. There is a mixture of bilateral and radial symmetry. Bilaterality is evident from the ventral position of the main nerve cord and the paired reproductive and excretory organs. The brain (a peribuccal nerve ring), the digestive system, the musculature, and several integumental structures show a radial symmetry. Some of the larvae exhibit more bilateral traits than the adults.

Radial symmetry, as well as the possession of an introvert, is connected with life in sediment. All species, including the fossil ones, are

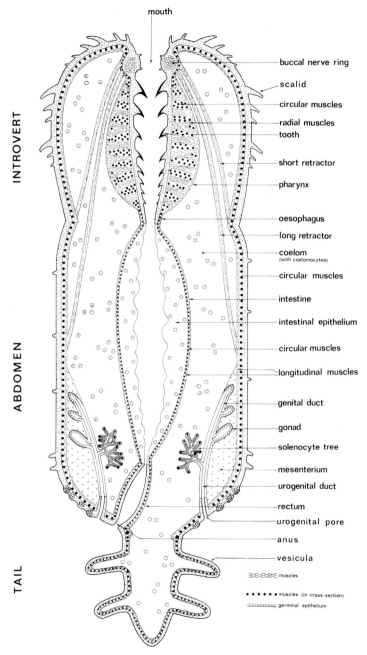

Fig. 17.1. Bodyplan of a priapulid (modified from van der Land 1970).

or were infaunal. Analogous evolutionary trends can be found in several other groups, which should be taken into account when comparisons are made with groups with similar life habits, such as the Sipuncula. Parasites such as Nematomorpha and Acanthocephala also show a tendency to radial symmetry, and make use of an introvert which could well be analogous.

Other important characters are the presence of a terminal mouth and anus; a single spacious body cavity, acting as a circulatory system and hydrostatic skeleton, with a special type of mesentery; an intra-epidermal nervous system; solenocytic protonephridia; and a chitinous cuticle which can be moulted.

Some morphological characteristics

1. Integument

The integument consists of two layers: a one-layered hypodermis and a structureless cuticle (Moritz 1972). Microvilii are rare, occurring only during the formation of the new cuticle before moulting. The cuticle is secreted as a continuous layer from the surface of the hypodermis. The fact that it is structureless indicates that the secretory process is continuous (Moritz and Storch 1970). It contains a chitin that very closely resembles α-chitin (Shapeero 1962).

In all true Spiralia (Platyhelminthes, Nemertea, Sipuncula, Echiura, Bryozoa, Annelida, and Arthropoda), cuticle formation is different, microvilli being numerous and fibrous material being deposited mostly in a criss-cross pattern between the microvilli. In Rotifera, Acanthocephala (Storch 1979), and Chaetognatha (Nørrevang unpublished) there is an 'intracellular' cuticle, which means that there is no extracellular layer on the body surface. Kinorhynchs have a dense, structureless cuticle, with no signs of microvilli (Merriman and Corwin 1973). Nematomorpha have a peculiar cuticle, consisting of layers of crossing fibres without sub-structures (Eakin and Brandenburger 1974). The Nematoda are the only group with a cuticle that (in some species) is similar to that of *Priapulus* (references in Moritz 1972). In Gastrotricha the epidermal cells have a few short microvilli, exceptionally penetrating the cuticle, which either consists of successive bilayers or is homogeneous; fibres are lacking.

2. Nervous system

The central nervous system of the Priapulida consists of a peribuccal nerve ring and a ventral longitudinal nerve cord, terminating in a

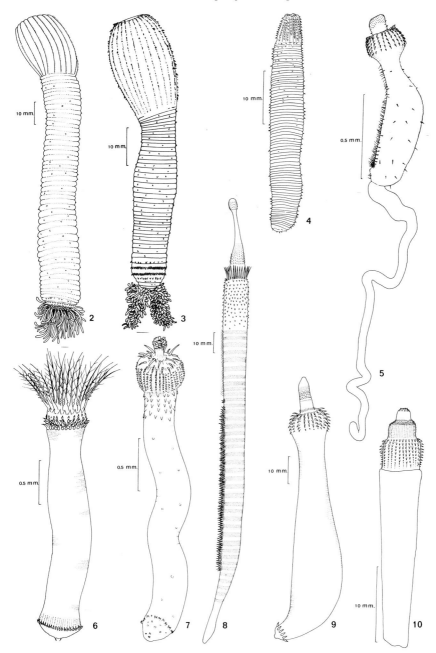

Figs 17.2-17.10. Priapulid types. Semi-diagrammatic drawings of: 2, *Priapulus*; 3, *Priapulopsis*; 4, *Halicryptus*; 5, *Tubiluchus* (male); 6, *Maccabeus*; 7, *Meiopriapulus*. Reconstructions of the Cambrian: 8, *Louisella*; 9, *Ottoia*; 10, *Selkirkia* (based on Conway Morris 1977).

ganglion-like swelling near the anus. It is intra-epidermal, both nerve ring and nerve cord being continuous with the hypodermal epithelium. The peripheral nervous system includes thirteen nerves from the nerve ring to the body wall and four nerves to the pharynx and gut. Sense organs and other epidermal structures (except those of the introvert) are innervated from the nerve cord. The tail is innervated from the caudal ganglion by two lateral nerves.

The intra-epidermal position of the nervous system is a primitive character of little phylogenetic significance. In the Kinorhyncha there is a circum-enteric nerve ring in a position similar to that of the nerve ring of Priapulida, and a ganglionated ventral nerve cord; the whole nervous sytem seems to be intra-epidermal. In Nematomorpha the nervous system is also essentially intra-epidermal, although the nerve cord has sunk below the body wall musculature. It consists of a circum-enteric nerve ring, a longitudinal ventral nerve cord, and an anal swelling. From an additional cloacal swelling two nerves extend into the caudal appendages. In gastrotrichs (Teuchert 1977) the brain consists of two halves with dorsal and ventral connections and the ventral nerve cord is paired, with the two branches widely separated. The organization of the nervous system in Acanthocephala is also different; there are a central ganglion, lateral nerves, and a number of other nerves running to muscles and sense organs (Kilian 1932).

3. Muscles and coelomic lining

Candia Carnevali and Ferraguti (1979) described the muscle type of priapulids as a variation on a basically smooth muscle pattern, unlike that of other types of transverse or obliquely striated muscle. They concluded that priapulid muscles have little in common with those of the Aschelminthes, but resemble those of the Sipuncula. Perhaps priapulids have conserved a primitive muscle system.

The presence and nature of the coelomic lining has always been crucial in discussions on the systematic position of the phylum. In the large species the presence of a membrane (peritoneum) is evident, but some observers described it as a flat epithelium with few nuclei, others as structureless. Little attention has been given to the structure of the different types of mesenteries (van der Land 1970).

Shapeero (1961) revived the discussion by describing a true coelomic lining, accompanied by convincing photomicrographs. He observed a cellular membrane lining the outside of muscles. Shapeero's findings resulted in the Priapulida being considered as coelomate by several authors. His statements remained unchallenged until Malakhov (1980) concluded from EM observations that there is no coelomic lining.

In order to explain these discrepancies the junior author made a detailed investigation of the interior body wall of *Priapulus* in different regions: the outside of the gut wall, the retractors, and the mesenteries. No epithelial lining or even epithelioid cells were found, but an explanation for earlier observations of a coelomic epithelium can now be given. According to Candia Carnevali and Ferraguti (1979) retractor muscle cells are of the circomyarian type, with the nucleus in the central axoplasm. However, we found that in the retractors the nucleus sometimes lies at the periphery of the cell, outside the myofilament regions. The muscle cells of the intestine have a peripheral nucleus, and this is also the case in several other muscle cell types, e.g. those found in the tail papillae (Fig. 17.12).

Muscle cells secrete extracellular membranes, which probably consist of the material normally found in glycocalyces, i.e. mucopolysaccharides and mucoproteins. The mesenteries illustrate this convincingly (Fig. 17.11). They are made up of a sheet of muscle cells, disorganized in adults, and an extracellular membrane on both sides. Between the limiting membranes there are other more or less coherent membrane sheets, sometimes up to four. The outer lining of retractor muscles has the same structure (Fig. 17.13). Apparently the extracellular membranes and the peripheral nuclei of muscle cells have given the impression of a coelomic epithelium in earlier light-microscopical observations. As far as we are aware the peculiar coelomic lining in priapulids is unique.

Reproduction

The spermatozoa of *Priapulus* are of a primitive type (Afzelius and Ferraguti 1978), and their features are of little use for phylogenetic considerations. The spermatozoa of *Tubiluchus* are atypical, and their structure is undoubtedly connected to internal fertilization in this genus. Alberti and Storch (1983) studied their fine structure and concluded that they are of a unique type.

The embryology of priapulids is imperfectly known. Zhinkin (1949) and Zhinkin and Korsakova (1953) described the early cleavage and concluded that it is radial, but somewhat atypical in that bilateral symmetry appears early. There are no elements of spiral cleavage. Cleavage ends quite early, which points to specialization. Gastrulation takes place as polar ingrowth.

According to Lang (1953, 1963), who also observed radial cleavage, the coeloblastula at the 64-cell stage is succeeded by an invagination gastrula. The mesoderm originates from two cell strings at the opening and completely fills the blastocoel. Consequently the mesoderm most

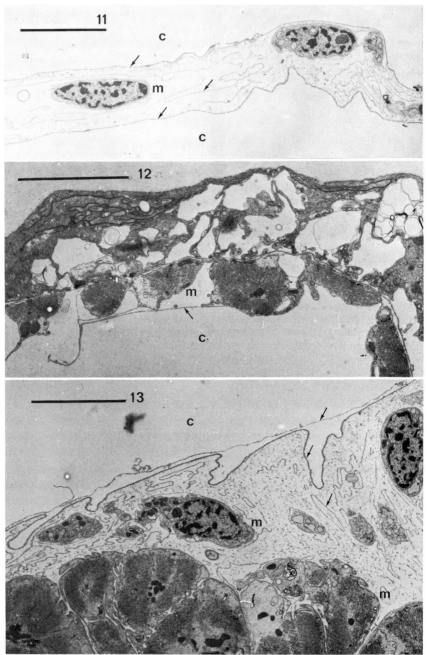

Figs 17.11-17.13. *Priapulus caudatus*, TEM images. 11, Mesentery between gut and retractor. 12, Epidermis and muscles of tail papilla. 13, Surface of retractor. Scale bars represent 10 μm. Abbreviations: c, coelom; m, muscle cell; arrows indicate extra-cellular membrane.

probably is of ectodermal origin. Lang (1953) also observed that the contractile, non-ciliated larva leaving the egg does not have a lorica at this stage. It consists of an ectoderm and an inner syncytial cell mass. Mouth, intestine, and anus are lacking, and other organs were not observed either.

The only definite conclusion is that the priapulids do not belong to the Spiralia.

Evolutionary radiation

1. Adults

The recent species can best be placed in three families (Table 17.1). The family Priapulidae (Figs 17.2-17.4) accommodates eight macro-benthic, mud-dwelling species, occurring in cold water (van der Land 1970). They are burrowing, carnivorous animals, mainly feeding on slow-moving prey, particularly polychaetes. The pharynx is provided with stout cuspidate teeth. *Halicryptus* (Fig. 17.4) is aberrant in that it has a very short introvert, with a single series of scalids (probably a primitive trait), and lacks a tail. *Acanthopriapulus*, with a single spiny tail, and *Priapulopsis* (Fig. 17.3), with two tails, appear to be more specialized than *Priapulus*.

The other families accommodate seven meiobenthic species. Because of the size of the largest specimens, none of them completely meets the morphometric meiofauna definition, but for all practical biological purposes they can be considered to belong to it. The Chaetostephanidae are semi-sessile tubicolous animals of muddy bottoms (Por and Bromley 1974; Salvini-Plawen 1974). They are carnivorous as shown by the strong pharyngeal armature, but catch their prey in a specialized fashion, the anterior scalids providing a trapping mechanism. The other scalids are also differentiated and the introvert cannot be used as a locomotory organ (Fig. 17.6). There is a posterior ring of hooks, but a tail is lacking. The Tubiluchidae live interstitially in carbonate sands of tropical waters (van der Land 1970, 1982; Morse 1982). They are characterized by a strongly differentiated pharyngeal armature, including delicate pectinate teeth, which means that they are likely to be microphagous. They also have a wide variety of external cuticular structures (Figs 17.5, 17.7). In this respect there are many similarities, but also differences between the two genera. *Tubiluchus* has internal fertilization and shows sexual dimorphism.

2. Fossil forms

Six fossil species have been described (Table 17.1). *Priapulites* from the Pennsylvanian does not differ significantly from the modern *Priapulus*,

Table 17.1. Known taxa of the Priapulida

Families and species	Distribution
Priapulidae:	
Priapulus caudatus Lamarck 1816	Circum-arctic
Priapulus tuberculatospinosus Baird 1868	Circum-antarctic
Priapulus abyssorum Menzies 1959	Cosmopolitan abyssal
Priapulopsis bicaudatus (Koren and Danielssen 1868)	Circum-arctic
Priapulopsis australis (de Guerne 1886)	Circum-antarctic
Priapulopsis cnidephorus Salvini-Plawen 1973	Adriatic
Acanthopriapulus horridus (Théel 1911)	South-Atlantic
Halicryptus spinulosus von Siebold 1849	Circum-arctic
*Priapulites konecniorum Schram 1973	Middle Pennsylvanian
Tubiluchidae:	
Tubiluchus corallicola van der Land 1968	Tropical West Atlantic
Tubiluchus remanei van der Land 1982	Red Sea
Tubiluchus sp. nov. (van der Land, unpublished)	Philippines
Tubiluchus sp. nov. (van der Land, unpublished)	Australia
Meiopriapulus fijiensis Morse 1981	Fiji islands
Chaetostephanidae:	
Maccabeus tentaculatus Por 1973	Mediterranean
Maccabeus cirratus (Malakhov 1979)	Indian Ocean
*Ottoidae:	
*Ottoia prolifica Walcott 1911	Middle Cambrian
*Selkirkiidae:	
*Selkirkia columbia Conway Morris 1977	Middle Cambrian
*Miskoiidae:	
*Louisella pedunculata Walcott 1911	Middle Cambrian
*Ancalagonidae:	
*Ancalagon minor (Walcott 1911)	Middle Cambrian
*Fieldiidae:	
*Fieldia lanceolata Walcott 1912	Middle Cambrian

* Fossil taxa, only known from North America.

but the five Cambrian species differ from all recent forms. Conway Morris (1977) made a detailed analysis of all available material and placed the species in separate families, which shows that considerable radiation had taken place. Observations on three species allow diagrammatic reconstructions (Figs 17.8-17.10). It is reasonable to include these animals in the phylum Priapulida, although the fossils lack certain decisive details and analogous developments can never be excluded completely. They certainly do not provide information which might aid in determining relationships with other phyla.

We may conclude that a variety of priapulid-like macrobenthic infauna worms occurred in Cambrian times and that they exhibited a comparable radiation to that observed in modern priapulids. *Selkirkia* (Fig. 17.10) is tubicolous like *Maccabeus*. *Ottoia* (Fig. 17.9) has a posterior ring of hooks like *Maccabeus* and *Meiopriapulus* (Figs 17.6, 17.7). In *Louisella* (Fig. 17.8) the scalids are not disposed in distinct rows, just as in *Acanthopriapulus* and *Maccabeus* (Fig. 17.6), and it has rows of ventral structures that we compare with the setae of *Tubiluchus* (Fig. 17.5).

3. Larvae

In all species of which the larvae are known, they are remarkably uniform when compared with the variation observed in adults. Thus, their size and general structure are the same. The hind part of the body is provided with a 'lorica', i.e. the cuticle is relatively thick and forms ridges. Between the introvert and the lorica there is a neck region. The introvert and the neck can be retracted into the lorica for protection. The scalids and cuspidate pharyngeal teeth exhibit little differentiation. It is evident that the larvae have a tendency to be more conservative than the adults.

Variable characters are the presence or absence and the number and distribution of tubuli on the lorica, the number of longitudinal ridges, the relative width of the areas between the ridges, the degree of flattening of the lorica, and the presence or absence of adhesive organs. We think the most primitive state is shown by the *Tubiluchus* larva, in which the lorica is not flattened dorso-ventrally. There are 10 primary ridges (with 10 incomplete secondary ridges in between), equally spaced around the body, and a large number of tubuli (up to about 20). All other known types are more or less flattened and the longitudinal ridges are not equally spaced so that they clearly exhibit bilateral symmetry. The number of ridges and tubuli is reduced. We envisage a development from the *Tubiluchus* type to the other three known types:

(*a*) *Maccabeus*. There are ten ridges, two dorsal and two ventral widened areas to form the dorsal and ventral 'plates' of the lorica, and two circlets of four tubuli.

(*b*) *Priapulus*. The larval lorica has eight ridges, small remnants of mid-dorsal and mid-ventral ridges and one circlet of four tubuli. The posterior end of the body carries an adhesive organ.

(*c*) *Halicryptus*. There are eight ridges. The cuticle of the dorsal and ventral fields is thickened to form a genuine plate, which is ornamented. Tubuli are lacking. The introvert is provided with a circlet of adhesive tubules.

Conclusions

The affinities of the Priapulida remain obscure, but at present we can at least be sure that they are not related to coelomate groups and do not belong to the Spiralia in general. They do show similarities with several groups traditionally assembled as Aschelminthes, possibly more with the Kinorhyncha than with the other groups. Within this assemblage the Priapulida are probably the least specialized.

The intraphyletic relationships are also enigmatic. Evidence frc.n the adults alone would suggest that *Halicryptus* is primitive and *Tubiluchus* specialized. However, the reverse is true for their larvae. Hence the interrelationships of the three modern families remain obscure, and connections with the five Cambrian families also are not obvious.

References

Afzelius, B. A. and Ferraguti, M. (1978). The spermatozoon of *Priapulus caudatus* Lamarck. *J. submicrosc. Cytol.* 10, 71-80.

Alberti, G. and Storch, V. (1983). Fine structure of developing and mature spermatozoa in *Tubiluchus* (Priapulida, Tubiluchidae). *Zoomorphologie* 103, 219-27.

Candia Carnevali, M. D. and Ferraguti, M. (1979). Structure and ultrastructure of muscles in the priapulid *Halicryptus spinulosus*: functional and phylogenetic remarks. *J. mar. biol. Ass. U.K.* 59, 737-44.

Conway Morris, S. (1977). Fossil priapulid worms. *Spec. Pap. Palaeont.* 20, 1-95.

Eakin, R. M. and Brandenburger, J. L. (1974). Ultrastructural features of a gordian worm. *J. Ultrastruct. Res.* 46, 351-74.

Kilian, R. (1932). Zur Morphologie und Systematik der Giganthorhynchidae (Acanthoceph.). *Z. wiss. Zool.* 141, 246-345.

Land, J. van der (1970). Systematics, zoogeography, and ecology of the Priapulida. *Zool. Verhandl. Leiden* 112, 1-118.

—— (1982). A new species of *Tubiluchus* (Priapulida) from the Red Sea. *Neth. J. Zool.* 32, 324-35.

Lang, K. (1953). Die Entwicklung des Eies von *Priapulus caudatus* Lam. und die systematische Stellung der Priapuliden. *Ark. Zool.* (2) 5, 321-48.

—— (1963). The relation between the Kinorhyncha and Priapulida and their connection with the Aschelminthes. In *The lower Metazoa, comparative biology and phylogeny* (ed. E. C. Dougherty) pp. 256-62. University of California Press, Berkeley, Ca.

Malakhov, V. V. (1980). Cephalorhyncha, a new type of animal kingdom uniting Priapulida, Kinorhycha, Gordiacea, and a system of Aschelminthes worms. *Zool. Zh.* 59, 485-99 (in Russian).

Merriman, J. A. and Corwin, H. O. (1973). An electron microscopical examination of *Echinoderes dujardini* Claparède (Kinorhyncha). *Z. Morph. Ökol. Tiere* 76, 227-42.

Moritz, K. (1972). Zur Feinstruktur integumentaler Bildungen bei Priapuliden (*Halicryptus spinulosus* und *Priapulus caudatus*). *Z. Morph. Ökol. Tiere* 72, 203-30.

—— and Storch, V. (1970). Über den Aufbau des Integumentes der Priapuliden und der Sipunculiden (*Priapulus caudatus* Lamarck, *Phascolion strombi* (Montagu)). *Z. Zellforsch. mikrosk. Anat.* 105, 55-64.

Morse, M. P. (1982). *Meiopriapulus fijiensis*, n. gen., n. sp.; An interstitial priapulid from coarse sand in Fiji. *Trans. Am. microsc. Soc.* 100, 239-52.

Por, F. D. and Bromley, H. J. (1974). Morphology and anatomy of *Maccabeus tentaculatus* (Priapulida: Seticoronaria). *J. zool. Soc., Lond.* 173, 173-97.

Salvini-Plawen, L. von (1974). Zur Morphologie und Systematik der Priapulidae: *Chaetostephanus praeposteriens*, der Vertreter einer neuen Ordnung Seticoronaria. *Z. zool. Syst. Evolutionsforsch.* 12, 31-54.

Shapeero, W. L. (1961). Phylogeny of the Priapulida. *Science, N.Y.* 133, 879-80.

—— (1962). The epidermis and cuticle of *Priapulus caudatus* Lamarck. *Trans. Am. microsc. Soc.* 81, 352-5.

Storch, V. (1979). Contributions of comparative ultrastructural research to problems of invertebrate evolution. *Am. Zool.* 19, 637-45.

Teuchert, G. (1977). The ultrastructure of the marine gastrotrich *Turbanella cornuta* Remane (Macrodasyoidea) and its functional and phylogenetical importance. *Zoomorphologie*, 88, 189-246.

Zhinkin, L. (1949). Early stages in the development of *Priapulus caudatus*. *Dokl. Akad. Nauk SSSR* 65, 409-12 (in Russian.)

—— and Korsakova, G. (1953). Early stages in the development of *Halicryptus spinulosus*. *Dokl. Akad. Nauk SSSR* 88, 571-3 (in Russian.)

18. Sipuncula: developmental evidence for phylogenetic inference

MARY E. RICE

Department of Invertebrate Zoology,
National Museum of Natural History,
Smithsonian Institution, Washington DC, USA

Abstract

Sipunculans are marine worms characterized by a unique combination of features which set them apart as a distinct coelomate phylum. A review of developmental features demonstrates clearly their similarities to annelids and molluscs. Cleavage is spiral, giving rise to a trochophore larva. Like molluscs, the cross cells of the apical plate are in the radial position. Close affinities to annelids are apparent in the prototrochal and metatrochal ciliary bands, the formation of the larval cuticle by transformation of the egg envelope, and especially in the development of the nervous system. They differ from annelids in that they lack segmentation. Thus, it is proposed that sipunculans are a primitive group which arose from a molluscan-annelidan stem.

Within the phylum the primitive developmental pattern is presumed to have included a moderately yolky egg covered by a thick egg envelope, a non-feeding trochophore encircled by the persistent egg envelope, and a planktotrophic pelagosphera larva. This developmental pattern is assumed to have arisen from an ancestral pattern with a planktotrophic trochophore and an egg having a simple envelope and low yolk content. Planktotrophy in sipunculans, apparently retarded by the development of the thick egg envelope and an associated moderate increase in yolk, was accordingly shifted to the specialized larval form, the pelagosphera. Based on these hypotheses two evolutionary sequences can be defined in extant developmental patterns of sipunculans: one toward an increase in yolk and decrease in the pelagic stage and the other toward a reduction in yolk and the prolongation of the pelagic stage as represented in long-lived planktotrophic pelagosphera larvae of the open ocean.

Introduction

Sipunculans are unsegmented, coelomate marine worms, characterized by a unique combination of features which have led to their recognition as a separate phylum (Sedgwick 1898; Baltzer 1931; Pickford 1947; Fisher 1952; Stephen and Edmonds 1972). The currently accepted name, Sipuncula, was proposed by Stephen (1964). Characters of the group include division of the body into two regions: a posterior thickened trunk and a narrow anterior retractable introvert usually terminating in a mouth surrounded by tentacles. Other definitory features include a spacious, undivided coelomic cavity, a recurved gut ending in a dorsal anus on the anterior trunk, a median unpaired ventral nerve cord, and usually two metanephridia located ventrolaterally in the anterior trunk.

The classification of Stephen and Edmonds (1972) recognized 4 families, 17 genera, and approximately 320 species. The genus *Golfingia*, containing by far the greatest number of species, has been presumed by many authors (Selenka, de Man, and Bülow 1883; Hérubel 1907; Gerould 1913; Åkesson 1958; Murina 1977) to represent the most primitive adult morphology. Primitive features are considered to be a ring of simple tentacles surrounding the mouth, four retractor muscles, a continuous layer of body wall muscles, two nephridia and a supra-oesophageal ganglion in a superficial position (cf. Clark, 1969). Other genera can be related to *Golfingia* on the basis of varying combinations of *Golfingia* characters or by specializations of these characters.

Although sipunculans are a homogeneous group, their phylogenetic position has long been an enigma, as shown by the widely divergent views that have been proposed for their relationships. In the early 1800s sipunculans were placed close to the holothurian echinoderms by Lamarck (1816) and Cuvier (1830). Subsequently, Quatrefages (1847) recognized their similarities to annelids and considered sipunculans as a link between holothurians and annelids. Thus he created the group 'Gephyrea' (from the Greek word, *gephyros*, meaning bridge) in which he included sipunculans, echiurans and priapulids. The Gephyrea became a convenient taxon for groups of uncertain affinities, and even after the relationships of sipunculans were better understood, the term persisted in the literature. Most frequently it was used to indicate a class of Annelida which included only sipunculans and echiurans. Another view which received considerable recognition was that of Lang (1888) who erected a group Prosopygii to contain sipunculans, phoronids, bryozoans, and brachiopods. This grouping was based on resemblances in the crown of tentacles, position of anus, unsegmented body, and the presence of one or two pairs of nephridia. A more recent proposal for the relationships of sipunculans, while retaining the accepted classification, has revived the ideas of Lamarck (1816), Cuvier (1830) and Meyer (1904), by again supposing an alliance between sipunculans and holothurians (Nichols 1967). Nichols has pointed to the similarities in general bodyplan, noting particularly the resemblance in structure and function of the water vascular system of holothurians and the coelomic tentacular system of sipunculans. Many other affinities for sipunculans have been proposed in the past (for review see Hyman 1959); however, all of these phylogenies have been based on morphological similarities with no reference to developmental biology.

Beginning in the late nineteenth century, studies on development of sipunculans suggested a relationship with annelids and to a lesser extent with molluscs (Selenka 1875; Hatschek 1883; Gerould 1903, 1906; Åkesson 1958, 1961a; Rice 1967, 1973, 1975a,b). The typical pattern of spiral cleavage found in sipunculan development and the formation of the trochophore larva have placed the sipunculans, along with the annelids and molluscs, in the Protostomia.

This paper briefly reviews the available information on development of sipunculans and compares it, where relevant, to that in other invertebrates, and considers its possible implications for phylogenetic interpretations, including evolutionary sequences within the phylum and affinities to other groups.

General developmental patterns

A wide diversity of developmental patterns is found within the Sipuncula, ranging from direct development with no pelagic stages to planktotrophic development with a long-lived larva which exists in the plankton for several months before undergoing settlement and metamorphosis into a juvenile worm. Table 18.1 gives information on the development of 20 species of sipunculans, representing all four families recognized in the classification of Stephen and Edmonds (1972). Based on this information, developmental patterns of sipunculans can be classified into four categories (Rice 1967, 1975a). Three are entirely lecithotrophic and the fourth includes a planktotrophic larva. Of the three lecithotrophic patterns, one is direct development, a second includes one larval stage, the trochophore, which transforms into a crawling vermiform stage, and the third includes a trochophore which metamorphoses into a second larval stage, the pelagosphera. The pelagosphera is a larval stage, unique to the Sipuncula, in which a prominent metatroch develops as the primary locomotory organ, and the prototroch is reduced or lost (Rice 1967). The fourth category is characterized by a lecithotrophic trochophore followed by a planktotrophic pelagosphera. The latter may live for a prolonged period in the plankton, attaining a length of 5-10 mm in some species (Åkesson 1961b; Damas 1962; Jägersten 1963). They are commonly found near the surface in warm oceanic current streams (Scheltema 1975).

Gametes, cleavage, and early development

Sipunculans, usually dioecious, spawn directly into the sea where fertilization occurs. This mechanism of discharging gametes is considered primitive among invertebrates (Jägersten 1972).

As in many other groups in which free spawning occurs, spermatozoa are of the primitive type (Franzén 1956).

All sipunculan eggs possess a thick egg envelope, comprising several layers and perforated by pores, but they show considerable variation in size, shape, pigmentation, and yolk content (Table 18.1). Although extremes in egg size represent two extremes in developmental pattern (i.e. direct and planktotrophic development) there is considerable overlap in size range within the various developmental patterns and no clear-cut correlation emerges.

Cleavage in sipunculans is spiral, holoblastic, and unequal. In a study of cell lineage of *Golfingia vulgaris* Gerould (1906) reported an alternating direction of spindles up to 48 cells and further in some regions of the egg. Differing from typical spiral cleavage, at the eight-cell stage the micromeres in the A, B, and C quadrants are larger than the macromeres—a size difference which is indicative of yolk content and is later reflected in the enormous yolk-laden cells of the prototroch. This peculiar feature of cleavage is found in the terebellid *Amphitrite* and some other polychaetes (Mead 1897; Gerould 1906) and in both cases the primary cells of the prototroch arise from cells la^2-ld^2. The 2d cell or somatoblast gives rise to the somatic plate and is the largest cell of the 16-cell stage. At the 48-cell stage in *Golfingia vulgaris* a radial division of the lq^{12} cells results in the formation of an apical cross. The cross cells, as described by Gerould, extend out from the tips of the rosette cells and lie in the sagittal and frontal planes of the future embryo. The arms of the cross are thus in the radial position of molluscs rather than the interradial of annelids. Some authors (Åkesson 1958; Hyman 1959) have mistakenly reported an annelidan cross in sipunculans because of confusion over Gerould's terminology. However, his description and illustration (Fig. 18.1) clearly indicate a molluscan cross. As in many other spiralians the third set of micromeres forms the

Table 18.1. A summary of developmental patterns of the Sipuncula*

Species	Egg size: diameter or length × width (μm)	Eight-cell stage relative size of micro- and macromeres in quadrants A, B, C	Gastrulation	Trochophore	Length of pelagic stage		
					Lecithotrophic	Pelagosphera	
						Lecithotrophic	Planktotrophic+
Category I							
Golfingia minuta[1]	260-280 × 214-230	?	Epiboly	0	0	0	0
Themiste pyroides[6]	190	Micromeres > macromeres	Epiboly	0	0	0	0
Phascolion cryptus[7]	136	Micromeres > macromeres	Epiboly	0	0		0
Category II							
Phascolion strombi[1]	125	Micromeres > macromeres	Epiboly	8 days	0		0
Phascolopsis gouldi[4]	150-180	Micromeres > macromeres	Epiboly	3 days	0		0
Category III							
Golfingia vulgaris[4]	150-180	Micromeres > macromeres	Epiboly	3 days	2 days		0
Golfingia elongata[2]	125	?	Epiboly + invagination	2 days	4 days		0
Golfingia pugettensis[6]	160	Micromeres = macromeres	Epiboly	8 days	13 days		0

Themiste alutacea[7]	138	?	Epiboly	2 days	6 days	0
Themiste lageniformis[11]	145	Micromeres > macromeres	Epiboly	0	8–12 days	0
Themiste petricola[3]	156	Micromeres > macromeres	Epiboly	2 days	5 days	0
Category IV						
Aspidosiphon parvulus[10]	139 × 107	Micromeres = macromeres	?	3 days	0	1 month
Golfingia misakiana[9]	108 × 77	Micromeres = macromeres	?	5 days	0	1 month
Golfingia pellucida[8]	105	Micromeres = macromeres	?	3 days	0	1 month
Paraspidosiphon fischeri[7]	103 × 94	?	?	2 days	0	1 month
Phascolosoma agassizi[6]	140 × 110	Micromeres = macromeres	Epiboly + invagination	8–10 days	0	1 month
Phascolosoma antillarum[7]	127 × 97	?	?	3 days	0	1 month
Phascolosoma perlucens[7]	112 × 91	Micromeres = macromeres	Epiboly + invagination	3 days	0	1 month
Phascolosoma varians[7]	104 × 90	?	?	3 days	0	1 month
Sipunculus nudus[5]	186	Micromeres < macromeres	Invagination	3 days	0	1 month

References: 1. Åkesson 1958; 2. Åkesson 1961a; 3. Amor 1975; 4. Gerould 1906; 5. Hatschek 1883; 6. Rice 1967; 7. Rice 1975a; 8. Rice 1975b; 9. Rice 1981; 10. Rice unpublished; 11. Williams 1972.

* Modified from Rice 1975b
† Time indicated is minimal period of survival in the laboratory. Metamorphosis to juvenile was not observed.

ectoderm, 3A-3D form endoderm and mesoderm. The 3D gives rise to 4d by a laeotropic division which in turn produces two daughter cells, the teloblasts of the mesoderm.

A survey of what is known of early cleavage in species other than *Golfingia vulgaris* reveals that the relative size of micromeres and macromeres at the eight-cell stage is related to the yolk content of the egg. Micromeres exceed the macromeres in size only in eggs with a high yolk content and in which the development is lecithotrophic. Those species with microlecithal eggs have micromeres equal to or smaller than the macromeres (Table 18.1).

Blastulation and gastrulation in sipunculans follow the pattern typical for other spiralians. Modifications are correlated with yolk content of the egg. A small blastocoel occurs in species with planktotrophic development. Gastrulation is accomplished entirely by epiboly in sipunculans having macrolecithal eggs, with the exception of *Golfingia elongata*, in which invagination plays a minor role (Åkesson 1958). Species which have planktotrophic development achieve gastrulation by a combination of epiboly and emboly. In *Phascolosoma perlucens* and *P. agassizi* gastrulation occurs mainly by epiboly, although a narrow archenteron is indicative of some invagination (Rice 1967, 1975a,b). In *Sipunculus nudus* gastrulation is accomplished primarily through invagination (Hatschek 1883).

The trochophore

Trochophores are essentially similar to those of polychaetes and molluscs. Typically the trochophore is top-shaped and characterized by pretrochal and post-trochal hemispheres, separated by a prominent

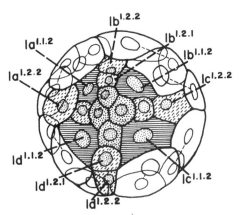

Fig 18.1 Apical view of the 48-cell stage of *Golfingia vulgaris*, showing molluscan cross and intermediate cells. Rosette cells are dotted, cross cells dashed, and intermediate cells barred. (Redrawn from Gerould 1906, after Rice 1975b).

band of ciliated prototrochal cells. Anteriorly an apical circle of rosette cells, encircled by the apical groove, bears the long cilia of the apical tuft. A pair of pigmented eyespots is located dorsolaterally in the pre-trochal hemisphere, and in a midventral position just posterior to the prototroch, there is a ciliated stomodaeum. The trochophore of sipun-culans is always completely enclosed by the thick egg envelope and thus is lecithotrophic or non-feeding, regardless of the content of yolk. Rudiments of the gut are present, and in the late trochophore of species with planktotrophic pelagospheras (Table 18.1) the gut may be fully formed, though not functional. The trochophore of sipunculans never develops a protonephridium, thus differing from many annelids and some molluscs.

Diversity among sipunculan trochophores is expressed in the pattern and degree of ciliation, the relative development of organs and the concentration of yolk (Fig. 18.2). In addition to the prototroch, pre-trochal and post-trochal (metatrochal) cilia occur in trochophores of *Golfingia elongata* (Åkesson 1961a) and *G. vulgaris* (Gerould 1906). *Sipunculus nudus* differs in that it is completely ciliated (Hatschek 1883; Gerould 1903). An equatorial band of ciliated cells, presumed to be homologous to the prototroch, grows posteriorly and anteriorly beneath the egg envelope, completely enclosing the embryo. The egg covering, thus composed of both prototroch cells and egg envelope, was referred to by Hatschek (1883) as the 'serosa'. Metatrochal cilia are present in the trochophore of *Sipunculus nudus*, but because they are enclosed by the egg envelope are non-functional. The embryos of directly develop-ing species lack ciliation, but have a broad equatorial band of large cells which are homologous to ciliated prototroch cells of other species. In *Golfingia minuta*, Åkesson (1958) reports rudimentary non-locomo-tory cilia on marginal cells above and below the prototroch cells. He interprets this condition as evidence that ancestors of this species were forms with a pelagic stage and that direct development is therefore secondary.

The end of the trochophore stage may occur in one of several ways, depending on the developmental category (Table 18.1). In categories 1 and 2 the non-swimming trochophore (or comparable stage) elongates, transforming into a small crawling vermiform stage. In other sipun-culans the end of the trochophore stage results in a metamorphosis to a pelagosphera larva—a process in which the prototroch is lost or reduced and the metatroch becomes the primary locomotory organ. In develop-mental category 3 the larva is lecithotrophic and in category 4 it is planktotrophic. The lecithotrophic larva swims for a short period before transforming into the vermiform stage. The planktotrophic larva, after a planktonic existence from one to several months, undergoes a second metamorphosis into the juvenile form.

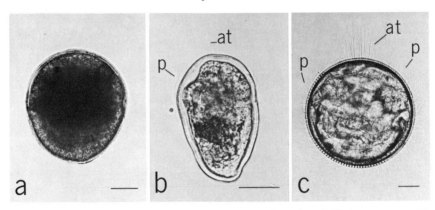

Fig. 18.2. Trochophores of sipunculans, illustrating diversity of ciliation patterns. (*a*) Trochophore of *Themiste lageniformis*, 2 days, lacking prototrochal ciliation (occasionally prototrochal cilia are weakly developed but non-functional in this species; the succeeding pelagosphera has a well-developed metatroch). (*b*) Late trochophore of *Phascolosoma varians*, 4 days. Note apical tuft and band of prototrochal cilia. (*c*) Trochophore of *Sipunculus* sp. collected from plankton. In this highly modified trochophore, prototrochal cilia surround the larva. An apical tuft is prominent anteriorly. Light micrographs of living larvae. Abbreviations: at, apical tuft; p, prototrochal cilia. Scale bars 40μm.

Developmental processes associated with the termination of the trochophore stage, regardless of developmental category, are formation or expansion of the coelom, dissolution of the prototroch, and formation of the larval cuticle either by transformation of the egg envelope or hatching from the egg coverings. An additional process, formation of the metatrochal band of cilia, occurs when metamorphosis results in a pelagosphera larva.

The coelom is formed in sipunculans, as in other protostomes, by splitting of mesoderm bands (Hatschek 1883; Gerould 1906; Rice 1967). The time of formation relative to other developmental events may vary among species, differences being related to the degree of yolk development. Coelom formation occurs in directly developing species at the time of dissolution of non-ciliated prototroch cells and elongation. In species with macrolecithal eggs and a lecithotrophic pelagosphera, coelom formation takes place simultaneously with the other events of trochophoral metamorphosis. In species with planktotrophic development the coelom is formed during the trochophore stage and is expanded at the time of metamorphosis of the trochophore to the pelagosphera stage.

Dissolution of the prototroch cells in species with macrolecithal eggs coincides with transformation of the trochophore into either the pelagosphera or the vermiform stage at the time of development of the coelom. In these species with relatively large micromeres at the eight-cell stage and large, yolk-rich prototroch cells, yolk material is released into the coelom and, when prototroch cells break down, it provides an important source of nutrition for the developing larva. Even in species with microlecithal eggs, the prototroch cells, although smaller, are relatively yolky and serve a nutritive function for the developing embryo. Granules are released into the prototrochal cavities of *Phascolosoma agassizi* and into the 'amniotic' cavity of *Sipunculus nudus* before metamorphosis of the trochophore. Thus it is evident that specialization of the prototroch to serve a nutritive function is characteristic of the phylum and is found throughout all of the developmental patterns, being most highly developed in species with lecithotrophic development.

At metamorphosis of the trochophore to the pelagosphera larva or transformation of the embryo to the vermiform stage, characteristic changes occur in the egg envelope as the cuticle of the larva or young worm is formed. The changes may involve shedding of the egg coverings or transformation of the egg envelope. Whether the egg envelope is shed or transformed has little relevance to the developmental pattern or amount of yolk. Trochophores of *Sipunculus nudus*, a species with planktotrophic development, shed the entire egg envelope along with the prototroch cells at metamorphosis of the trochophore (Hatschek 1883). Other species with lecithotrophic development, *Golfingia vulgaris* and *Phascolopsis gouldi*, shed only the egg envelope and develop a new larval cuticle (Gerould 1906). *Phascolion cryptus*, a species lacking a pelagic larval stage, loses the prototrochal and pretrochal portions and retains the post-trochal egg envelope (Rice 1975b). Egg envelopes of other species appear to be transformed into the larval cuticle either entirely or in part (Rice 1967, 1973, 1975a,b). The process includes a loss of porosity and lamellation of the envelope and, as the larva elongates, the envelope increases in elasticity. A similar transformation has been reported in many polychaetes; for example in *Phragmatopoma lapidosa* by Eckelbarger and Chia (1978).

Pelagosphera larvae

Pelagosphera larvae, whether planktotrophic or lecithotrophic share basic features (Figs 18.3, 18.4). Three regions of the body are clearly distinguished: head, metatrochal collar, and posterior trunk. Dorsally the head bears a U-shaped prototroch comprised a short, weakly-developed cilia. On the inner side of the prototroch there is a pair of dorsolateral pigmented eye spots. Ventrally the head is ciliated, both above and below the region of the mouth and may be bifurcated by a median channel leading to the mouth. Formation of the ventral head occurs at trochophoral metamorphosis when the egg envelope overlying the stomodaeum ruptures and the latter opens outward to the exterior to form the entire ventral ciliated surface of the head. At the same time the prototrochal band is modified by loss of the ventral portion and a reduction in size of cells and length of cilia. In planktotrophic pelagospheras the ciliated portion of the ventral head posterior to the mouth is expanded into the ventral lip, a lobe which usually extends out perpendicularly to the head, or, when the larva is feeding, can be flattened against a substratum. Organs associated with the mouth and presumed to be used in feeding are the buccal organ and lip gland. The buccal organ is a protrusible muscular sac which can be extended through a transverse slit just below the mouth. The lip glands (one or two pairs) are pendulous lobes hanging down into the anterior coelom and opening to the exterior through a pore on the lip.

The metatrochal collar or 'thorax' (Jägersten 1963) bears a prominent band of long, active cilia which serve as a means of locomotion. The metatrochal collar may be considerably expanded during swimming. Both head and metatroch can be retracted into the trunk.

The trunk is by far the largest region of the larval body and encloses a spacious coelom which in lecithotrophic larvae is filled with yolk granules originating from prototrochal and other cells. A terminal attachment organ is found at the posterior tip of most lecithotrophic and planktotrophic larvae although the degree of development may vary. When highly developed it is used for temporary attachment to a

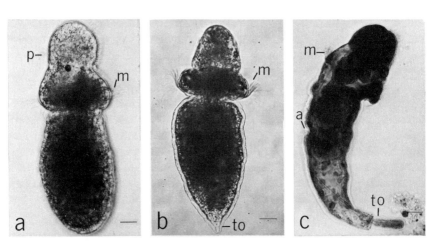

Fig. 18.3. Young pelagosphera larvae, reared from eggs in the laboratory. (*a*) *Themiste alutacea*, 3 days; a lecithotrophic larva lacking a terminal organ. (*b*) *Golfingia pugettensis*, 13 days; a lecithotrophic larva with a terminal organ. (*c*) *Phascolosoma perlucens*, about 7 days; a planktotrophic larva showing a prominent terminal organ. Light micrographs of living larvae. Abbreviations: a, anus; m, metatroch; p, prototroch; to, terminal organ. Scale bars 25 μm. (From Rice 1975a, b).

substratum, but if relatively small may be primarily sensory (Hatschek 1883).

The internal organs of a pelagosphera larva consist of a central nervous system including brain, circumoesophageal connectives and ventral nerve cord; digestive system comprised in planktotrophic larvae of oesophagus, stomach, intestine, and rectum; a pair of nephridia with internal and external openings; retractor muscles which serve to withdraw the head and metatrochal region into the trunk.

Metamorphosis of the pelagosphera to the juvenile form is marked by a loss of metatrochal cilia in both lecithotrophic and planktotrophic forms. In the former there is a transition through a vermiform stage which may last for several weeks during which time the gut is completed, yolk absorbed, introvert and body elongated, and the tentacular lobes formed. Metamorphosis of planktotrophic larvae is more rapid, the adult habitus being acquired usually within 1-3 days of loss of the metatroch (Rice 1978). The major changes occur in the region of head and metatroch and involve movement of the mouth from a ventral to a terminal position, formation of the tentacles and elongation of the introvert. Most of the organ systems of the larva are retained in the adult, including central nervous system, digestive system, retractor muscles and nephridia. During growth and differentiation of the juvenile, however, there may be some loss or specialization of organs.

Some authors have suggested resemblances of the pelagosphera larva of sipunculans to larvae of molluscs and entoprocts. Gerould (1906) pointed to the similarity of the buccal organ and lip glands of planktotrophic pelagospheras to the rudiments of the radular sac and pedal gland in chiton larvae. Jägersten (1972) further compared the lip of sipunculans to the ventral creeping lobe or foot that occurs between mouth and anus in larvae of molluscs and entoprocts. Other authors (Åkesson 1958; Rice 1973) have presumed that these organs of the mouth of planktotrophic pelagospheras may represent adaptations for feeding activity with little phylogenetic significance.

The pelagosphera of sipunculans represents a later developmental stage than that of the trochophore of polychaetes. Features considered to be more advanced in the pelagosphera are the well developed trunk, spacious coelom, and metanephridia. Evidence of segment rudiments is entirely lacking in the pelagosphera. Papillae, bearing bristles, and sometimes arranged in transverse rows, are frequently found in young sipunculan larvae, but they are generally not considered as evidence of metamerism, nor are the bristles homologous to polychaete chaetae (Åkesson 1958, 1961a; Clark 1969). The pelagosphera is seen as a specialized form without demonstrable parallels in other groups.

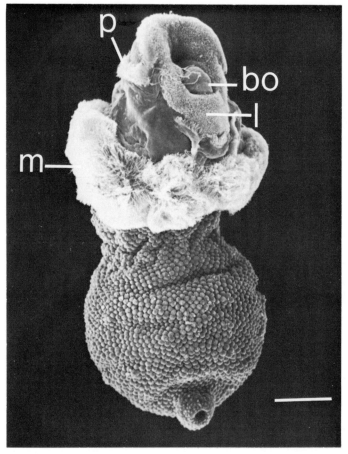

Fig. 18.4. Scanning electron micrograph of an oceanic planktrophic pelagosphera larva, *Aspidosiphon* sp. The terminal organ at the posterior extremity is retracted. Abbreviations: bo, buccal organ; l, lip; m, metatroch; p, prototroch. Scale bar 150 μm. (After Rice 1981).

Organogenesis

Development of the nervous system and the nephridia in sipunculans is of particular concern when making phylogenetic comparisons. The central nervous system in sipunculans and polychaetes has many striking similarities both in anatomy and development. The ventral nerve cord develops in both as a ventral ectodermal thickening which is typically double and segmented in annelids, but in sipunculans is generally single, median, and unsegmented. An exception is found in the development of the nerve cord in *Phascolosoma agassizi* which, although unsegmented, is double at its inception but unites to form a single cord in the older larva of two months (Rice 1973). Similarly the nerve cord in the young larva of *Golfingia pellucida* throughout much of its length, though not completely divided, is distinguished by two rows of nerve cell nuclei joined medially by the neuropile. In the nerve cords of all other larval species examined, the nerve cell nuclei form a continuous band on the ventral side of a median cord and the fibres are dorsal and central. Gerould (1906) reported a transitory metamerism in the nerve cord of developing larvae, but later repudiated this statement (Hyman 1959).

The phylogenetic significance of the development of the nerve cord has been interpreted variously by different authors. Gerould (1906) suggested that the single, unpaired nerve cord in the development of species which he studied related sipunculans to annelids such as *Polygordius*. Åkesson (1958), on the other hand, took the position that the orthogonal nervous system of turbellarians should be considered primitive for molluscs and annelids, and that the single nerve cord of sipunculans was therefore a derived feature. He considered this as evidence that sipunculans split off from the annelids after the orthogonal system was reduced. The rudimentary paired nerve cord found in the development of *Phascolosoma agassizi* is interpreted by Rice (1973) to be reminiscent of a double cord that occurred in the ancestors of extant sipunculans, thus relating sipunculans more closely to annelids.

A protonephridium, found in the trochophores of many annelids and some molluscs, is absent in sipunculans. However, it has been noted by Anderson (1966) and others that protonephridia in annelids are characteristic of planktotrophic trochophores, but not yolky larvae. As the trochophore of sipunculans is lecithotrophic or non-feeding, the significance of the lack of a protonephridium may not be as great as supposed earlier (Gerould 1906; Åkesson 1958; Hyman 1959). The larval nephridia of sipunculans are U-shaped structures opening at one end into the coelom and at the other to the exterior. Usually occurring as a pair, they are located ventrolaterally in the trunk posterior to the metatroch. These are also the nephridia of the adult sipunculan, undergoing little or no change at metamorphosis of the pelagosphera into the juvenile. Goodrich (1945) classified the paired sipunculan nephridia as mixonephridia, similar to those found in most polychaetes, but not in molluscs.

Functional morphology and behaviour of larvae

Ciliary bands of sipunculan larvae show considerable diversity in degree of development and function, even though, at least in the case of the prototroch, they are homologous. Information is not sufficient to determine whether the metatroch has the same lineage (Gerould 1906). If we compare the function of the cilia in the planktotrophic trochophore of a polychaete such as *Pomatoceros* (Segrove 1941) or *Eupomatus* (Shearer 1911) with that in the planktotrophic pelagosphera of a sipunculan we find some striking differences. In the polychaete the prototroch produces a downward current for both swimming and feeding whereas the more weakly developed metatroch evokes an upward current. The latter moves particles upward to an intermediate band of cilia (adoral) continuous with the mouth which directs food particles toward the mouth. This method of feeding, known as the opposed-band mechanism (Jägersten 1972; Strathmann 1978) does not occur in sipunculans in which the function of the cilia is quite different. Here the strongly beating metatroch functions primarily in locomotion, and the prototroch, reduced to a weak partial band on the head, is of unkown function. Feeding may be accomplished in swimming or resting pelagospheras by the movement of suspended particulate matter into the mouth by means of the short but numerous cilia on the ventral head and median ventral groove. Presumably the lip glands participate by secreting an adhesive substance for trapping food particles.

Another more specialized mode of feeding, not known to occur in polychaete larvae, involves feeding on a substratum by application of the ventral head to a surface. With the head in this position, particles can be moved into the mouth through activity of the ventral cilia alone or with the assistance of the protrusible buccal organ. This organ may also aid in swallowing larger particles, scraping material from the substratum or ejection of particles from the mouth. Feeding on a substratum can be accomplished either while the larva moves along with posterior end directed upward or, if attached by the terminal organ, the larva can bend over and graze the surrounding surfaces. This latter method of feeding is particularly characteristic of young benthopelagic pelagospheras.

The functional significance of the terminal organ may vary with age in planktotrophic pelagospheras (Rice 1978, 1981; Ruppert and Rice 1983). All young pelagospheras that have been studied in the laboratory spend much of their time near the bottom of culture dishes, either attached, resting or swimming. Yet older pelagospheras are found in the surface waters of the open ocean, presumed to survive there for many months (Scheltema and Hall 1975). In the older larvae the terminal organ may be proportionately reduced, and even when in contact with a substratum under laboratory conditions is rarely used for attachment. Young larvae, adapted for a benthic existence, do not appear able to metamorphose, but seem to represent a stage of growth and differentiation. Older larvae, on the other hand, are highly adapted for a planktonic existence, and are able to delay metamorphosis until a substratum is contacted. They serve as a highly specialized means of dispersal for the species (Rice 1978, 1981).

Evolutionary sequences of development within the phylum

Previous papers on sipunculan development have presented arguments in favour of the primitiveness of lecithotrophic development in the phylum (Gerould 1906; Åkesson 1958; Rice 1967, 1975a). But if we assume, along with other authors (Jägersten 1972; Strathmann 1978; Chia 1974; Hermans 1979) that planktotrophy is the primitive mode of development among marine invertebrates, then it would follow that planktotrophy in sipunculans must be a derived condition and the planktotrophic pelagosphera must have re-acquired the feeding habit. In view of studies made since the previous hypotheses another interpretation is now suggested.

The previous rationale for the primitiveness of lecithotrophy was based on the premise that the genus *Golfingia*, having the most primitive adult features and being the largest genus with characters from which the other genera could be derived (see Introduction) would include the most primitive developmental patterns. At the time these arguments for lecithotrophy were developed, all species of *Golfingia* for which developmental studies had been made did indeed exhibit lecithotrophic development. Now, in view of additional information on other species of *Golfingia* we know that nearly every developmental category is represented in the genus *Golfingia* (as defined by Stephen and Edmonds 1972). A directly developing species, *Golfingia minuta*, has exceptionally large, yolky eggs which it broods within its tube, epibolic gastrulation, and a lecithotrophic developmental period of 8 weeks. *Golfingia vulgaris*, *G. elongata*, and *G. pugettensis* have a lecithotropic development which includes a lecithotrophic pelagosphera of a relatively short pelagic existence of 5, 6, and 21 days respectively. Two other species studied more recently, *G. pellucida* and *G. misakiana*, have a planktotrophic pelagosphera which in the former species lives for 1 to 2 months in the laboratory before metamorphosis of the pelagosphera to the juvenile worm (Rice 1975b, 1978, 1981). The pelagosphera of *G. misakiana* has a much more prolonged pelagic existence and is commonly collected in the surface waters of the component currents of the Gulf Stream System where they are believed to live for several months (Hall and Scheltema 1975; Rice 1978).

If we assume with Jägersten (1972) and others that a planktotrophic pelagic larva is common to the development of the ancestors of metazoans, and that a planktotrophic development similar to that found in extant *Golfingia pellucida* is primitive for sipunculans, we can then speculate on the derivation of sipunculan developmental patterns. We can hypothesize that the ancestral larva common to sipunculans, annelids, and molluscs was a feeding trochophore with prototrochal and metatrochal ciliary bands and an opposed-band mechanism of feeding, and that this larva metamorphosed by processes of elongation and reduction of cilia to a benthic adult (cf. Jägersten 1972). The ancestral egg which gave rise to this larva was presumably low in yolk and covered by a thin egg envelope. In contrast, in the presumably primitive development of *G. pellucida* the eggs are moderately yolky and covered by a thick egg envelope which persists in the trochophore, so precluding feeding. The non-feeding trochophore is followed by a feeding larva, the pelagosphera, which then metamorphoses into the benthic adult. To derive the supposed primitive development of sipunculans from that of an ancestor, we can propose an increase in yolk of the egg associated with an increase in the thickness of the egg envelope. The additional yolk, mostly stored in the prototroch cells, served to provide nutriment for the developing trochopore while enclosed by the envelope. The feeding stage, thus retarded, occurred later in development when the envelope overlying the anus and stomodaeum ruptured and the gut became functional. The prototroch of the ancestral trochophore was retained in the trochophore of sipunculans for locomotion. The metatroch, used for feeding in the ancestral trochophore, did not develop in the non-feeding sipunculan trochophore, but appeared in the pelagosphera in which it assumed the function of locomotion rather than feeding. The prototroch was reduced in the pelagosphera and a new method of feeding adopted. The ciliated ventral head and lower lip, derived from the stomodaeum, along with the accessory organs of the mouth, buccal organ, and lip gland, took over the process of feeding. According to Jägersten (1972) the ciliation of the head and buccal organ were ancient adult characters, used by the adult for crawling and feeding, and were shifted to the larval stage.

If then, we assume that a development similar to that of *Golfingia pellucida*, characterized by a moderately yolky egg and a relatively short-lived planktotrophic pelagosphera, is primitive for the phylum, then two evolutionary sequences become apparent. One sequence would diverge toward a direct development with no pelagic stage, and the other toward a development with a planktotrophic larva specialized for a prolonged existence in the plankton (Fig. 18.5). The first sequence would evolve towards development such as that exhibited by *Themiste pyroides* and *Phascolion cryptus* in which the increase in yolk correlated with the large size of the non-ciliated prototroch cells and an increased significance in their nutritive function. Beginning this sequence would be a development such as that of *Golfingia vulgaris, G. pugettensis*, and *Themiste alutacea*, in which there is a trochophore as well as a lecitho-

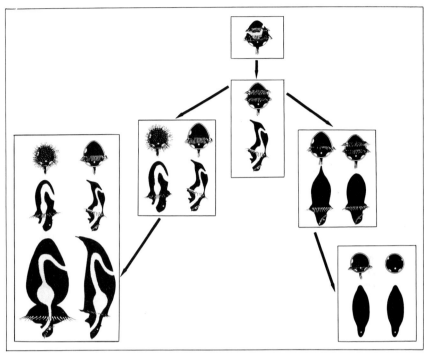

Fig. 18.5. Hypothetical scheme for evolution of developmental patterns within the Sipuncula. The ancestral development, represented at the bottom, is presumed to have had an egg with simple envelope and a planktotrophic trochophore with prototrochal and metatrochal bands of cilia. Derived from this, the primitive development of sipunculans probably had an egg with the thick egg envelope characteristic of all present-day species, a non-feeding trochophore and a newly evolved, feeding pelagosphera similar to that of *Golfingia pellucida*. From this, two evolutionary sequences are proposed, based on extant developmental patterns. The left branch shows a tendency to an increase in yolk and a reduction in pelagic stages via an intermediate lecithotrophic pelagosphera. On the right the sequence is towards a reduction in yolk and a prolongation of the pelagic phase by the large pelagospheras of the open ocean. See text for detailed explanation.

trophic pelagosphera. Intermediate in this sequence is the development of species such as *Phascolion strombi* and *Phascolopsis gouldi* which show a pelagic trochophore but no pelagosphera stage. The highly modified development of *Golfingia minuta*, occurring within the protective tube of the parent, is considered to be derived from this line. The second sequence would evolve towards a reduction in yolk in the egg, a highly modified trochophore as occurs in *Sipunculus nudus*, and the long-lived planktotrophic pelagosphera of the open ocean. The greatest modification in this sequence is found in the trochophore of *S. nudus* in which the prototroch grows over the entire larva.

All species of planktotrophic larvae studied thus far have terminal attachment organs, except those of species of *Sipunculus*. Assuming *Sipunculus* to have the most modified development, we can conclude that the terminal organ is a primitive feature for sipunculan pelagospheras. In the presumably primitve development of *Golfingia pellucida*, the terminal organ remains prominent and functional until metamorphosis into the juvenile. However, in *G. misakiana*, a species with a prolonged planktonic larva, the terminal organ is highly developed in the young pelagosphera as an organ of attachment, whereas in the older oceanic pelagosphera it is considerably reduced and rarely extended. Based on these observations, and on the adaptations previously mentioned for bottom feeding in pelagosphera larvae, it can be assumed that primitively the pelagosphera was 'bentho-pelagic', i.e. it remained close to the substratum, whereas the long-lived planktotrophic pelagosphera evolved more recently. Terminal organs are also present in some but not all lecithotrophic pelagospheras. In species in which the terminal organ is missing, such as *Themiste alutacea*, terminal glands produce a substance by which the lecithotrophic larva may adhere to a substratum. Thus, in lecithotrophic larvae the presence of the terminal organ is considered to be a derived feature.

Phylogenetic affinities

Developmental evidence for phylogenetic affinities of sipunculans has been discussed and reviewed by many authors (Hatschek 1883; Gerould 1906; Åkesson 1958; Rice 1967, 1975a; Clark 1969; Jägersten 1972). The typical spiral cleavage in sipunculans relates them by definition to annelids and molluscs as members of the Protostomia. As in other protostomes, the stomodaeum forms at the site of the blastopore and the coelom develops by schizocoely. Cleavage, gastrulation, and closure of the blastopore in sipunculans, polychaetes and many molluscs result in a trochophore larva characterized by an apical plate bearing a tuft of cilia, an equatorial band of ciliated prototroch cells, a ventral median stomodaeum below the prototroch, and lateral bands of mesoderm on either side of the gut. The anus of sipunculans does not form at the posterior blastopore as in annelids, but arises later as a new formation in a dorsal position. Another difference is that sipunculan trochophores, being enclosed by a thick egg envelope, are always non-feeding, even in species with low yolk content. A protonephridium, found in the trochophores of many annelids and some molluscs, is lacking in sipunculans, but this is true also in non-feeding larvae of polychaetes.

Resemblances of sipunculan development to that of polychaetes in particular include the similarity of prototrochal and metatrochal bands of cilia, the retention of the egg envelope in some species to form the larval cuticle, and the development of the nervous system. Although the ventral nerve cord, usually double in polychaetes, is single in the development of most sipunculans, exceptions have been noted for *Phascolosoma agassizi* and *Golfingia pellucida* which relate the sipunculans more closely to the annelids. The former has a double nerve cord in early larval stages and the latter shows an incipient splitting in the young pelagosphera larva.

Differing from annelids, sipunculans lack segmentation during both developmental and adult phases. Larval epidermal papillae appearing in pairs or transverse rows in some species are not considered indicative of segmentation as it occurs in polychaetes nor are their bristles homologous to polychaete chaetae (Clark 1969; Rice 1975a).

As in molluscs, the apical cross at the 48-cell stage of sipunculans is in the radial position, representing the sagittal and frontal planes of the future embryo rather than in the interradial position of polychaetes. The highly modified and hypertrophied prototroch of *Sipunculus nudus* has been proposed as homologous to the velum of molluscs. Gerould (1906) noted that the prototroch is shed in this species as is the velum of molluscs, but as pointed out by Åkesson (1958) the velum is shed at the end of the pelagic stage and the prototroch of *Sipunculus* at an early stage of development after which another ciliary band, the metatroch, takes over the function of locomotion for the larva. Less convincing is the comparison of the lip of the planktotrophic pelagospheras to the foot of veliger larvae and the supposed homology of the lip gland and buccal organ of pelagospheras to the pedal gland and radular sac of veligers of chitons.

Trochophores of sipunculans have been compared to those of both polychaetes and molluscs. Gerould (1906) noted the resemblance of the trochophore of *Golfingia vulgaris* to that of the polychaete *Amphitrite* and also the likeness to trochophores of *Chiton, Patella*, and other molluscs. However, there is no larval stage of either polychaetes or molluscs that is comparable to the pelagosphera—a specialized larval form unique to the Sipuncula.

The differences in developmental features from both molluscs and annelids suggest that the Sipuncula is a distinctive group within the Protostomia. The lack of segmentation and the retention of larval organs in the adult as well as the simple organization of the adult are assumed to be primitive features. Thus it is concluded that sipunculans are a primitive group derived from the annelidan-molluscan stem and closely related to the common ancestor of annelids and molluscs.

Acknowledgements

The able technical assistance of Julianne Piraino and Hugh Reichardt is appreciated. Carolyn B. Gast made valuable contributions to the artwork, and June Jones provided efficient secretarial services. This is contribution number 121 of the Smithsonian Marine Station at Link Port.

References

Åkesson, B. (1958). A study of the nervous system of the Sipunculoideae with some remarks on the development of the two species *Phascolion strombi* Montagu and *Golfingia minuta* Keferstein. *Unders. Oresund* 38, 1-249.

—— (1961a). The development of *Golfingia elongata* Keferstein (Sipunculidea) with some remarks on the development of neurosecretory cells in sipunculids. *Ark. Zool.* 13, 511-31.

—— (1961b). Some observations on *Pelagosphaera* larvae. *Galathea Rep.* 5, 7-17.

Amor, A. (1975). El desarrollo de *Themiste petricola* (Amor, 1964) (Sipuncula, Golfingiidae). *Physis* A 34, 357-70.

Anderson, D. T. (1966). The comparative embryology of the Polychaeta. *Acta zool., Stockh.* 47, 1-42.

Baltzer, F. (1931). Sipunculida. In *Handbuch der zoologie* (eds W. Kükenthal and T. Krumbach) Vol. 2, Pt 9, pp. 15-16. De Gruyter, Berlin.

Chia, F. S. (1974). Classification and adaptive significance of developmental patterns in marine invertebrates. *Thalassia jugosl.* 10, 121-30.

Clark, R. B. (1969). Systematics and phylogeny: Annelida, Echiura, Sipuncula. In *Chemical zoology* (eds M. Florkin and B. T. Scheer) Vol. 4, pp. 1-68. Academic Press, New York.

Cuvier, G. (1930). *Le regne animal distribué d'après son organisation, pour servir de base a l'histoire naturelle des animaux et d'introduction a l'anatomie comparée*, Ser. 2, Vol. 3, pp. 1-244. Deterville, Paris.

Damas, H. (1962). La collection de *Pelagosphaera* du 'Dana'. *Dana Rep.* 59, 1-22.

Eckelbarger, K. J. and Chia, F. S. (1978). Morphogenesis of larval cuticle in the polychaete *Phragmatopoma lapidosa*. *Cell Tissue Res.* 186, 187-201.

Fisher, W. K. (1952). The sipunculid worms of California and Baja California. *Proc. U.S. natn. Mus.* 102, 371-450.

Franzén, Å. (1956). On spermiogenesis, morphology of the spermatozoan, and biology of fertilization among invertebrates. *Zool. Bidr. Upps.* 31, 355-482.

Gerould, J. H. (1903). Studies on the embryology of the Sipunculidae, I. The embryonal envelope and its homologue. *Mark Anniversary Volume* (ed. G. H. Parker) No. 22, pp. 439-52. New York.

—— (1906). Studies on the embryology of the Sipunculidae, II. The development of *Phascolosoma*. *Zool. Jb. Abt. Anat.* 3, 77-162.

—— (1913). The sipunculids of the eastern coast of North America. *Proc. U.S. natn. Mus.* 44, 373-437.

Goodrich, E. S. (1945). The study of nephridia and genital ducts since 1895. *Q. Jl microsc. Sci.* 86, 115-392.

Hall, J. R. and Scheltema, R. S. (1975). Comparative morphology of open-ocean pelagosphaera. In *Proceedings of the international symposium on the biology of the Sipuncula and Echiura* (eds M. E. Rice and M. Todorovic) Vol. 1, pp. 183-97. Naucno Delo Press, Belgrade.

Hatschek, B. (1883). Ueber Entwicklung von *Sipunculus nudus*. *Arb. zool. Inst. Univ. Wien* 5, 61-140.

Hermans, C. O. (1979). Polychaete egg sizes, life histories, and phylogeny. In *Reproductive ecology of marine invertebrates* (ed. S. E. Stancyk) pp. 1-9. University of South Carolina Press, Columbia, SC.

Hérubel, M. A. (1907). Recherches sur les Sipunculides. *Mém. Soc. zool. Fr.* 20, 107-419.

Hyman, L. H. (1959). *The invertebrates*. Vol. 5, *Smaller coelomate groups*. McGraw-Hill, New York.

Jägersten, G. (1963). On the morphology and behaviour of Pelagosphera larvae (Sipunculoidea). *Zool. Bidr. Upps.* 36, 27-35.

—— (1972). *Evolution of the metazoan life cycle*. Academic Press, London.

Lamarck, J. P. B. A. de M. (1816). *Histoire naturelle des animaux sans vertébrés* Vol. 3, pp. 76-9. Verdière, Paris.

Lang, A. (1888). *Lehrbuch der vergleichenden Anatomie der wirbelosen Theire*. Gustav Fischer, Jena.

Mead, A. D. (1879). The early development of marine annelids. *J. Morph.* 13, 229-326.

Meyer, E. (1904). Theoretische Betrachtunger über die ersten Anfänge des ambulacralen Wassergefässsystems der Echinodermen und die Abstammung ihrer bilateralen Vorfähren. *Zool. Jb. Abt. Anat.* 21, 339-78.

Murina, V. V. (1977). *Sipunculid marine worms of the arctic and boreal waters of Eurasia*. Nauka Publishing House, Leningrad (in Russian).

Nichols, D. (1967). The origin of echinoderms. *Symp. zool. Soc. Lond.* 20, 209-29.

Pickford, G. E. (1947). Sipunculida. *Encyclopedia Britannica* 20, 719-23.

Quatrefages, A. de (1847). Mémoire sur l'échine de Gaertner. *Annls Sci. nat., Zool.* Ser. 3, 221.

Rice, M. E. (1967). A comparative study of the development of *Phascolosoma agassizii*, *Golfingia pugettensis*, and *Themiste pyroides* with a discussion of developmental patterns in the Sipunculida. *Ophelia* 4, 143-71.

—— (1973). Morphology, behavior, and histogenesis of the pelagosphera larva of *Phascolosoma agassizii* (Sipuncula). *Smithson. Contr. Zool.* 132, 1-51.

—— (1975a). Observations on the development of six species of Caribbean Sipuncula with a review of development in the phylum. In *Proceedings of the international syposium on the biology of the Sipuncula and Echiura* (eds M. E. Rice and M. Todorovic) Vol. 1, pp. 141-60. Naucno Delo Press, Belgrade.

—— (1975b). Sipuncula. In *Reproduction of marine invertebrates* (eds A. Giese and J. Pearse) Vol. 2 , pp. 67-127. Academic Press, New York.

—— (1978). Morphological and behavioral changes of metamorphosis in the Sipuncula. In *Settlement and metamorphosis of marine invertebrate larvae* (eds. F. S. Chia and M. E. Rice) pp. 83-102. Elsevier-North Holland Press, New York.

—— (1981). Larvae adrift: patterns and problems in life histores of sipunculans. *Am. Zool.* 21, 605-19.

Ruppert, E. E. and Rice M. E. (1983). Structure, ultrastructure and function of the terminal organ of a pelagosphera larva (Sipuncula). *Zoomorphologie* 102, 143-63.

Scheltema, R. S. (1975). The frequency of long-distance larval dispersal and the rate of gene-flow between widely separated populations of sipunculans. In *Proceedings of the international symposium on the biology of the Sipuncula and Echiura* (eds M. E. Rice and M. Todorovic) Vol. 1, pp. 199-210. Naucno Delo Press, Belgrade.

—— and Hall, J. R. (1975). The dispersal of pelagosphera larvae by ocean currents and its relationship to geographical distribution of sipunculans. In *Proceedings of the international symposium on the biology of the Sipuncula and Echiura* (eds M. E. Rice and M. Todorovic) Vol. 1, pp. 103-15. Naucno Delo Press, Belgrade.

Sedgwick, A. (1898). *A student's textbook of zoology.* Vol. 1. MacMillan, New York.

Segrove, F. (1941). The development of the serpulid *Pomatoceros triqueter* L. *Q. Jl. microsc. Sci.* 56, 543-90.

Selenka, E. (1875). Eifurchung und Larvenbildung von *Phascolosoma elongatum* (Kef.). *Z. wiss. Zool.* 25, 442-50.

——, Man, J. G. de, and Bülow, C. (1883). Die Sipunculiden. In *Reisen im archipel der Philippenen* (ed. C. F. Semper) Vol. 4, Pt 1, pp. 1-131. C. W. Kriedel, Wiesbaden.

Shearer, C. (1911). On the development and structure of the trochophore of *Hydroides uncinatus* (*Eupomatus*). *Q. Jl microsc. Sci.* 56, 543-90.

Stephen, A. C. (1964). A revision of the classification of the phylum Sipuncula. *Ann. Mag. nat. Hist.,* Ser. 13, 7, 457-62.

—— and Edmonds, S. J. (1972). *The phyla Sipuncula and Echiura.* British Museum (Natural History), London.

Strathmann, R. R. (1978). The evolution and loss of feeding larval stages of marine invertebrates. *Evolution* 32, 894-906.

Williams, J. (1972). Development of a rock burrowing sipunculid inhabiting stony coral. *Am. Zool.* 12, 723 (Abstr.).

19. Functional constraints in the evolution of the Annelida

C. METTAM

Department of Zoology, University College, Cardiff, UK

Abstract

The bodyplan of the Annelida is distinguished by a coelomic cavity surrounding the gut, which is supported by transverse septa and mesenteries. The septa divide the body into segments, often visible externally as annulations of the body wall or from the serial occurrence of cilia or chaetae projecting through the cuticle. The functional correlates of this organization are well established for large worms, less so for small, meiofaunal ones.

Several scenarios have described functional stages in the evolution of the annelid state. Attempts to define a stem form for the phylum which is polychaete-like or oligochaete-like have been unsatisfactory either because the selective advantage proposed between one stage and another is mechanically unsound or because the steps lack functional continuity. The protoannelid could have been an interstitial form, developing chaetae for grip during simple contraction, or a pelagic larva with chaetae as flotation devices. It is possible that increasing size, associated with increasing muscular activity, has brought about the characteristic features of the phylum as adaptations to burrowing, sinuous writhing or parapodial swimming modes of locomotion. No single, unique sequence of events can be claimed to explain the emergence of annelid characteristics from a simpler, acoelomate organization.

Introduction

The phylum Annelida holds together and is set apart from other soft-bodied worms by being both coelomate and segmented, the segments characteristically with protruding chaetae. In traditional phylogeny it occupies a position between unsegmented worms and higher invertebrates, notably the Arthropoda, with which annelids are sometimes combined to form the Articulata.

Annelids exist in a splendid diversity of size and form. For this discussion an arbitrary distinction will be made between macro-annelids and meio-annelids. A further category of mega-annelids could be invented to accommodate the giants of the phylum for which respiratory demands may impose an upper constraint on size (Alexander 1971). The classes Hirudinea, Oligochaeta, and Polychaeta are conventionally recognized within the macro-annelids. The Hirudinea are widely accepted as a derivation from oligochaete stock and both may be included in the taxon Clitellata. The relationship between the major classes Oligochaeta and Polychaeta is ambiguous: they can be difficult to tell apart although they normally have distinctive chaetae and reproductive characters. The Myzostomaria are specialized ectoparasites, sometimes referred to the Polychaeta.

Among the smaller annelids, morphological distinctions between the classes, and indeed between the Annelida and other phyla, are obscured (e.g. Giere and Riser 1981; Rieger 1980). Many meio-annelids are characterized by structural simplicity and often show only rudimentary annelidan features; indeed some were first identified as Turbellaria. Hermans (1969), reviewing a century of debate on the phylogenetic significance of this assemblage of worms, concluded that some are oligochaetes (Aeolosomatidae) and some are referable to known polychaete families, leaving a cluster of families with polychaete features for which the name Archiannelida (reflecting their traditional position as primitive annelids) is retained (see Westheide, this volume).

Probably no discussion of annelid classification avoids a similarly prejudiced terminology; annelids may be called 'primitively simple' or 'secondarily simplified' according to the need to bolster a particular view of phylogeny, rather than using neutral, descriptive terms. In fact nothing definite can be said either about the precise relationships of the Annelida to other phyla or about relationships within the phylum. The fossil record gives little guidance on the history of annelids, beyond demonstrating their immense antiquity. Fauchald (1974) viewed polychaete evolution as a Precambrian radiation and considered that perhaps the only evolutionary trend in the class has been cephalization, 'increasing the morphological distance from the ancestral polychaete'. Brinkhurst (1982) reviewed classification schemes for the Oligochaeta, which follow a traditional emphasis on the form and disposition of the reproductive organs in different segments. He argued that the stem form had a plesioporous, octogonadal reproductive system and is represented by the living family Haplotaxidae, a widely distributed group of moderately sized worms with earthworm-like chaetae.

Generations of zoologists have agonized over the transformations required in various organ systems to produce the annelid state from a simpler precursor and sought to identify primitive taxa. Much of the dispute, within the Oligochaeta and the Polychaeta alike, has turned on the position of the meio-annelids: are they primitively simple or secondarily simplified from macro-annelids? Species can be arranged in a linear phyletic series, or a more complex branching one, based on morphological similarity. The problem is to determine the direction in which to read such a series.

Clark (1964) proposed a resolution of this problem in his book *Dynamics in Metazoan Evolution* and subsequently in a series of reviews (Clark 1981 and references therein). He proposed that the coelom arose as a burrowing adaptation, that segmentation arose as an adaptation for sustained burrowing and that a macro-oligochaete body form best represents the ancestral annelid condition. The theory provides the basis for Fauchald's oligochaetoid ancestral polychaete (Fauchald 1974) and Brinkhurst's ancestral oligochaete (Brinkhurst 1982). Clark's argument is succinctly and powerfully expressed in the following extract:

Taking the Metazoa as a whole it cannot be too strongly emphasised that the only consistent correlation between the condition of the coelom and of segments and function is in their mechanical properties: the coelom as a hydrostatic skeleton and its subdivision by septa as a mechanical device limiting the transmission of hydrostatic pressure in the system.

These are mechanical requirements in macroscopic animals: they are totally irrelevant in small animals. It follows that no theory which postulates that the earliest metamerically segmented worms were very small can account for the existence of a patent coelom, and still less of segments, in them. (Clark 1969)

The influence of this theory has been due to the correlation of structure with function and the consequent identification of selective advantages in key annelid characters; the extensive literature survey (and catchy title) of the book which made accessible to modern readership a debate that had become unfashionable; the implications of the theory which showed the difference between annelid segmentation (primarily a division of the coelom) and vertebrate segments (a structuring of the musculature for swimming). Although Clark has considered problems associated with this theory, his most recent evolutionary scenario is not fundamentally changed (Clark 1981).

Two different views of annelid evolution might also claim to follow the stricture quoted above. According to Bonik, Grasshoff, and Gutmann (1976) the primary reason for the (simultaneous) origin of the coelom and segments is to improve a mechanism of locomotion by lateral undulation. Canals in the body matrix give rise to serially repeated coelomic chambers. Parapodial lobes develop to increase grip (the authors argue that these should be ventral but that their lateral position is the result of a compromise between several functions). The lobes become stiffened with chaetae as an improvement of the worm's swimming ability and there is the protoannelid: a versatile animal that can enjoy all annelid modes of locomotion and which is essentially a polychaete. Oligochaetes are derived by adaptation to peristaltic burrowing, not possible in the early stages of this scenario.

Pilato (1981) considered the protoannelid to have a complex, turbellarian-like arrangement of musculature. The change, with increasing size, from ciliary to muscular locomotion meant that 'with the aid of the cuticle, coelom and metameres, it could perform the various movements even better than the Turbellaria and Nemertea' (Pilato 1981). The Hirudinea therefore retain the primitive musculature and only keep their association with the Oligochaeta by diverging very close to the ancestral stem.

Methods and Approaches

Traditionally, zoologists have supposed the morphology of living animals to be a combination of their ancestral characters and their adaptive specialities. It is possible to disregard adaptation and retain a set of precepts for phylogenetic modelling (embracing cladistics, punctuated equilibria, vicariance biogeography). In this model, speciation events (which may be random) provide raw material for selection. Evolutionary pathways are reconstructed as dichotomous branching patterns (hence clades). Animals are the residuals of their ancestry and functional morphology can be irrelevant. A difficulty with this view is its demand that homologous and derived characters (and, by implication, primitive ones) in an animal must be identified with confidence. A parsimonious cladogram tends to reconstruct a monophyletic evolution. Thus it is often assumed that the whole assemblage of the Articulata is monophyletic. Why such evolutionary changes with clear functional advantages should have happened only once is not explained.

Equally, it is possible to follow an alternative adaptationist philosophy which disregards ancestry. This is the initial hypothesis of the experimental functional morphologist. Its assumptions are that animals are not relics of previous speciation events but are the best adaptation possible: there are no neutral characters. Morphology is under constant surveillance by natural selection; enduring features within a species population must be maintained by selection even if their selective advantage is not apparent. In this philosophy no character is necessarily either primitive or advanced; the terms become irrelevant. Functional demands lead to unlimited convergence and animals exist as organizational grades rather than clades. Evolution would be expected to proceed between grades on a broad front rather than by monophyly; for a discussion around this point see Gould and Lewtontin (1979) and reply by Cain (1979). Evolutionary inferences can be made from both these approaches but since they depend on different assumptions, there is a risk, in blending the two, of presenting an argument that shifts its base expediently from one set of premises to the other. The rest of this paper is concerned with evolutionary inferences from functional morphology. Relating structure to function is not without its own difficulties and a limited survey may give a misleading impression of the constraints imposed by structure on habit: domestic goats in north Africa and the Gray Fox of North America have tree-climbing habits but are not morphologically different from their non-scansorial relatives.

Since we may assume a general pattern of increasing complexity in the course of evolution, upon which reverses may be imposed, it is perfectly proper to establish grades or types of organization and pathways of transformation between them. There are few practical rules for establishing evolutionary pathways:

(i)   Totally new organisms do not arise out of the blue, even sister groups must be similar. This is a matter of degree and significant exceptions may arise through developmental causes, e.g. heterochrony and homeoesis (Raff and Kaufman 1983).

(ii)   The terminus of a phyletic series of existing species must be a derived, not primitive, state if it disobeys rule (i), since it can have no ancestor except one within the same series. Mettam (1971) suggested that *Aphrodita* held this position. Storch (1968), who held that the complex musculature of *Aphrodita* was primitive, did not explain how it came into being.

(iii)   Functional continuity must exist between stages. Manton (1977) demonstrated that leg mechanisms of various Hexapoda could not be derived one from another without first dismantling their specialized joint articulations, so any common ancestor must have possessed unjointed, lobopodal limbs and each hexapod group is a separate evolutionary line. Boudreaux (1979) reworked the same ground from the point of view of a monophyletic cladist but disappointingly did not confront the critical issue—is there functional continuity or not?

Functional continuity is less easy to determine in annelids than in arthropods, which exhibit a precision of articulation in their joints and in their locomotory gaits. We might also suppose the musculature of soft-bodied worms to be labile and of limited taxonomic value. However, the musculature of the stomodeum (Dales 1962, 1977) and the body wall musculature (Storch 1968) appear to be determined more by taxonomic position than by habit and habitat in the Polychaeta. Unfortunately we can say little more than this since the functional morphology of polychaete musculature is little known and poorly understood. I shall therefore return to a consideration of the position of meio- and macro-annelids as grades of organization in an evolutionary framework.

Critique of theories

1. The meio-annelids as ancestors

Hermans (1969), excluding the oligochaetous archiannelid, *Aeolosoma*, concluded that 'the archiannelids are neither primitive nor degenerate but are polychaetes primarily adapted for interstitial life'. Indeed, the adaptationist philosophy expects that their small size, poorly defined segmentation, poorly formed septa and coelom, and other features best fit them for life among the sand grains. Some features such as a well developed surface ciliation, are clearly explicable in these terms. For others, the adaptive value is not readily apparent, which may be a limitation of the zoologist, not the worm. If the feeble coelom and segmentation are functional adaptations, they could have originated in these small worms.

Not all adaptations are morphological of course and Gould (1977) pointed out the adaptive value of precocious sexual development (progenesis) in *r*-selected, opportunist species. The short-term advantage of high reproductive rate achieved by a rapid turnover of generations sacrifices morphological adaptation (slow, progressive adaptation in evolution characterizes K-selection) and gives a justification for the view of archiannelids as 'reduced' or 'degenerate'—their morphological adaptation is poor because their real adaptation lies in an *r*-selected life history strategy in an unstable habitat (Gould 1977). It has been suggested many times that the archiannelids are paedomorphic and progenesis provides a logic for these views.

If meio-annelids are morphological degenerates, their status as an ancestral grade is diminished. But are they morphologically well adapted or not? If *r*-selection is the dominant force there should be poor correlation between morphology and habit or niche specialization and this is open to practical investigation, not merely speculation. If not maladapted degenerates, the meio-annelids can provide a functional continuity to the macro-annelids by increase in size and complexity.

The claim that meio-annelids do not represent a primitive grade of organization (Clark 1969) depends on the truth of the contention that the coelom and segmentation are irrelevant features for such small animals. This is based retrospectively on evidence that they are effective in large ones, but the converse is not proven. Gray (1969) argued for a functional role for the structural organization of the archiannelid *Saccocirrus* which uses muscles, chaetae and mucous glands in its locomotion and there does seem to be a correlation between coelomic development and muscular activity in small annelids (Fransen 1980).

The notion that the coelom necessarily arose in large animals cannot be upheld even if its primary function was mechanical, as a hydroskeleton. Other functions served by the body cavity, whether mechanical (e.g. as a lubricant between the gut and body wall) or physiological (e.g. as a transport system or as a store for metabolites or gonads) may have been equally important in its evolution.

2. Macro-annelids as ancestors

Clark (1964) envisaged evolutionary changes as a progressive improvement of burrowing ability. With increasing muscular activity came the coelom. The condition compares with the present day Echiura. Next the coelom is divided by transverse muscular septa into hydraulic units, resembling the earthworm condition, where septa localize muscular forces. The observation that septate coelomates are more active burrowers then either unsegmented coelomates or annelids that have incomplete septa and live semi-sedentary lives, is the basis for the assertion that segmentation has its primary function in burrowing.

Corroborative evidence comes from Hirudinea which, if they are derived from Oligochaeta, must have exchanged their burrowing habit for surface living and a characteristic looping locomotion (a form of peristalsis) and correspondingly lost all functional need for a spacious coelom, septa, and chaetae (Clark 1964). These are lacking in typical leeches.

There is an exception to prove (i.e. test) this rule: the erpobdellid leech *Trocheta subviridis* is almost as large as the oligochaete *Lumbricus terrestris* and both burrow deeply in heavy clay soil, penetrating subsoils. The leech burrows by multiple, retrograde peristalsis, like the earthworm, using especially the front third of the body where there are no coelomic chambers and no septa: the long, triradiate oesophagus is narrow and the body is almost 'solid muscle'. The absence of oligochaete features does not prevent this annelid from being a most effective burrower.

A more direct obstacle to Clark's hypothesis is the lack of functional continuity between an open coelom and one divided by septa into separate hydraulic units: to be effective the septa must be complete, muscular and leakproof. The evolutionary gap could be bridged if the segment formation is part of an ancestral developmental and growth process persisting into adult morphology, but then the primary reason for the existence of segments would be embryological.

Functional continuity is maintained if the segmented condition with septal barriers to fluid pressure changes originated through serial cavitation in a mass of tissue which would already have a similar damping effect. Multiple retrograde peristaltic waves occur in acoelomate nemertean worms, indicating local isolation of pressure differences. Chapman (1975), however, was of the opinion that 'the selective advantage of a body filled with liquid over a mass of deformable cytoplasm probably lies, not in its being better able to transmit pressure, but in needing less force to deform it.' A large body cavity may then be associated with the need to operate, at low pressures, gentler activities than driving a path through resistant soil, such as irrigatory peristalsis.

Muscular movements that pass along the body in waves are said to be retrograde if the leading end (normally the head) reaches forwards to a new position and the wave of longitudinal elongation passes back down the body. If, instead, the trailing end is pulled forward by longitudinal contraction, the advancement of the body begins posteriorly and the wave of shortening moves forwards as a direct wave. In the time since Clark's theory was presented in detail (Clark 1964) more information has become available on peristalsis, recently summarized by Elder (1980). Direct peristalsis (tail to head) which Clark (1964) thought inappropriate for burrowing, is now known to be used by unsegmented worms and those polychaetes in which septa are not barriers to bulk movement of fluid. The open coelom permits locomotion by both direct and retrograde waves as well as stationary irrigation, while peristaltic irrigation is possible in fully septate worms only with special parapodial modifications (Mettam 1969). The greater versatility of peristalsis in aseptate worms is well illustrated in the echiuran *Bonellia viridis* which readily adjusts its irrigatory waves to the variable dimensions of its burrow and is capable of squeezing through minute holes and crevices (Schembi and Jaccarini 1977). In other worms, lack of septa permits high pressures to be developed and an even application of pressure to the burrow in homogeneous sediments (Seymour 1970); it is also an adaptation for life in soft, flocculent sediments where direct waves and low fluid pressures are appropriate (Elder 1980). It is no longer possible to regard the absence of septa as a simple condition or as a degeneration associated with a less active habit. It is a more versatile body form with its own ecological potential, adopted by many polychaetes which acquired septa in their normal growth process: there is no reason to think of the unsegmented coelomates as primitive, and no need to interpolate this organization into the evolution of the annelids.

Since the evolving annelids may bypass the unsegmented coelomate stage there is no reason to assume a simple arrangement of complete muscular layers in the body wall and no reason to suppose that the earliest segmented and septate stage resembled the burrowing oligochaetes. Few, if any, Precambrian trace fossils can be attributed to large peristaltic burrowers (Runnegar 1982).

Bonik *et al.* (1976) consider the Polychaeta, with elaborated parapodia, to be the ancestral annelids but their supposed origin of parapodial lobes, without chaetae, seems to be a device to incorporate a preconceived lobopod ancestor for all the Articulata and is difficult to justify on grounds of functional morphology. The subsequent strengthening of parapodia with chaetae as an aid to swimming is also unlikely to have occurred (Clark 1979): stiffened parapodia and body undulations are more likely to work against each other and produce opposing thrusts.

Reconstructing evolutionary pathways in the Annelida from functional morphology

In this final discussion I shall suggest alternative ways to derive the annelid condition from a simpler acoelomate one by gradual adaptation with functional continuity, ignoring for the moment the possibilities for evolutionary diversions contained in the developmental process (Raff and Kaufman 1983). The locomotion of few living annelids has been described in detail but some of the properties of a worm-like animal can be predicted from a model of the worm as a cylinder (Chapman 1950).

Chapman (1950) emphasized the mutual antagonism of circular and longitudinal muscles in cylindrical worms but this does not necessarily represent a primitive organization. The relative importance of each muscle layer depends on its thickness as measured in a transverse section of the worm, assuming that a muscle produces a force proportional to its cross-sectional area. Since stresses in the circumference of a cylinder are twice the longitudinal stresses (Chapman 1950) we might expect the circular muscle to be the thicker of the two layers but the opposite is usually the case. Circular muscles are often poorly developed or even absent in small worms and the longitudinal muscle is usually much thicker, even in burrowing earthworms, indicating its greater importance in exerting forces against the ground.

These differences in stresses in the circumferential and longitudinal dimensions apply to a cylinder of membrane thickness. If the muscular body wall is thick, the difference is accentuated by the greater effective radius for the circular muscle layer which encloses the other muscle tissues as part of the hydraulic skeleton. A balance between the antagonistic muscle pair is more easily achieved with a thin body wall and a spacious cavity, such as occurs in many irrigating worms. In the primitive annelid, longitudinal muscles may have been the main locomotory agent or simply a device for retracting the body quickly.

An earthworm such as *Lumbricus* is a classic realization of a segmented cylindrical animal. The front end is advanced by circular muscle contraction as retrograde waves pass down the body and, indeed, most of the locomotory force is developed by the anterior segments: the body of earthworms is partly dragged along behind the advancing end (Piearce 1983). Body segments that are merely shunted forwards to keep pace with the advancing front end can largely dispense with circular muscles since the segments will be stretched and pulled forwards by the shortening of segments further along the line. A feeble development of circular muscles throughout the body would, however, require some alternative mechanism for advancing the leading end. In small worms this could be achieved by ciliary creeping. Larger annelids, using muscular movements could progress by writhing undulations, or parapodial stepping, or hydraulic eversion of anterior segments or proboscis, without necessarily putting any demand on the circular muscles. *Polyphysia crassa*, which advances its body by direct waves, excavates a path with the anterior parapodia (Hunter, Moss, and Elder 1983). In some other polychaetes, for example *Scololepis*, the thoracic segments have recurved chaetae which act as scoops, sometimes in conjunction with lateral body undulations, to dig through sediment and the abdominal segments keep pace by retrograde peristalsis.

Let us consider a simple protannelid wriggling through a coarse sediment with movements that approximate to sinusoidal waves. Gray (1939) pointed out that in direct, lateral waves of undulation, used by many polychaetes, each section of the body contacts the ground in turn and is temporarily stationary where the crest of the wave pushes out laterally. Lateral attachment organs would therefore be appropriately developed at intervals along the body. If the attachment is by protruding chaetae, contact is improved by appropriate tilting of the chaetae and, with further elaboration, the chaetae can contribute directly to the forward thrust of the body. Unlike the retrograde lateral undulations of vertebrates, which are elaborated at the expense of limbs in lizards and snakes, the direct undulations of polychaetes would have encouraged the progressive evolution of parapodia (Mettam 1971). There is an unbroken functional continuity in the evolution of parapodia associated with direct waves of muscle contraction and the elaboration of the coelom could be associated with the increasing importance of muscular body movements.

A possible origin for the chaetae in annelids can be traced through the existing meio-annelids where chaetae may be present or absent even in members of the same genus. Several kinds of gland are found in the meio-annelids (Martin 1978). Apart from the secretions associated with the ciliary tracts, there are adhesive glands (with associated releaser secretions) in the pygidium and, in some species, also in segmental bands. Perhaps in some ancestral form, the adhesive glands, needed in the fast flow of interstitial water, were replaced by chaetae as mechanical attachments, either associated with axial contraction in peristalsis or the ditaxic contraction of undulating progress.

If the annelid features of segmentation and enlargement of the coelomic cavity accompanied parapodial development, they could have arisen in a swimming worm in which parapodia developed as the primary means of locomotion. Polychaete larvae often have special chaetae as flotation devices, subsequently replaced by adult chaetae. If serially repeated larval chaetae can produce locomotory thrust, perhaps augmented by body undulations such as those seen in phyllodocid larvae, a pelagic polychaete would result by appropriate reorganization of the skeletomuscular system. The pelagic Alciopidae have a musculature like that of the Phyllodocidae, suggestive of a benthic ancestry, but the Tomopteridae, with lobopodal parapodia, have a unique system of dorsal and ventral oblique muscles bending the parapodia and betray no trace of a benthic origin.

The burrowing habit, leading to a particular oligochaete form with short chaetae, makes a convenient but not essential ancestor for the Oligochaeta. Polychaetes do not need to pass through this stage, indeed the extreme muscular simplification of the oligochaetes makes it difficult to derive either Polychaeta or Hirudinea from them (Pilato 1981). It is not possible to exclude the meio-annelids from their role as an ancestral grade of organization, and the characteristics of the phylum Annelida can be seen as emergent features in worms that straddle the meio-/macrofaunal boundary.

References

Alexander, R. McN. (1971). *Size and shape*. Edward Arnold, London.

Bonik, K., Grasshoff, M., and Gutmann, W.F. (1976). Die Evolution der Tierkonstruktionen IV. Die Entwicklung der Ringelwurmer und ihre Aufgliederung in Vielborster, Wenigborster und Egel. *Naturw. Mus. Frankf.* 106, 303-16.

Boudreaux, H. B. (1979). *Arthropod phylogeny with special reference to insects*. John Wiley, New York.

Brinkhurst, R. O. (1982). Evolution in the Annelida. *Can. J. Zool.* 6, 1043-59.

Cain, A. J. (1979). Introduction to the general discussion. *Proc. R. Soc.* B205, 599-604.

Chapman, G. (1950). Of the movement of worms. *J. exp. Biol.* 27, 29-39.
—— (1975). Versatility of hydraulic systems. *J. exp. Zool.* 194, 249-70.
Clark, R. B. (1964). *Dynamics in metazoan evolution: the origin of the coelom and segments.* Clarendon Press, Oxford.
—— (1969). Systematics and phylogeny: Annelida, Echiura, Sipuncula. In *Chemical zoology* (eds M. Florkin and B. T. Scheer) Vol. 4, pp. 1-68. Academic Press, New York.
—— (1979). Radiation of the Metazoa. In *The origin of the major invertebrate groups* (ed. M. R. House), Systematics Association Special Vol. 12, pp. 55-102. Academic Press, London.
—— (1981). Locomotion and the phylogeny of the Metazoa. *Boll. Zool.* 48, 11-28.
Dales, R. P. (1962). The polychaete stomodeum and the interrelationshiops of the families of Polychaeta. *Proc. zool. Soc. Lond.* 1939, 289-328.
—— (1977). The polychaete stomodeum and phylogeny. In *Essays on polychaetous annelids in memory of Dr Olga Hartman* (eds D. J. Reish and K. Fauchald) pp. 525-45. Allan Hancock Foundation, Los Angeles.
Elder, H. Y. (1980). Peristaltic mechanisms. In *Aspects of animal movement* (eds H. Y. Elder and E. R. Trueman) No. 5, pp. 71-92. Cambridge University Press, Cambridge.
Fauchald, K. (1974). Polychaete phylogeny: a problem in protostome evolution. *Syst. Zool.* 23, 493-506.
Fransen, M. M. (1980). Ultrastructure of coelomic organisation in annelids: 1. Archiannelids and other small polychaetes. *Zoomorphologie* 95, 235-49.
Giere, O. W. and Riser, N. W. (1981). Questidae—polychaetes with oligochaetoid morphology and development. *Zool. Scr.* 10, 95-103.
Gould, S. J. (1977). *Phylogeny and ontogeny.* Harvard University Press, Cambridge, Mass.
—— and Lewontin, R. C. (1979). The spandrells of San Marco and the Panglossian paradigm: a critique of the adaptationist programme. *Proc. R. Soc. Lond.* B205, 581-98.
Gray, J. (1939). Studies in animal locomotion VIII. *Nereis diversicolor. J. exp. Biol.* 16, 9-17.
Gray, J. S. (1969). A new species of *Saccocirrus* (Archiannelida) from the west coast of North America. *Pacif. Sci.* 23, 238-51.
Hermans, C. O. (1969). The systematic position of the Archiannelida. *Syst. Zool.* 18, 85-192.
Hunter, R. D., Moss, V. A., and Elder, H. Y. (1983). Image analysis of the burrowing mechanism of *Polyphysia crassa* (Annelida: Polychaeta) and *Priapulus caudatus* (Priapulida). *J. Zool., Lond.* 99, 305-23.
Manton, S. M. (1977). *The Arthropoda: habits, functional morphology, and evolution.* Clarendon Press, Oxford.
Martin, G. G. (1978). An ultrastructural study of muco-ciliary locomotion and adhesive mechanisms in lower inverebrates. Ph.D. thesis, University of California, Berkeley, Ca.
Mettam, C. (1969). Peristaltic waves of tubicolous worms and the problem of locomotion in *Sabella pavonina. J. Zool., Lond.* 158, 341-56.

—— (1971). Functional design and evolution of the polychaete *Aphrodite aculeata. J. Zool., Lond.* 63, 489-514.

Piearce, T. G. (1983). Functional morphology of lumbricid earthworms, with special references to locomotion. *J. nat. Hist.* 17, 95-111.

Pilato, G. (1981). The significance of musculature in the origin of the Annelida. *Boll. Zool.* 48, 209-26.

Raff, R. A. and Kaufman, T. C. (1983). *Embryos, genes and evolution.* Collier MacMillan, London.

Rieger, R. M. (1980). A new group of interstitial worms, Lobatocerebridae nov. fam. (Annelida) and its significance for metazoan phylogeny. *Zoomorphologie* 95, 41-84.

Runnegar, B. (1982). Oxygen requirements, biology and phylogenetic significance of the late Precambrian worm *Dickinsonia*, and the evolution of the burrowing habit. *Alcheringa* 6, 223-39.

Schembi, P. J. and Jaccarini, V. (1977). Locomotory and other movements of the trunk of *Bonellia viridis* (Echiura, Bonellidae). *J. Zool., Lond.* 182, 477-94.

Seymour, M. K. (1970). Skeletons of *Lumbricus terrestris* L. and *Arenicola marina* (L.). *Nature, Lond.* 228, 383-5.

Storch, V. (1968). Zur vergleichenden Anatomie der segmentalen Muskelsysteme und zur Verwantschaft der Polychaeten-Familien. *Z. Morph. Tiere* 63, 251-342.

20. The systematic position of the Dinophilidae and the archiannelid problem

WILFRIED WESTHEIDE

Spezielle Zoologie, Universität Osnabrück,
Osnabrück, Federal Republic of Germany

Abstract

Six families (Polygordiidae, Saccocirridae, Protodrilidae, Nerillidae, Dinophilidae, and Diurodrilidae) are still assigned in modern classificatory systems to the taxon Archiannelida, erected in the last quarter of the 19th century for a number of small annelids having a seemingly primitive organization. With the exception of some larger Polygordiidae these are typically interstitial species. However, synapomorphic characters which can substantiate the monophyly of the Archiannelida are not present. Recent ultrastructural investigations of the organization of the foregut indicate homologies only between the pharyngeal organs of saccocirrids, protodrilids, and nerillids. In particular, the pharynx of dinophilids is considered to have an independent origin within the annelids.

There are numerous indications that several of these taxa, especially the dinophilids, have undergone neotenous evolution. The genera of the Dinophilidae can easily be included in a morphological series of dorvilleid polychaetes with increasingly pronounced larval characters. These relationships are also supported by the discovery of the neotenous genus *Apodotrocha* possessing a mosaic of characters from both the Dorvilleidae and Dinophilidae. The evaluation of this latter family as a primitive (basally positioned) taxon has to be excluded; their characters cannot be considered as retained primitive annelid features but rather belong to a secondarily simplified larval organization with several highly-derived non-larval adaptations to the interstitial biotope. Characters that most probably were present in the 'true archiannelid' (i.e. the stem species of the Annelida) were homonomous segmentation, numerous chaetae, and moderate or even large size.

The taxon 'Archiannelida' must therefore be eliminated. Polygordiida (Fam. Polygordiidae), Protodrilida (Fam. Protodrilidae, Saccocirridae), Nerillida (Fam. Nerillidae) and Diurodrilida (Fam. Diurodrilidae) should be considered as separate orders in the Polychaeta; the Dinophilidae are included in the Eunicida but a fusion with the Dorvilleidae is for the time being rejected.

The archiannelid concept

The concept of a taxon Archiannelida may be traced back to Hatschek. He considered the achaetous genus *Polygordius* (1878), and later also *Protodrilus* and *Dinophilus*, as 'der ursprünglichsten Stammform der Anneliden noch am nächsten stehend' (Hatschek 1891). The genus *Saccocirrus* with its stump-like segmental appendages ('protopods') and its simple chaetae forms in his System der Anneliden (1893) a group 'Protochaeta' which he placed as the first order within the Chaetopoda beside Polychaeta and Oligochaeta. It is evident from his paper of 1878 (p. 65) and his phylogenetic tree published in 1893 that the Protochaeta should connect the Archiannelida with all other annelids. Hermans (1969) gives a comprehensive review of the confusing history of archiannelid descriptions and their classifications. The archiannelid theory was based on the popular assumptions that stem annelids were small and had a primitive organization. Hatschek (1878) regarded a trochophore-like organism as the stem species from which the Annelida among other taxa have developed. This was the immediate reason why primitive annelids were sought only among small forms. In general, however, it was taken for granted that the stem species of a large taxon, such as the Annelida, could only be an organism of small body dimensions. The results of palaeontological research with the discovery of numerous phyletic lines, beginning with small forms and leading to, in part, gigantic species (Cope's rule) may well have added weight to this supposition. The second assumption of the archiannelid concept is based on the idea that simple and primitive organizations are equivalent, so implying an evolutionary pathway within the annelids from forms lacking parapodia and chaetae to those possessing chaetigerous parapodia, and from homonomously segmented to heteronomously segmented species. The most primitive characters were those found in the developmental stages of typical polychaetes. The large 'true polychaetes' thus recapitulate stages of phylogeny in their ontogeny, and these stages are still represented by the Archiannelida.

It is therefore not surprising that a series of forms characterized by small size and supposedly simple organization were placed in the

Archiannelida. But the lack of characters shared with well-known annelid families was also a reason for grouping them into the archiannelids. The Archiannelida can therefore also be seen as a dumping ground for taxonomically difficult Annelida. When relationships with other taxa were subsequently recognized this led in most cases to rapid reclassification. The Histriobdellidae are an impressive example which, on account of their small size and lack of chaetae, were considered as characteristic archiannelids (Foettinger 1884; Shearer 1910; Heider 1922) until careful analysis of their jaw apparatus unequivocally demonstrated their inclusion in the Eunicemorpha (Mesnil and Caullery 1922). Other taxa whose archiannelid membership has been disputed are Ctenodrilidae (Reisinger 1925), Parergodrilidae (Reisinger 1925), Poeobiidae (Berkeley and Berkeley 1960), Aeolosomatidae (Reisinger 1925), and Oweniidae (Bubko 1973).

Recent supporters of the archiannelid concept include the following families: Polygordiidae, Protodrilidae, Saccocirridae, Nerillidae, and Dinophilidae. Kristensen and Niilonen (1982) have, with good reason, excluded *Diurodrilus* from the Dinophilidae; thus a sixth family Diurodrilidae exists (Fig. 20.1). The phylogenetic position of these families is still controversial, and their systematics is handled in very different

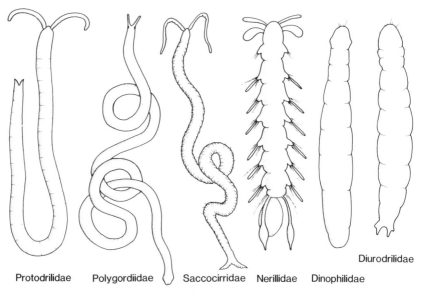

Protodrilidae   Polygordiidae   Saccocirridae   Nerillidae   Dinophilidae   Diurodrilidae

Figure 20.1. Genera of the so-called archiannelid families: Protodrilidae: *Protodrilus*; Polygordiidae: *Polygordius*; Saccocirridae: *Saccocirrus*; Nerillidae: *Nerillidium*; Dinophilidae: *Trilobodrilus*; Diurodrilidae: *Diurodrilus*. Scale of the individual representatives greatly different. Largest species up to 10 cm or more in the genus *Polygordius*; smallest animals in the genus *Diurodrilus* (about 400 μm). Redrawn from various sources.

ways. Reviews of annelid morphology or physiology generally separate Archiannelida from Polychaeta, Oligochaeta, and Hirudinea. Original contributions on individual representatives of the above-named families almost always identify them as 'Archiannelida'. Some textbooks and keys still place them at the base of the Annelida (e.g. Fretter and Graham 1976); others indicate their doubt in the archiannelid concept but do not in the end discard the grouping (e.g. Hartmann-Schröder 1982). An increasing number of authors, however, reject this taxon completely and give the individual families ordinal rank (e.g. Fauchald 1977; Pettibone 1982; Westheide 1985). This contradictory situation may be examined further by posing two questions and attempting to find the answers to them: do the six archiannelid families form a monophyletic taxon and was their organization originally simple or has it been secondarily simplified?

Do the archiannelid families form a monophyletic taxon?

The individual subgroups of a monophyletic taxon must exhibit at least one common derived (= synapomorphous) character. Orrhage (1974) proved convincingly that the characters discussed so far are either limited to some taxa (e.g. circular muscles absent) or are widespread amongst the so-called 'true polychaetes' (e.g. absence of parapodia and chaetae, a ventral ciliary tract, dorsal ciliary rings, a 'simple' system of blood vessels, a nervous system closely connected with the epidermis). Ultrastructural examinations of the cuticle (Brandenburg 1970; Rieger and Rieger 1976; Jouin 1978b; Purschke 1984) and eyes (Eakin, Martin, and Reed 1977; Brandenburger and Eakin 1981) have similarly not established a synapomorphy. The small body dimension is unsuitable as a shared character: not only are there considerable differences in size (*Polygordius lacteus* c. 100 mm, *Diurodrilus westheidei* 400 μm), but within the mesopsammon there are also numerous 'true polychaetes' with body dimensions corresponding to those of the above mentioned families (Westheide 1971, 1984).

Only one character requires special attention: the ventral pharyngeal muscle organ (Fig. 20.2) which is present in all families with the exception of the Polygordiidae, some species of *Saccocirrus*, the gutless protodrilid *Astomus taenioides* and *Apharyngtus punicus*. Jägersten's (1947) comparative light microscopical investigation seemed to prove its homology within all so-called archiannelids, so that Hermans (1969) stated that '. . . until further study proves otherwise, it must be concluded that the structure of the stomodaeum apparently links the families of archiannelids.' The first TEM analysis by Rieger and

Rieger (1975), examined the pharyngeal plate muscle cells of *Trilobo-drilus* (Dinophilidae) and tried to explain by ingenious hypotheses poss-ible evolutionary pathways for both the formation of the pharyngeal bulb and the plate muscle cells. Jouin (1978a), however, rejected any idea of a structural homology of this organ in the families Protodrilidae and Dinophilidae. Kristensen and Niilonen (1982) showed that the pharyngeal organ of the Diurodrilidae undoubtedly differs from that of the other families. Finally, Purschke's (1984) comprehensive compara-tive TEM examination of this organ in seven annelid families also tends to the view that the dinophilid pharynx differs anatomically, cytologically, and functionally from the bulb of any other annelid (see Fig. 20.2(d)). In addition to the Polygordiidae, which lack a bulb, the

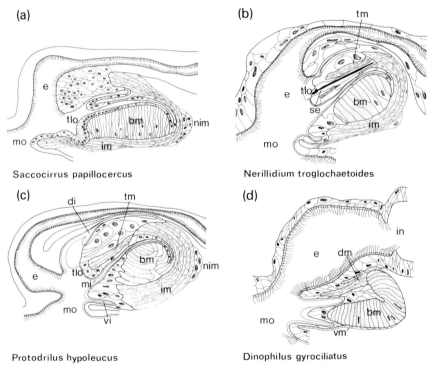

Fig. 20.2. Sagittal sections of the ventral pharyngeal organ of (a) *Saccocirrus papillocercus;* (b) *Nerillidium troglochaetoides*; (c) *Protodrilus hypoleucus*; (d) *Dinophilus gyrociliatus*. (c-d) Reconstructions from a series of electron micrographs; (a) redrawn after Goodrich (1901); (b) after Jouin (1978a) from Purschke (1984). Abbreviations: bm, transversely orientated bulbus muscles; dm, dorsal (= retractor) longitudinal muscle; di, dorsal in-vagination; e, ciliated ectodermal oesophagus; im, investing muscles of pharyngeal organ; in, endodermal intestine; mi, median invagination; mo, mouth opening; nim, nuclei of the investing muscles; se, one of the four intracellular elements in the tongue-like organ; tlo, tongue-like organ; tm, tongue muscles; vi, ventral invagination; vm, ventral (= protractor) longitudinal muscles.

Diurodrilidae and Dinophilidae cannot be joined in a monophyletic taxon either because of this organ. There remain the Saccocirridae, Protodrilidae, and Nerillidae, whose pharynges indicate a high degree of similarity, certainly according to Purschke's (1984) examinations. For the saccocirrids and protodrilids a proposed close relationship is supported by corresponding habitus, the head tentacles, and the contractile head cavity. Indeed, Goodrich (1901) stated that 'Wherever we put *Protodrilus* in our system of classification, there also we must place *Saccocirrus*.' The inclusion of the nerillids, which differ greatly in their habits, in a monophyletic group consisting of these three families is, however, only feasible if their pharyngeal bulbs—or substructures—can be considered not only as homologous, but also as derived synapomorphic characters. According to Purschke's (1984) investigations, derived structures are indeed present in the pharyngeal organ of Protodrilidae and Nerillidae: cross-striated stylet-like structures in the tongue-like organ, circomyarian muscle cells of the hirudinean type, specific tongue muscles in the tongue-like organ. A taxon Protodrilida, comprising Saccocirridae, Protodrilidae and Nerillidae, appears therefore to be realistic, but needs to be confirmed by additional synapomorphies.

Synapomorphies between the pharyngeal organs of all so-called archiannelids and corresponding organs in different polychaete families, e.g. Ctenodrilidae, Parergodrilidae, Orbiniidae, Flabelligeridae, Sabellidae, are not present according to Purschke (1984). There is also no convincing proof for the primitiveness of this type of construction and, in any case, it would be of little importance for recognition of phylogenetic relationships (see Orrhage 1973, 1974; Wilfert 1973). The assumption of convergent evolution of the ventral pharyngeal organs (at least three times in the archiannelid families discussed here) is therefore the most likely explanation (see e.g. Bunke 1967). The causes of the convergence are presumably the combined tongue functions of food gathering and swallowing (Reisinger 1925; Jones 1954; Jennings and Gelder 1969; Schmidt and Westheide 1972; Rieger and Rieger 1975; Ivanov and Tzetlin 1981).

Retention of a primitive annelid organization or a secondarily simplified one?

Even if the concept of a monophyletic taxon Archiannelida comprising the above-mentioned families is refuted, it is nonetheless feasible that these families could have separated independently from the evolutionary pathway leading to the 'true polychaetes' at a very early date and thus have retained a particularly large number of primitive characters. Such characters would include: small body dimensions correlated, in part, with a small number of segments; homonomy of segments; absence of parapodia; few simple chaetae or no chaetae at all; a ventral ciliary tract, segmental ciliary rings, and sometimes connected with this a swimming capacity; restriction of the pharynx to the region behind the mouth opening; close connection of the nervous system with the epidermis; presence of dorsal and ventral mesenteries; simple cuticle without prominent grids of collagen fibres, but with numerous well-developed microvilli; typical protonephridia with distal flame cells. These features are characteristic also of larval or juvenile polychaetes, but these forms halt at a stage which is more or less recapitulated by the more highly developed 'true polychaetes' in their ontogeny, or they represent progenetic (= neotenous) forms which mature at a larval or juvenile level of organization. According to Osche (1982) neoteny can be recognized only when not all phenes of an organism are affected by fetalization. These so-called archiannelids certainly possess a series of highly derived non-larval characters, particularly in their reproductive biology and their reproductive organs: highly modified sperm in Protodrilidae, Saccocirridae, Nerillidae, Dinophilidae, and Diurodrilidae; complex male copulatory organs in Saccocirridae and Dinophilidae; female receptive organs in Saccocirridae and Nerillidae; copulation-supporting structures in Saccocirridae and Nerillidae; direct transmission of sperm by pseudocopulation, transmission of spermatophores, direct hypodermic injection or true copulation in Protodrilidae, Saccocirridae, Nerillidae and Dinophilidae; hermaphroditism in Nerillidae and Dinophilidae; development without planktonic larvae in Protodrilidae, Saccocirridae, Nerillidae, Dinophilidae and Diurodrilidae (?); eggs spawned in cocoons in Protodrilidae and Dinophilidae; brooding and associated supporting organs in Nerillidae (see Westheide 1984 for references). Further derived characters may include a body wall consisting solely of longitudinal muscles in Polygordiidae, additional nerve cords between the main ventral nerve stems in *Dinophilus* (Kotikova 1973), and intraovarian oogenesis with nurse cells in dinophilids (Eckelbarger 1983). Most of these above-mentioned characters are typical adaptations of the mesopsammon that have developed convergently in many other interstitial taxa. This mosaic of larval and highly derived characters, therefore, strongly supports interpretation of at least some of these so-called archiannelids as neotenous species or species with partly-fetalized organization which are adapted to the interstitial habitat (see also Hermans 1969). This hypothesis receives support from the typical life-cycle of many macrobenthic polychaetes: fertilization is external, the larvae live in the water column; at a certain stage of development they enter the bottom sediment, their small dimensions enabling them at first to settle in the pore system between the sediment particles. They are then part of the mesopsammon forming the 'temporary meiofauna' (McIntyre 1969; Kudenov 1979). Developmental stages of many polychaetes therefore have direct and easy access to the interstitial habitat; as Rieger (1980) stated their life cycle offers the potential of specialization for interstitial existence. They could occupy this biotope permanently by maturing at a larval or juvenile stage (Gould 1977; Westheide 1984). Schminke (1981) has developed a very similar concept for the evolution of the interstitial bathynellacean Crustacea, which are considered as having originated by neoteny from a zoea-like ancestor. Other interstitial taxa for which progenesis or neoteny have been discussed include psammodrilid polychaetes, the new annelid taxon Lobatocerebridae, opisthobranch molluscs, copepods, and mystacocarids (see Westheide 1982 for references).

The neotenous nature of the Dinophilidae and their systematic position within Annelida

An indication of neotenous evolution of a taxon is given when conformity exists with the larval stages of another taxon. Such is the case with the dinophilids which externally compare well with the polytrochous larvae of the polychaete order Eunicida. Genera of the eunicid family Dorvilleidae (*Pettiboneia, Exallopus, Ophryotrocha, Gymnodorvillea, Parophryotrocha, Ikosipodus, Parapodrilus*) can be arranged with the dinophilid genera *Dinophilus* and *Trilobodrilus* into a morphological series with increasingly neotenous characters. This series shows striking similarities with a series of developmental stages of larger, non-interstitial dorvilleid genera, e.g. *Schistomeringos* (Westheide 1984). Mecznikow (1866) already recognized the great similarity of *Dinophilus* to eunicids long before Hatschek's archiannelid theory: '. . . daß *Dinophilus* als eine stationäre Annelidenlarve zu betrachten ist, und mithin zu den Anneliden ebenso, wie Appendicularia zu den Ascidien sich verhält.' Nelson (1904) recognized that the ontogeny of this taxon is in no way primitive. It is therefore only logical when he wrote 'Thus in the light of cleavage . . . Metschnikoff's conjecture appears almost prophetic. *Dinophilus* probably should be regarded as a 'stationäre Annelidenlarva', but one in which the development ends and which has become correspondingly modified'.

Neoteny and the close relationship to the Eunicida has been repeatedly discussed (see Hermans 1969 for references), Bubko (1973) taking an extreme stand by arguing against the neotenous origin with the proposal that the organization of the dinophilids goes far beyond a larva in many characters and thus is an exact reversal of the criterion used by Osche (1982) for neoteny. Åkesson (1977) was the first to add new evidence for a close relationship between dinophilids and eunicid polychaetes with the demonstration of reciprocal infection with coelomic coccidia of the genus *Grellia*, parasites that are generally extremely host specific. He drew attention to a morphological character joining dorvilleids and dinophilids: the unpaired median ventral appendage of the pygidium (Fig. 20.3), characteristic of most *Ophryotrocha* species, *Parapodrilus psammophilus*, and *Dinophilus* species.

The most recent argument which supports a close relationship between eunicid polychaetes and dinophilids is the discovery of a taxon that may be a 'missing link': *Apodotrocha progenerans* Westheide and Riser, 1983 (Fig. 20.3). This 600 μm long species with six trunk segments possessing neither parapodia nor chaetae, looks like a typical

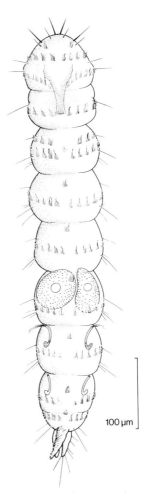

Fig. 20.3. *Apodotrocha progenerans* Westheide and Riser 1983; female, ventral side. From Westheide and Riser (1983).

Ophryotrocha larva (Westheide and Riser 1983). If the characters it has in common with the dorvilleids are considered as plesiomorphous (e.g. ciliary rings, sensory cilia, three caudal appendages, the specific micro-villar structure of the gut) and the characters which conform with the dinophilids are considered as synapomorphies (e.g. lack of parapodia, chaetae and head appendages, a trunk with six segments, pattern of the body wall musculature, presence of protonephridia), a phylogenetic dendrogram can be constructed that places the genera *Dinophilus* and *Trilobodrilus* as extremely derived taxa of the Dorvilleidae and as the sister group of the genus *Apodotrocha* (Fig. 20.4). This dendrogram is only absolutely convincing, however, if it is assumed that the taxa followed each other step by step with increasingly neotenous organi-zation in the course of evolution, developing one from the other. The other extreme in the evolution of very similar neotenous taxa is, how-ever, also feasible: independent evolutionary pathways from one or

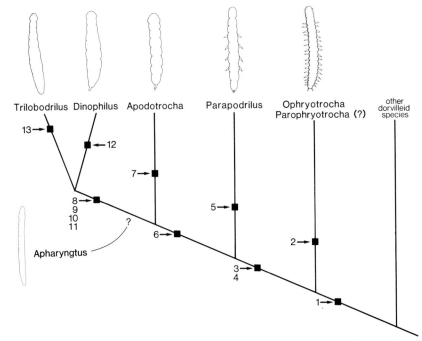

Fig. 20.4. Dendrogram indicating the possible phylogenetic-systematic relationships of Dinophilidae and Dorvilleidae. 1, Median unpaired caudal appendage; 2, specific structure of jaws; 3, loss of jaws and ventral muscular pharyngeal apparatus; 4, loss of head appendages; 5, constant number of four setigerous segments; 6, loss of parapodia and chaetae; 7, two pairs of seminal receptacles; 8, loss of paired anal appendages; 9, ventral muscular bulb that lacks sagittal investing muscles, but with presence of two opposing myofilament systems in the individual cell; 10, unpaired copulatory organ; 11, specific type of spermatozoon; 12, ocelli with lenses; 13, loss of median caudal appendage. Modified from Westheide and Riser (1983).

more larger dorvilleid species to various larval or juvenile stages with
early maturity. Although this latter possibility in no way alters the supposed close relationship between the taxa discussed here, it is important
for elucidation of the systematics of the groups involved. Corroboration
of the results seen in the tentative dendrogram would necessitate a fusion
of the two families (as Dinophilidae for nomenclatural reasons). Parallel
development, however, retains open a sister group relationship Dino-
philidae-Dorvilleidae. Elimination of the family Dorvilleidae (including
the genus *Apodotrocha*) is, therefore, for the present being rejected. It is
likely that new taxa from this group (e.g. a new achaetous genus with
jaws found by Dr. Kristensen, personal communication) will be
described shortly, making one or other evolutionary possibility more
probable.

Such a clear connection to 'true polychaetes' cannot be drawn for
the other five families. Hatschek's (1893) phylogenetic tree, however,
already indicated a suspected close relationship between Saccocirridae
and Spionidae. This idea acquired support from Orrhage's (1974)
investigations of the head anatomy of the Spiomorpha.

What did the stem species of the annelids look like?

If the six families do not have any particular connection with the stem
species of the annelids and are not to be considered as 'Archiannelida',
then what is the 'true archiannelid'? I will not attempt to select taxa
from the recent fauna which appear to be especially primitive, but
rather restrict myself to a theoretical consideration of the polychaete
stem species. As I accept the sister group relationship Annelida-
Arthropoda, the stem species of both taxa must have possessed a segmented body, probably homonomously, and segmented ventral nerve
cords. The chaetae of annelids have a similar fine structure and method
of formation to those of Echiura and Pogonophora, the chaetae of the
mantle margin of Brachiopoda and the organs of Kölliker in Cephalo-
poda (George and Southward 1973). These structures are regarded as
homologous, and annelid chaetae must be derived from a structure that
was present in the common stem species of annelids and arthropods.

The question regarding the parapodia is more difficult to answer.
Three possibilities can be discussed (Fig. 20.5):

1. The articulate stem species already had parapodia and the extremities of the arthropods developed from these. The parapodia within
the annelids then reduced in the stem species of the Clitellata. The frequently quoted homologization of parapodium and oncopodium-
arthropodium (e.g. Lauterbach 1978) is principally based on their

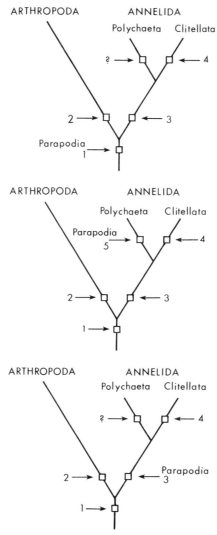

Fig. 20.5. Three possible dendrograms of phylogeny of Articulata. Rectangles indicate the origin of features in the evolutionary pathway that characterize the individual group. 1, Stem species of Articulata: homonomous segmentation, segmental ventral nerve cord, parapodia with chaetae. 2, Stem species of Arthropoda: arthropod extremity that originated from a parapodium (also other characters). 3, Stem species of Annelida: specific obliquely striated musculature and specific cuticle. 4, Stem species of Clitellata: clitellum, specific ontogenesis, hermaphroditism, gonads restricted to a few segments etc., parapodia reduced to dorsal and ventral bundles of chaetae. 5, Stem species of Polychaeta: biramous parapodia. ?, Autapomorphies cannot be demonstrated for the stem species of Polychaeta.

approximate similarity of position on the ventro-lateral sides of the segments.

2. The Articulata stem species did not possess any parapodial appendages. The arthropod extremities are new formations from the arthropod stem species. Not until the stem species of the polychaetes evolved was it characterized by parapodia. According to this concept absence of parapodia in the Clitellata is a primitive feature.

3. The third possibility considers the parapodia as having originated independently of arthropod limbs, but as being already present in the stem species of the annelids. In the case of the Clitellata the parapodia must therefore have been reduced. According to the present stage of knowledge there is for the polychaetes no apomorphous character clearly defining the group. This concept includes the possibility that the Clitellata do not form the sister group of all Polychaeta but are more closely related to a certain group of them.

There are no unequivocal arguments in favour of any of these three theories. The third possibility is nevertheless regarded as the most probable at present. It avoids the difficult homologization of parapodium and arthropod extremity (Siewing 1978), and considers the simple chaetae in the oligochaetes as structures reduced as an adaptation to a terrestrial-limnic life.

In summary, the stem species of the annelids possessed a metameric body with parapodia and chaetae. As the primary function of the chaetae was probably protective (e.g. Storch 1968), it is supposed that their numbers were originally relatively high and their dimensions large. Further, it may be supposed that there were probably not numerous strong chaetae in very small species. Segmentation, septation of the coelom and a hydroskeleton (Clark 1964) are unlikely to have originated in small species. It is more probable that the stem species of the annelids was part of the macrofauna.

According to Fauchald (1974) the stem species was probably a shallow water marine form that was gonochoristic, spawning into the sea and producing larvae that were planktotrophic.

Conclusions

It can be concluded that the so-called Archiannelida are not a monophyletic group and are not primitive annelids. It is quite probable that the Dinophilidae are neotenous Dorvilleidae belonging to the polychaete order Eunicida. The position of the other families cannot be determined conclusively. At the moment it is best to consider them as separate orders in the Polychaeta: Polygordiida (Fam. Polygordiidae), Protodrilida (Fam. Protodrilidae, Saccocirridae), Nerillida (Fam.

Nerillidae), Diurodrilida (Fam. Diurodrilidae) (see also Pettibone 1982); the Protodrilida may be placed in the vicinity of the Spionida. The old taxon 'Archiannelida' should be eliminated and the term no longer used when referring to the above families.

Acknowledgements

I am indebted to Dr Margaret Fransen for kindly reading the English manuscript, Dr Günter Purschke and Henning von Nordheim for valuable discussions. Support of the Deutsche Forschungsgemeinschaft is gratefully acknowledged.

References

Åkesson, B. (1977) Parasite-host relationship and phylogenetic systematics. The taxonomic position of dinophilids. In *The meiofauna species in time and space* (eds W. Sterrer and P. Ax), Workshop Symposium, Bermuda 1975. *Mikrofauna Meeresbod.* 61, 19-28.

Berkeley, E. and Berkeley, C. (1960). Some further records of pelagic Polychaeta from the northeast Pacific north of latitude 40°N and east of longitude 175°W, together with records of Siphonophora, Mollusca, and Tunicata from the same region. *Can. J. Zool.* 38, 787-99.

Brandenburg, J. (1970). Die Cuticula des *Dinophilus* (Archiannelida). *Z. Morph. Tiere* 68, 300-7.

Brandenburger, J. L. and Eakin, R. M. (1981). Fine structure of ocelli in larvae of an archiannelid, *Polygordius* cf *appendiculatus. Zoomorphology* 99, 23-36.

Bubko, O. V. (1973). On systematic position of Oweniidae and Archiannelida (Annelida). *Zool. Zh.* 52, 1286-96 (in Russian).

Bunke, D. (1967). Zur Morphologie und Systematik der Aeolosomatidae Beddard 1895 and Potamodrilidae nov. fam. (Oligochaeta). *Zool. Jb. Abt. Syst.* 94, 187-368.

Clark, R. B. (1964). *Dynamics in metazoan evolution: the origin of the coelom and segments.* Clarendon Press, Oxford.

Eakin, R. M., Martin, G. G., and Reed, C. T. (1977). Evolutionary significance of fine structure of archiannelid eyes. *Zoomorphologie* 88, 1-18.

Eckelbarger, K. J. (1983). Evolutionary radiation in polychaete ovaries and vitellogenic mechanisms: their possible role in life history patterns. *Can. J. Zool.* 61, 487-504.

Fauchald, K. (1974). Polychaete phylogeny: a problem in protostome evolution. *Syst. Zool.* 23, 493-506.

—— (1977). The polychaete worms. Definitions and keys to the orders, families and genera. *Nat. Hist. Mus. Los Angeles County, Science Series* 28, 1-190.

Foettinger, A. (1884). Recherches sur l'organization de *Histriobdella homari*,

P.-J. van Beneden rapportée aux archiannelides. *Archs Biol., Paris* 5, 435-516.

Fretter, V. and Graham, A. (1976). *A functional anatomy of invertebrates.* Academic Press, New York.

George, J. D. and Southward, E. C. (1973). A comparative study of the setae of Pogonophora and polychaetous Annelida. *J. mar. biol. Ass. U.K.* 53, 403-24.

Goodrich, E. S. (1901). On the structure and affinities of *Saccocirrus. Q. Jl microsc. Sci.* 44, 413-28.

Gould, S. J. (1977). *Ontogeny and phylogeny.* Harvard University Press, Cambridge, Mass.

Hartmann-Schröder, G. (1982). Stamm Annelida. In *Lehrbuch der Speziellen Zoologie, Bergründet von A. Kaestner* (4th edn) (ed. H.-E. Gruner) Vols. 1,3, pp. 276-469. Fischer, Stuttgart.

Hatschek, B. (1878). Studien über Entwicklungsgeschichte der Anneliden. Ein Beitrag zur Morphologie der Bilaterien. *Arb. zool. Inst. Univ. Wien* 1, 277-404.

—— (1891). *Lehrbuch der Zoologie, 3. Lfg.* Fischer, Jena.

—— (1893). System der Anneliden, ein vorläufiger Bericht. *Lotos* 13, 123-6.

Heider, K. (1922). Über Archianneliden. *Sber. preuss. Akad. Wiss.* 39-44.

Hermans, C. O. (1969). The systematic position of the Archiannelida. *Syst. Zool.* 18, 85-102.

Ivanov, D. L. and Tzetlin, A. B. (1981). Origin and evolution of the cuticular pharyngeal armature in trochophore animals having the ventral pharynx. *Zool. Zh.* 60, 1445-54 (in Russian).

Jägersten, G. (1947). On the structure of the pharynx of the Archiannelida with special reference to there-occurring muscle cells of aberrant type. *Zool. Bidr. Upps.* 25, 551-70.

Jennings, J. B. and Gelder, S. R. (1969). Feeding and digestion in *Dinophilus gyrociliatus* (Annelida: Archiannelida). *J. Zool., Lond.* 158, 441-51.

Jones, E. R. (1954). The operation of the pharyngeal bulb in the archiannelid *Dinophilus. Q. J. Fl. Acad. Sci.* 19, 2-3.

Jouin, C. (1978a). Anatomical and ultrastructural study of the pharyngeal bulb in *Protodrilus* (Polychaeta, Archiannelida). I. Muscles and myoepithelial junctions. *Tissue Cell* 10, 269-87.

—— (1978b). Anatomical and ultrastructural study of the pharyngeal bulb in *Protodrilus* (Polychaeta, Archiannelida). II. The stomodeal epithelium and its cuticle. *Tissue Cell* 10, 289-301.

Kotikova, E. A. (1973). New data concerning the nervous system of Archiannelida. *Zool. Zh.* 52, 1611-5 (in Russian).

Kristensen, R. M. and Niilonen, T. (1982). Structural studies on *Diurodrilus* Remane (Diurodrilidae fam. n.) with description of *Diurodrilus westheidei* sp. n. from the Arctic interstitial meiobenthos, W. Greenland. *Zool. Scr.* 11, 1-11.

Kudenov, J.-D. (1979). Post-larval polychaetes in sandy beaches of Tomalis Bay, California. *Bull. S. Calif. Acad. Sci.* 78, 144-7.

Lauterbach, K.-E. (1978). Gedanken zur Evolution der Euarthropoden-Extremität. *Zool. Jb. Abt. Anat.* 99, 64-92.

McIntyre, A. D. (1969). Ecology of marine meiobenthos. *Biol. Rev.* 44, 245-90.

Mecznikow, E.(1866). *Apsilus lentiformis*, ein Räderthier. *Z. wiss. Zool.* 16, 346-56.

Mesnil, F. and Caullery, M. (1922). L'appareil maxillaire d'*Histriobdella homari*; affinités des Histriobdellides avec les Euniciens. *C. r. hebd. Séanc. Acad. Sci., Paris* 174, 913-7.

Nelson, J. A. (1904). The early development of *Dinophilus*: a study in cell-lineage. *Proc. Acad. Nat. Sci. Philadelphia* 56, 687-737.

Orrhage, L. (1973). Two fundamental requirements for phylogenetic-scientific works as a background for an analysis of Dale's (1962) and Webb's (1969) theories. *Z. Syst. Evolutionsforsch.* 11, 161-73.

—— (1974). Über die Anatomie, Histologie und Verwandtschaft der Apistobrachidae (Polychaeta Sedentania) nebst Bemerkungen über die systematische Stellung der Archianneliden. *Z. Morph. Tiere* 79, 1-45.

Osche, G. (1982). Rekapitulationsentwicklung und ihre Bedeutung für die Phylogenetik—Wann gilt die 'Biogenetische Grundregel'? *Verh. naturw. Ver. Hamb.* 25, 5-31.

Pettibone, M. H. (1982). Annelida. In *Synopsis and classification of living organisms* (ed. S. P. Parker) Vol. 2, pp. 1-43. McGraw-Hill, New York.

Purschke, G. (1984). Vergleichende anatomische und ultrastrukturelle Untersuchungen ventraler Pharynxapparate bei Polychaeten und ihre phylogenetische Bedeutung. Ph.D. thesis, Universität Göttingen.

Reisinger, E. (1925). Ein landbewohnender Archiannelide. Zugleich ein Beitrag zur Systematik der Archianneliden. *Z. Morph. Ökol. Tiere* 13, 197-254.

Rieger, R. M. (1980). A new group of interstitial worms, Lobatocerebridae nov. fam. (Annelida) and its significance for metazoan phylogeny. *Zoomorphologie* 95, 41-84.

—— and Rieger, G. E. (1975). Fine structures of the pharyngeal bulb in *Trilobodrilus* and its phylogenetic significance within Archiannelida. *Tissue Cell* 7, 267-79.

—— (1976). Fine structure of the archiannelid cuticle and remarks on the evolution of the cuticle within the Spiralia. *Acta zool., Stockh.* 57, 53-68.

Schmidt, P. and Westheide, W. (1972). *Dinophilus gyrociliatus* (Polychaeta). Nahrungsaufnahme und Fortpflanzung. *Encycl. Cinemat.* 1750/1971, 1-16.

Schminke, H. K. (1981). Adaptation of Bathynellacea (Crustacea, Syncarida) to life in the interstitial ('Zoeatheory'). *Int. Rev. ges. Hydrobiol. Hydrogr.* 66, 575-637.

Shearer, C. (1910). On the anatomy of *Histriobdella homari*. *Q. Jl. microsc. Sci.* 55, 287-359.

Siewing, R. (1978). Zur mutmaßlichen Phylogenie der Arthropoden-extremität. *Zool. Jb. Abt. Anat.* 99, 93-8.

Storch, V. (1968). Zur vergleichenden Anatomie der segmentalen Muskelsysteme und zur Verwandtschaft der Polychaeten-Familien. *Z. Morph. Tiere* 63, 251-342.

Westheide, W. (1971). Interstitial Polychaeta (excluding Archiannelida). *Smithson. Contr. Zool.* 76, 57-70.

—— (1982). *Ikosipodus carolensis* gen. et sp. n., an interstitial neotenic poly-chaete from North Carolina, U.S.A., and its phylogenetic relationships within Dorvilleidae. *Zool. Scr.* 11, 1117-26.

—— (1984). The concept of reproduction in polychaetes with small body size: adaptations in interstitial species. *Fortschr. Zool.* 29. In *Polychaete reproduction* (eds A. Fischer and H. D. Pfannenstiel) pp. 265-87.

—— (1985). Annelida. In *Lehrbuch der Zoologie,* Bd II, Systematik (ed. R. Siewing). Fischer, Stuttgart.

—— and Riser, N. W. (1983). Morphology and phylogenetic relationships of the neotenic interstitial polychaete *Apodotrocha progenerans* n. gen., n. sp. (Annelida). *Zoomorphology* 103, 67-87.

Wilfert, M. (1973). Ein Beitrag zur Morphologie, Biologie und systematischen Stellung des Polychaeten *Ctenodrilus serratus. Helgoländer wiss. Meeresunters.* 25, 332-46.

21. Vestimentiferan pogonophores: their biology and affinities

MEREDITH L. JONES

Department of Invertebrate Zoology,
National Museum of Natural History,
Washington DC, USA

Abstract

The presumed coelomic cavities, as reflected in body segmentation, of an obturate pogonophoran, *Rifta pachyptila* Jones, are compared with those (*sensu* Southward) of the perviate pogonophorans. The latter are concluded to possess at least two such cavities anterior to the opisthosome, and the former, four cavities in the same region. It is concluded that the relationship of the Pogonophora with the Annelida is more distant than previously thought and that similarly the relationship between the Obturata and the Perviata is more distant. Further, it is concluded that the Perviata departed from the annelid line earlier than the Obturata.

Introduction

In the course of a description of *Lamellibrachia barhami* from off the coast of California, Webb (1969a) recognized four regions in the body of this vestimentiferan pogonophore. The first of these, the 'tentacular crown', comprises 'tentacles' (perhaps, more properly, branchiae) surrounding a central 'lophophoral organ', now known as the obturaculum. Both the obturaculum and the plume of branchiae are provided with coelomic cavities, the former as perivascular cavities in which the obturacular vessels are situated, the latter as the elongate cavities of the branchial filaments. Webb found no extensive coelomic cavities in the second body region, the vestimentum, but noted that the paired excretory ducts of this region open to the exterior by an unpaired medial excretory pore. He did not observe any connection between the coelom(s) of the branchial crown and the excretory tubules/ducts. The third region, the trunk, is provided with a single, extensive pair of coelomic cavities separated by a medial mesentery. The gonads are situated in this third region and the genital pores of both males and females are located near the extreme anterior end of the trunk coeloms where they extend into the posterior third of the vestimentum. The fourth region, the opisthosome, is not described in detail. Webb (1969a and subsequent papers) considered the nerve cord to be dorsal whereas I regard it as ventral and thus consider that the medial excretory pore, as well as the genital pores, open on the dorsal surface of *L. barhami*.

In a paper describing the excretory system of *L. barhami*, Webb (1975) confirmed his previous observation of no communication between the excretory system and the coeloms of the branchiae and obturaculum, and further stated that the coelomic canals of the branchiae and obturaculum are separate. In other papers Webb (1969b,c, 1977; Webb and Ganga 1980) has commented at length on the four body regions of the pogonophorans, both in general terms, as well as specifically on the extent of the coelomic cavities of the vestimentiferans.

On the basis of a single incomplete specimen, van der Land and Nørrevang (1975) described *Lamellibrachia luymesi* from the western Atlantic. In addition to a general description of the anatomy and morphology of the new species, they proposed that the Pogonophora and Vestimentifera be assigned as separate classes to the phylum Annelida. Later van der Land and Nørrevang (1977) presented a detailed description of all the organ systems and anatomical features of *L. luymesi*, and, of particular relevance to this paper, described the coelomic cavities. Furthermore, they proposed the general term 'obturaculum' for the anterior structure bearing the branchial plume and pointed out that it is actually a pair of structures, the obturacula.

Following the discovery of a series of hydrothermal vents on or associated with the East Pacific Rise, and the collection of members of the exotic and unique faunal community, Jones (1981a) described a third species of vestimentiferan, *Riftia pachyptila*. Morphological features in common with both *Lamellibrachia barhami* and *L. luymesi* were such that there was no problem in assigning *Riftia* to the Vestimentifera, but differences in morphology were deemed sufficient to separate the two genera at the familial level. Overall similarities were such that the lamellibrachiid and riftiid vestimentiferans were retained within the phylum Pogonophora, but important morphological differences dictated that the three vestimentiferan species be set apart (as the subphylum Obturata) from other pogonophores (subphylum Perviata). It should be noted that throughout the description of *Riftia* (Jones 1981a), the term 'coelom' was used in reference to the spacious cavity of the third (trunk) region of the body, but was also applied, with some question, to the elongate cavity of the branchial filaments. Subsequently Jones (1981b) persisted in his cautious application of the term 'coelom' and having described the histological characteristics of each of the five types of cavity found in *Riftia*, noted that only the perivascular cavity of the paired, blind-ending, obturacular vessels would qualify as 'coeloms', in the sense of Hyman (1951).

The purposes of this paper are to re-examine the morphology and histology of the vestimentiferans, especially of *Riftia* and an undescribed species, in order to establish the extent and nature of their coelomic cavities and to comment on the systematic relationships of the pogonophoran subphyla Obturata and Perviata.

Material and methods

The present study is based on specimens of vestimentiferan Pogonophora in the collections of the Division of Worms, Department of Invertebrate Zoology, National Museum of Natural History (USNM), Smithsonian Institution. Certain specimens have been embedded in paraffin, sectioned at 5 μm, and subjected to standard histological staining procedures, i.e. Masson's trichrome, Mayer's haematoxylin and eosin, or chlorazol black.

The material examined in detail is as follows:

(i) *Riftia pachyptila* Jones 1981 (see Jones 1981a for geographical coordinates and depths):

USMN 59953, paratype, collected at 'Garden of Eden' hydrothermal vent, Galapagos Rift, *Alvin* Dive 773, 16 March 1977;

USNM 59958, paratype, 'Rose Garden' vent, Galapagos Rift, *Alvin* Dive 889, 14 February 1979;

USNM 81950, 'Rose Garden' vent, Galapagos Rift, *Alvin* Dive 990, 7 December 1979.

(ii) Undescribed vestimentiferan from 13°N:

USNM 81951, 11° 30′N; 103° 48′W, *Cyana* Dive 07, January 1982, 2540 m depth.

Observations

The paired, coelomic cavities of the *Riftia* trunk are separated medially by a mesentery in which are located the dorsal and ventral vessel, as well as the trophosome and gonads (Fig. 21.19 and Jones 1981a,b). Posteriorly the cavities are delimited by the septum separating the

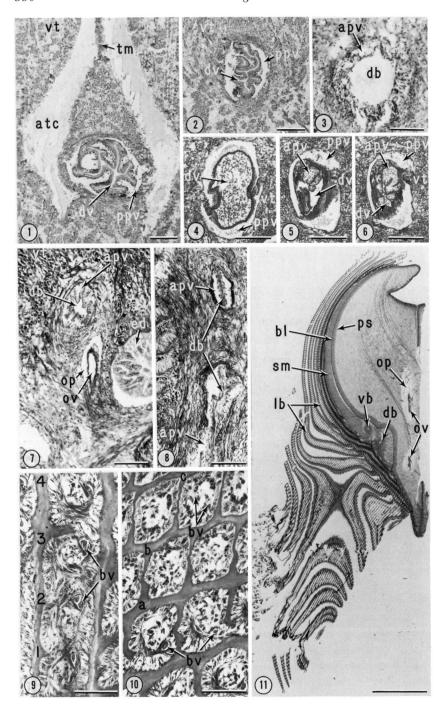

trunk from the opisthosome (Figs 21.17, 21.18). Anteriorly, although ultimately bordered by the muscle and connective tissue that comprises the bulk of the vestimentum, the coelomic cavities of the trunk, and the trunk itself, extend over the posterior one-fifth to one-fourth of the dorsal surface of the vestimentum (Figs 21.1, 21.22). Internally, the anterior extensions of the trunk coelom carry the gonoducts forward to their external openings, and then continue for a short distance on either side of the medial mesentery which carries the dorsal vessel (Fig. 21.1). The mesentery is quite stout and is composed mainly of numerous longitudinal muscle bundles bound in a connective tissue matrix. The dorsal vessel is carried in a perivascular cavity and is supported by another mesentery (Fig. 21.1 and Jones 1981a). Immediately posterior to the beginning of the perivascular cavity, the dorsal vessel is surrounded by strong circular musculature constituting a sphincter. As has been noted by Jones (1981a) the dorsal vessel at this level is composed of an innermost thin cellular layer of muscle, a more peripheral thick layer of connective tissue, and a thick surrounding layer of striated circular muscles. This structure is similar to that of the obturacular vessels (Fig. 21.12). More anteriorly the anterior extensions of the trunk coelom are lost and the dorsal vessel and its surrounding perivascular cavity are embedded in the tissue of the vestimentum (Fig. 21.2). Just

---

Figs 21.1-21.11 (Figs 21.1-21.3, 11, *Riftia pachyptila* Jones; Figs 21.4-21.10, undescribed vestimentiferan from 13°N). Fig. 21.1 Dorsal vessel in posterior perivascular cavity, isolated in anterior portion of mesentery separating trunk coeloms. Fig. 21.2. Dorsal vessel in posterior perivascular cavity surrounded by vestimental tissue. Fig. 21.3. Branch of dorsal vessel at level of brain in poorly defined anterior perivascular cavity. Fig. 21.4 Dorsal vessel in posterior perivascular cavity just posterior to site of formation of anterior perivascular cavity; ventral is to the right. Fig. 21.5. Same but with anterior cavity beginning to form in wall of dorsal vessel; ventral is to the right. Fig. 21.6. Same but with fully established anterior cavity (20 μm anterior to Fig. 21.5); ventral is to the right. Fig. 21.7. Branch of dorsal vessel near junction with obturacular vessel in its perivascular cavity. Fig. 21.8. Sections of branch of dorsal vessel moving ventrally in transverse plane. Fig. 21.9. Cross sections of bases of adjacent branchiae of same lamella; due to oblique plane of sectioning, base at 1 is more basal than level 4. Fig. 21.10. Cross section of bases of branchiae of adjacent lamellae; due to oblique plane of sectioning, base at level a is more basal than level c; level 4 of Fig. 21.9 is comparable to level a, here. Fig. 21.11. Cross section of right half of obturaculum with associated lamellae, muscles, and blood vessels. Figs 21.1-21.3, Mayer's haematoxylin and eosin; Figs 21.4-21.10, Masson's trichrome; Fig. 21.11, chlorazol black. Scale bar represents 200 μm in Figs 21.1, 21.2, 21.4-21.6; 100 μm in Figs 21.3, 21.7, 21.8; 50 μm in Figs 21.9, 21.10; 2 mm in Fig. 21.11. Abbreviations: a-c, successively more distal lamellar sections; apv, anterior perivascular cavity; atc, anterior extensions of trunk coelom; bl, basal lamina; bv, blood vessel; db, anterior branch of dorsal vessel; dv, dorsal vessel; ed, excretory duct; lb, lamellae of branchial filaments; op, obturacular perivascular cavity; ov, obturacular vessel; ppv, posterior perivascular cavity; ps, parasagittal muscles; sm, supportive muscle sheath; tm, trunk medial mesentery; vb, anterior branch of ventral vessel; vt, vestimental tissue; 1-4, successively more distal sections of branchiae.

posterior to the brain the dorsal vessel divides into two lateral branches
and within a short distance each of these gives rise to anterior, medially situated branches, the obturacular vessels (Figs 21.22, 21.23).
More anteriorly and laterally the branches of the dorsal vessels undergo
a considerable change (Fig. 21.3). In particular, the thick layer of connective tissue becomes thin and little remains of the thick layer of
circular muscles apart from a few muscle bundles at some distance
from the lumen of the branch of the dorsal vessel. Between the inner
surface of the wall of the dorsal vessel and the loosely arranged circular muscle bundles, there appears to be a space, albeit obscured by
a loose matrix of cells (Fig. 21.3). Unfortunately, the anterior fate of
this cavity cannot be determined in the available histological sections of
Riftia.

Studies on histological sections of the dorsal vessel at a similar level
of the vestimentum as it occurs in an undescribed species of vestimentiferan from 13°N on the East Pacific Rise, reveal the course of the dorsal blood vessel and the fate of its perivascular cavity. The structure of
the dorsal vessel and its disposition in the surrounding cavity are
comparable to that of *Riftia* (cf. Figs 21.2 and 21.4). Somewhat posterior
to the branching of the dorsal vessel a new layer appears, interposed
between the heavy connective tissue lining of the dorsal vessel and the
surrounding layer of circular muscles (Fig. 21.5). Within 20 μm the
new layer completely encloses the dorsal vessel, separating it from the
layer of circular muscles (Fig. 21.6). As the dorsal vessel proceeds
anteriorly and undergoes branching there is a continual diminution of
the perivascular cavity until it disappears. There is, however, an
accompanying development of a *new* perivascular cavity that lacks
mesenteries, but is provided with a loose cellular material around the
external surface of the dorsal vessel and its anterior branches (Figs.
21.7 and 21.8).

It should be noted that in *Riftia* there is a different disposition of the
anterior circulatory elements, relative to those of both species of *Lamellibrachia*, and the undescribed species from 13°N. In *Riftia* the branches
of the dorsal vessel proceed along the length of each of the obturacula
(Fig. 21.11), giving rise to transversely oriented lamellar vessels at the
base of each of the branchial lamellae (see Jones 1981b); these in turn
give rise to branchial vessels, one to each branchial filament. In the
other vestimentiferans the branches of the dorsal vessel remain at the
base of the obturacular plume in an approximately transverse plane
(similar to the lamellar vessels of *Riftia*) and give off a succession of
lamellar vessels at the base of each branchial lamella; these in turn give
rise to branchial vessels, one to each branchial filament.

Throughout the length of this portion of the circulatory system the
blood vessels are enclosed in a perivascular cavity. This applies to the

primary branches of the dorsal vessel (Fig. 21.7), the subsequent transversely oriented region of the vessel (Fig. 21.8), and the individual branchial vessels (Figs 21.9, 21.10). All these extensions of the anterior
perivascular cavity of the dorsal vessel are confluent with the cavities of
the branchial filaments (Fig. 21.10).

As mentioned above, the obturacular vessels arise from the branches
of the dorsal vessel at a point close to the branching of the latter, end
blindly near the anterior end of the obturacula, and are situated in a
perivascular cavity supported by mesenteries (Fig. 21.12). When traced
posteriorly through the obturacula the vessels are seen to undergo
a number of dorso-ventral loops (Fig. 21.13) before emerging from the
bases of the obturacula adjacent to the brain through two narrow pass-
ageways in the connective tissue (basal lamina) covering the obturacula.
The obturacular vessels, still in their respective perivascular cavities,
pass into the tissue of the brain (Fig. 21.13), undergo an anterior loop,
then pass dorso-posteriorly to emerge from the brain into a medial,
anterior extension of the excretory organ (Fig. 21.14). Here, the perivascular cavities may fuse, but the obturacular vessels retain their
integrity. In this region it has been noted that a number of ciliated
excretory organ tubules open into the perivascular cavities of the
obturacular vessels (Figs 21.14, 21.15). Subsequent to the emergence
of the obturacular vessels from the excretory organ, still in their perivascular cavities, the vessels join the branches of the dorsal vessel (Figs
21.7, 21.16). An examination of the region immediately before and
after the section illustrated in Figure 21.16 shows that although the
perivascular cavities of the obturacular vessel and of the branch of the
dorsal vessel are in close proximity as the vessels join, there is no connection between them (Fig. 21.16).

Earlier observations on the segmentation of the fourth body region of
Riftia (Jones 1981a) have been confirmed in histological sections of
a juvenile (Fig. 21.17). Each of the segments, marked externally by
nearly circumferential rows of chaetae in the more anterior segments, is
set apart anteriorly and posteriorly by septa (Figs 21.17, 21.19). The
segments are paired by virtue of a medial mesentry (Figs 21.17, 21.18,
21.20). The septa are composed of two layers of muscle bundles separated by a basal lamina (Fig. 21.18). There appears to be a sequential
development of certain features of the opisthosome. In this specimen
(Figs 21.17, 21.18) there are 30 septa delimiting 30 segments (the most
posterior segment has the tissue of the opisthosomal tip as its posterior
boundary). The anterior six (right) and seven (left) segments are provided with rows of chaetae; the posterior 24 and 25 segments have
none. The anteriormost seven septa are provided with well-developed
layers of muscle bundles on both faces (Fig. 21.18); on the next four
septa the two sets of muscle bundles are poorly developed; the posterior

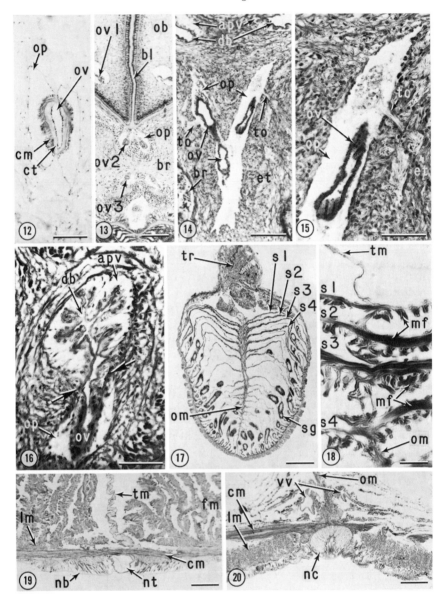

Figs 21.12-21.20. (Figs 21.12, 21.13, 21.17-21.20, *R. pachyptila*; Figs 21.14-21.16, undescribed vestimentiferan from 13°N). Fig. 21.12. Cross section of obturacular vessel in perivascular cavity. Fig. 21.13. Cross section of base of obturaculum showing loops of obturacular vessels in obturaculum and brain tissue. Fig. 21.14. Longitudinal section of obturacular vessels in perivascular cavities surrounded by ciliated tubules of excretory organ, two of which open into cavity. Fig. 21.15. Detail of Fig. 21.14. Fig. 21.16. Junction of branch of dorsal vessel and obturacular vessel, each in own cavity, separated by tissue at points indicated by arrows. Fig. 21.17. Frontal section of

20 septa appear to lack muscle bundles, at least as seen with light microscopy.

Some observations of the ventral nerve cord are pertinent to the following discussion. Throughout the length of the trunk the ventral nerve has the configuration of a longitudinal band (Fig. 21.19). The medial portion is occupied by a single neurular tube. The band is associated with the epithelial layer of the trunk wall, and is separated from a well-developed layer of circular muscles by a basal lamina. At the anterior margin of the opisthosome, the nerve band becomes a more compact, rounded nerve cord (Fig. 21.20). It is single and medial throughout the length of the opisthosome and lacks an obvious neurular tube.

Discussion

A number of points may be made concerning the coelomic cavities of *Riftia* and the unidentified vestimentiferan from 13°N. I am confident that these comments are equally applicable to other vestimentiferans.

First, the trunk coelom is paired by virtue of its medial mesentery and is confined to the trunk *per se*. This includes both the anterior extensions of the trunk cavities and the dorsal trunk wall that together overlie the posterior 20-25 per cent of the vestimentum. Webb (1977) stated that in *Lamellibrachia barhami* 'the trunk coelom is continuous with the restricted coelom of the vestimental region which lies on either side of the ventral [= dorsal] blood vessel.' This would seem to imply that there has been a fusion of the trunk coelom and a (?posterior) vestimental coelom. I do not think this is the case. From a functional standpoint the anterior trunk coelom extensions allow the gonopores to be closer to the tube opening and, because of the dorsal overlap of the

opisthosome of juvenile, at level of centrally located dorsal vessel. Fig. 21.18. Frontal section of medial mesentery of trunk and opisthosome and first four septa of opisthosome. Fig. 21.19. Cross section of ventral nerve 'band' of trunk region. Fig. 21.20. Cross section of ventral nerve 'cord' of opisthosome. Figs 21.12, 21.13, 21.19, 21.20; Mayer's haematoxylin and eosin Figs 21.14-21.18, Masson's trichrome. Scale bar represents 100 μm in Figs 21.12, 21.14; 200 μm in Figs 21.13, 21.17; 21.19, 21.20; 50 μm in Figs 21.15, 21.16; 20 μm in Fig. 21.18. Abbreviations: apv, anterior perivascular cavity; bl, basal lamina; br, brain; cm, circular muscles; ct, connective tissue; db, anterior branch of dorsal vessel, et, ciliated excretory tubules; fm, feather muscles; lm, longitudinal muscles; mf, muscle fibres; nb, ventral nerve band; nc, ventral nerve cord; nt, neurular tube; ob, obturaculum; om, opisthosomal medial mesentery; op, obturacular perivascular cavity; ov, obturacular vessel; ov1-3, successive loops of obturacular vessels; S1-4, first to fourth opisthosomal septa; sg, septal glands; tm, trunk medial mesentery; to, opening of ciliated tubule; tr, trophosome; vv, ventral vessel.

vestimental 'wings' and the so-called vestimental canal formed when the animal is in place in its tube, the genital products always have ready access to the surrounding sea water.

Second, I consider the perivascular cavity of the dorsal vessel through-out its length, provided as the vessel is with a heavy inner layer of con-nective tissue (the 'elastica' of van der Land and Nørrevang 1977) and a thick outer layer of circular muscles, to be another coelom which is also paired by a medial mesentery. This pair of coelomic cavities extend over 50-60 per cent of the middle vestimental length and are isolated from the trunk coeloms and anterior coeloms. Van der Land and Nørrevang (1975, 1977) suggest that there is a single coelom in the vestimentiferan *Lamellibrachia luymesi* and that all apparent coelomic cavities are interconnected. On the basis of what I have observed in the vestimentiferans studies here, I cannot agree.

Third, beginning in the anterior region of the vestimentum, there is yet another perivascular cavity enveloping the dorsal vessel. This associ-ation continues anteriorly along the branches of the dorsal vessel, the lamellar vessels, the individual branchial vessels, and ultimately is con-tinuous with the coelomic spaces of the individual branchial filaments. I view this entire perivascular complex as an elaborate coelomic space.

Fourth, the pair of blind-ending obturacular vessels traverse the length of the obturacular region, each enclosed in perivascular cavities, and again paired by the mesentery that supports the vessels in the cavity. The presence of ciliated excretory openings into this space strongly suggests not only that it should be considered coelomic, but that the excretory ducts should be regarded as coelomoducts. A similar occurrence of ciliated tubules opening into the obturacular cavities has been observed in an undescribed vestimentiferan from hydrothermal vents at 21°N on the East Pacific Rise. This is a different taxon from that at 13°N. Webb (1969a, 1975) found no communication between the excretory organ and either the tentacular or obturacular coelomic systems; this is not surprising for, more often than not, sections in the critical region are equivocal and those shown in Fig. 21.14 and 21.15 are rather fortuitous. It might be added that the observation of van der Land and Nørrevang (1977) that the mesenteries of the obturacular coeloms are probably not continuous seems not to be the case in the material examined here.

Fifth, the vestimentiferan opisthosome is provided with septa that set apart each of the opisthosomal segments. There is also a medial mesen-tery that makes these segmental coelomic spaces paired. This is most probably the case in *Lamellibrachia barhami*, *L. luymesi*, and all other vestimentiferans.

Southward (1980) reported that in the case of *Siboglinum fiordicum* there appears to be an 'embryonic growth zone' at the posterior end of the opisthosome. This observation provides confirmation of the findings of Nørrevang (1970) on a species of *Siboglinum*, in which he stated that '. . . rather large cells are here [in the posterior part of the opisthosome] arranged in arcs with more or less distinct lumina between each layer.' These observations, suggestive of an annelid-like pygidial formation of posterior segments, stand in contrast to the interpretation by Ivanov (1975) of the formation of opisthosomal 'metameres' in the perviate pogonophores, i.e. that the septa are '. . . simple folds of the somatopleura growing inwards from the walls of the body.' Ivanov also states that the metameres so formed '. . . are not independent segments but constitute parts of the unique telosomal region and are unlike the postlarval segments of Annelida.' Although I have not been able to observe larval development in *Riftia* or any other vestimentiferans, I would suggest that the restriction of chaetal rows, and of septa provided with sheets of muscles, to the anterior part of the opisthosome in *Riftia* provides evidence that segmental development is from posterior to anterior, and that the oldest segment is that immediately following the trunk.

Sixth, I would suggest tentatively that the obturaculum (excluding the surrounding branchial complex) represents a first coelomic segment (Fig. 21.22, bf1); the branchial complex and the anterior portion of the dorsal vessel, a second (Fig. 21.22, bf2); the vestimentum and the length of the dorsal vessel throughout which the vessel is comprised of a thick connective tissue layer and a heavy covering of striated circular muscles ('heart') within a paired perivascular cavity, a third segment (Fig. 21.22, bf3); the trunk, with the reproductive organs and endosymbiotic bacteria (trophosome), a fourth segment (Fig. 21.22, bf4); and the posterior section, the opisthosome, which comprises an accruing number of segments (Fig. 21.22, bf5, 9 etc).

From a functional standpoint, the first segment serves to seal off the tube when the worm is withdrawn and to support the branchial complex when the plume is extended from the tube. The second segment bears the branchial plume, which apart from serving the usual respiratory functions probably also takes up materials, such as sulphide for the endosymbiotic bacteria on which the worm appears to rely for most of its nutrition. The third segment not only secretes and applies tube material and extends the length of the tube, but also serves as a point of temporary attachment to hold the plume beyond the opening of the tube. By virtue of a blood sinus system just internal to the ventral ciliated field, this segment probably serves as a secondary site of respiration; the channel formed by the overlapping of the 'wings' provides a means of egress for genital products and for water trapped in the tube when the worm withdraws into the tube. The fourth segment, in addition to being the site of gamete formation and of indirect nutrition, secretes

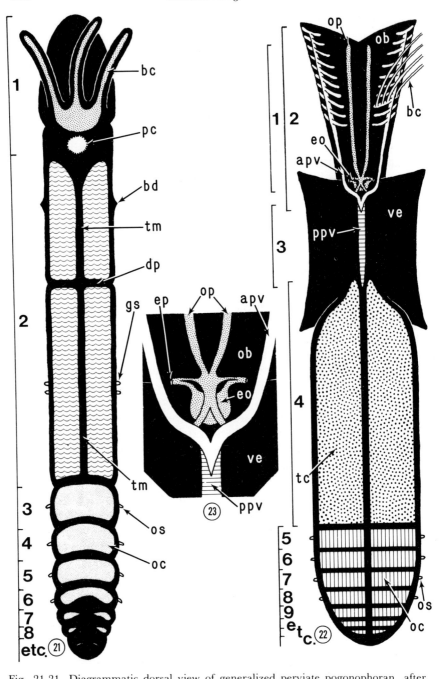

Fig. 21.21. Diagrammatic dorsal view of generalized perviate pogonophoran, after Southward (1980), showing apparent segmentation based on coelomic spaces (segment 1, cephalic region; 2, first metamere; 3 *et seq.*, second and succeeding metameres).

tube material that serves to thicken the main part of the tube. The fifth and succeeding segments serve as the basal holdfast, the circumferential rows of chaetae holding this terminal portion of the worm's body securely to the inner surface of the tube when the longitudinal muscles of the trunk contract during withdrawal into the tube.

I believe that this present account is reasonable, but not thoroughly adequate. I am concerned that I have observed only one opening of the ciliated tubules of the excretory organ, and have not been able to decipher the main mass of tubules. Webb (1975) also was unable to determine the nature and structure of the ciliated tubules that comprise the excretory organ of *Lamellibrachia barhami*. Van der Land and Nørrevang (1975, 1977) likewise were unable to determine the nature of the inner ends of the ciliated tubules of *L. luymesi*. They suggested that there is an 'interstitial' coelom in the vestimentum, i.e. a finely divided coelomic space between the muscles of the region. This is a tempting, but as yet unproven, possibility.

In considering the generalities that emerge from the foregoing discussions a comparison of the obturate and the perviate Pogonophora is useful (Figs 21.21-21.23). If we accept the interpretation of Southward (1980) and restrict ourselves to a consideration of 'apparent segments', rather than regions, it transpires that the first segment of the perviates bears the branchiae and, if it proves to be a consistent structure, the pericardium (Fig. 21.21). The second segment, secondarily divided by the diaphragm, bears the bridle in the anterior part, the girdle externally, and the genital system internally, in the posterior part (Fig. 21.21). It is divided into right and left halves by a medial mesentery. The subsequent segments, comprising the opisthosome, are single and are traversed by three ventral nerve trunks that are provided with what appear to be ganglia (Southward 1980). The coelom of the first segment is unpaired, the coeloms of both parts of the second are paired by their medial mesentery, and each of the opisthosomal segments is unpaired, at least when fully formed (Fig. 21.21).

In the Obturata, as represented by *Riftia pachyptila*, the first segment is the obturaculum *sensu stricto*, i.e. the central part of the obturacular plume, circumscribed by a basal lamina and lined internally by parasagittal

---

Fig. 21.22. Diagrammatic dorsal view of an obturate pogonophoran (*R. pachyptila*), showing apparent segmentation based on coelomic spaces; branchial coelomic spaces on the left and all but one series on the right have been omitted. Fig. 21.23. Detail of Fig. 21.22, at junction of obturaculum and vestimentum. Common patterns of stippling do not imply homology of coeloms. Abbreviations: apv, anterior perivascular cavity; bc, branchial coelom; bd, bridle; dp, diaphragm; eo, excretory organ; ep, excretory pore; gs, girdle chaetae, ob, obturaculum; oc, opisthosomal coelom; op, obturacular perivascular cavity; os, opisthosomal chaetae; pc, pericardium; ppv, posterior perivascular cavity; tc, trunk coelom; tm, trunk medial mesentery ve, vestimentum.

muscles (Fig. 21.11). Coelomic cavities of this segment are the peri-vascular cavities of the obturacular vessels. Because of its position relative to the obturacular vessel (Fig. 21.13), I believe that the brain should be considered a part of the first segment. The second segment has undergone considerable modification, in that it extends anteriorly from behind the base of the first segment to near its anterior end. The second segment is comprised of the branchial complex and a dorsally tapering sheet of muscle just external to the obturaculum whose func-tion appears to be that of supporting the obturaculum and/or the res-piratory complex. Further observations are required to determine whether the excretory organ is part of the first or second segment. The third segment includes the portion of the dorsal vessel that is provided with such a well-developed coat of circular muscles that it can be con-sidered as a 'heart'. In addition, the third segment includes the vestimentum and its lateral 'wings'. The fourth segment is the elongate trunk with the trophosome and gonads. The subsequent segments, comprising the opisthosome, are paired and are traversed by a single nerve trunk that appears to have no ganglia. The coeloms of all segments of *Riftia* are paired, either by virtue of a medial mesentery or by lateral branching. The true nature of the coelom(s?) of the first seg-ment, with its right and left cavities, each with a pair of cavities sep-arated by a mesentery, poses a problem in interpretation which requires embryological/larval material for its resolution.

Earlier, I suggested (Jones 1981a) that the Obturata and the Perviata, in spite of differing in a number of morphological characters, are suf-ficiently related to be retained in the same phylum. For the present I maintain this position, but, on the basis of this comparison of seg-mentation and of the extent of coelomic cavities, I view the two subphyla as being even more distantly related than previously thought. In addi-tion, I have stated that Pogonophora are '. . . most closely related to the phylum Annelida' (Jones 1981a) and that they '. . . stand in close relationship to the Annelida . . .' (Jones 1981b). Once again, on the basis of the segmentation pattern and the nature of the coeloms, this relationship seems on present evidence to be more tenuous than I previously imagined.

The presence of segmentation in the Perviata and the Obturata pro-vides a link with the Annelida. The lack of a medial mesentery in the opisthosome of the Perviata suggests not only a basic morphological difference from the Annelida, but also an embryological one in that given schizocoely, how can single, unpaired coelomic cavities be formed? In the case of the Obturata, the opisthosomal coeloms are not the source of difficulty in relating them to the Annelida. It is the configuration of the anterior coelomic cavities that pose the problem. There is no prece-dent among the Annelida for a schizocoelous coelomate first segment in adults; likewise there is no precedent for a branchial second segment.

In summary, I suggest that the Perviata departed from the annelid line before the Obturata, perhaps before schizocoely was established and gave consistent paired coelomic segments, or alternatively, perhaps the perviates arose entirely independently of the Annelida. I would suggest that the Obturata departed from the annelid line after such schizocoely was established and then evolved independently. I conclude that whatever departures took place, they occurred early in the evolution of the Annelida.

Acknowledgements

I thank K. Fauchald and W. D. Hope of the National Museum of National History, and S. L. Gardiner of the University of North Carolina for fruitful discussions, K. Fauchald for reviewing the manuscript, D. Desbruyeres, CNEXO, Brest and M. Segonzac, CENTOB, Brest, for making available specimens from the French expeditions to 13°N, and S. D. Ripley, Secretary, Smithsonian Institution, for travel funds.

This paper is contribution No. 50 of the Galapagos Rift Biology Expedition, supported by the National Science Foundation.

References

Hyman, L. H. (1951). *The invertebrates*. Vol. 2. *Platyhelminthes and Rhynchocoela, the acoelomate Bilateria*. McGraw-Hill, New York.

Ivanov, A. V. (1975). Embryonalentwicklung der Pogonophora und ihre systematische Stellung. *Z. zool. Syst. Evolutionsforsch*. Sonderheft *1, 10-44*.

Jones, M. L. (1981a). *Riftia pachyptila*, new genus, new species, the vestimentiferan worm from the Galapagos Rift geothermal vents (Pogonophora). *Proc. biol. Soc. Wash*. 93, 1295-313.

—— (1981b). *Riftia pachyptila* Jones: Observations on the vestimentiferan worm from the Galapagos Rift. *Science, N.Y.* 213, 333-6.

Land, J. van der, and Nørrevang, A. (1975). The systematic position of *Lamellibrachia* [sic] (Annelida, Vestimentifera). In *The phylogeny and systematic position of Pogonophora* (ed. A. Nørrevang) *Z. zool. Syst. Evolutions-forsch*. Sonderheft 1, 86-101.

—— (1977). Structure and relationships of *Lamellibrachia* (Annelida, Vestimentifera). *K. danske Vidensk. Selsk. biol. Skr*. 21, 1-102.

Nørrevang, A. (1970). On the embryology of *Siboglinum* and its implications for the systematic position of the Pogonophora. *Sarsia* 42, 7-16.

Southward, E. C. (1980). Regionation and metamerisation in Pogonophora. *Zool. Jb. Abt. Anat. Ontog*. 103, 264-75.

Webb, M. (1969a). *Lamellibrachia barhami*, gen. nov., sp. nov. (Pogonophora), from the northeast Pacific. *Bull. mar. Sci*. 19, 18-47.

—— (1969b). An evolutionary concept of some sessile and tubicolous animals. *Sarsia* 38, 1-8.

—— (1969c). Regionation and terminology of the pogonophoran body. *Ibid.* 38, 9-24.

—— (1975). Studies on *Lamellibrachia barhami* (Pogonophora). I. The excretory organs. In *The phylogeny and systematic position of Pogonophora.* (ed. A. Nør-revang) *Z. zool. Syst. Evolutionsforsch.* Sonderheft 1, 86-101.

—— (1977). Studies on *Lamellibrachia barhami* (Pogonophora). II. The repro-ductive organs. *Zool. Jb. Abt. Anat. Ontog.* 97, 455-81.

Webb, M. and Ganga, K. S. (1980). Studies on *Lamellibrachia barhami* (Pogo-nophora). III. Plaques, glands and epidermis. *Annale Univ. Stellenbosch,* Ser. A2 (Soologie) 2, 1-27.

22. Non-skeletalized lower invertebrate fossils: a review

S. CONWAY MORRIS

Department of Earth Sciences, University of Cambridge, Cambridge, UK

Abstract

The fossil record of those lower invertebrates lacking substantial hard skeletal parts is briefly reviewed. Much of the evidence is dependent on exceptional examples of soft-part preservation, but in a number of cases trace fossils provide additional evidence. Examples are discussed from amongst the cnidarians (especially those from the late Precambrian Ediacaran fauna), ctenophores, platyhelminths, nemerteans, nematodes, nematomorphs, rotifers, priapulids, pogonophores, echiurans, and sipunculans. The available fossil record of these groups is unlikely to revolutionize our understanding of lower invertebrate phylogeny, but it does give some indications on first appearances and the subsequent evolution of lower invertebrates, especially with respect to adaptive radiations.

Introduction

There is no precise agreement on the number of metazoan phyla; most estimates arrive at a figure of about 35 and of these some 20 are generally included in the lower invertebrates. Hard parts capable of fossilization occur in a variety of these lower invertebrate phyla, but the majority are effectively soft-bodied and have left a poor fossil record. The aim of this paper is to review briefly what can be salvaged from such palaeontological debris. No attempts is made to address the extensive information on well-skeletized lower invertebrates; such exclusion embraces groups with spiculate skeletons that have a relatively well documented fossil record, e.g. sponges and octocorals. Much of the evidence in this review is thus based on soft-bodied fossils that normally have a minimal preservation potential and require exceptional conditions for fossilization. Trace fossils left by these animals may occasionally be sufficiently diagnostic to identify the maker, but I believe the greatest future potential may lie in the search of micropalaeontological samples for distinctive but hitherto unrecognized hard parts.

Despite the antiquity of the Earth (*c.* 4550 Ma) the fossil record indicates that metazoans only became abundant in the late Precambrian and thereafter diversified in a series of major adaptive radiations during which most, and perhaps all, the major bodyplans evolved. Indeed it seems clear that a number of major bodyplans became extinct during the early Palaeozoic, and although extinct phyla may have played an important role in early metazoan evolution, the paucity of information does not yet allow their inclusion in a reliable phylogenetic framework. The initial metazoan radiation spanned a period of approximately 150 Ma (*c.* 680-530 Ma), and is best known from the relatively abrupt appearance of hard parts near the base of the Cambrian (*c.* 570 Ma). The time of appearance of the first metazoans is, however, far from established and may not coincide with the onset of the late Precambrian adaptive radiations. Equivocal records of trace (e.g. Kauffman and Steidtmann 1981) and body fossils (e.g. Towe 1981), together with indirect evidence based on the supposed constancy of molecular clocks (Runnegar 1982a) could indicate that metazoans evolved hundreds of millions of years before their abundant appearance in the stratigraphical column.

Cnidarians and ctenophores

Late Precambrian (*c.* 680-570 Ma) metazoan fauna are generally referred to as Ediacaran, the name stemming from early discoveries of these fossils in the Edicara Hills of the Flinders Ranges, South Austrialia. Ediacaran faunas are now known to have an almost world-wide distribution (Glaessner 1979a), and typically are dominated by forms interpreted as cnidarians. A number of genera have been placed in extant groups, with representatives of scyphozoans (Wade 1969, 1972a; Palij, Posti, and Fedonkin 1979; Glaessner 1979a; Fedonkin 1981), chondrophores (Glaessner and Wade 1966, Wade 1971, 1972a), possibly other hydrozoans (Wade 1972a; Fedonkin 1981) and pennatulaceans (Glaessner and Wade 1966; Jenkins and Gehling 1978) all being identified. Such assignments, however, are not without their problems. For example, Wade (1969, 1971) clearly indicates that placing certain medusoids with an apparently scyphozoan grade of organization in the

Scyphozoa is a legitimate exercise, although unresolved patterns of convergence and polyphyletic origins suggest that valuable phylogenetic information may be obscured by such a procedure. The vicissitudes surrounding the correct taxonomic placement of a variety of frond-like fossils are also instructive. Earlier accounts regarded them as pennatulaceans (Glaessner and Wade 1966), but more recently the concept of a distinctive group of Precambrian leaf-like fossils, sometimes folded into saccate forms, and generally known as the petalonamids (Pflug 1972; Glaessner and Walter 1975; Glaessner 1979a) has won increasing acceptance with only a few genera e.g. *Charniodiscus* still being accepted by some workers as pennatulaceans (Jenkins and Gehling 1978). Unfortunately, a full understanding of petalonamid anatomy and affinities is still awaited, not least because of disagreements on the interpretation of fossilized structures and apparently extensive synonymy between supposedly separate genera (Glaessner and Walter 1975; Jenkins, Plummer, and Moriarty 1981). Jenkins *et al.* (1981) have presented a reconstruction of the petalonamid *Ernietta*, showing it to be a sac-like organism anchored or embedded in the sea floor and possessing a flared aperture. The body wall was evidently lamellate and divided by approximately radial partitions into a series of canals. Further work could show the petalonamids to be separate from the cnidarians, representing a distinct phylum, albeit of diploblastic organization.

Even if existing proposals of certain Ediacaran fossils representing scyphozoans and other extant groups continue to command acceptance, this still leaves a considerable variety of medusoids and other cnidarians (Glaessner and Wade 1966; Glaessner 1979a, Palij *et al.* 1979; Fedonkin 1981) which have defied ready classification, but could represent crucial stages in early cnidarian evolution. Fedonkin (1982) has hinted at the recognition of 'a separate taxon of high rank' to accommodate a variety of enigmatic medusoids, but its characterization is still fluid. Fedonkin (1982) also stressed the unusual occurrence of a tri-radiate symmetry in certain medusoids, e.g. *Albumares* and *Skinnera*, and noted further that *Tribrachidium*, long regarded as an early echinoderm with three-fold symmetry, may represent a cnidarian.

The stratigraphic position of Ediacaran cnidarians should in principle make them of immense value in providing direct palaeontological evidence of early stages of cnidarian evolution and the origin of major taxa within the phylum. Certainly the diversity of forms is consistent with their evolution being dominated by an early series of adaptive radiations. Glaessner (1971) indicated a central role for *Conomedusites* in the ancestry of the Conulata and probably Scyphozoa, but it must be admitted that present evidence otherwise adds comparatively little to neontological speculation. Wade (1972a) suggested that if the medusoid *Cyclomedusa* represented a hydrozoan then it may have shared a common ancestor with the chondrophores. In general, however, interrelationships between known Ediacaran fossils are not well established, in part because the respective species have a relatively simple morphology, an assessment of which is further confused by vagaries in the style of preservation. Indeed, Wade (1972a) indicated that some major steps in cnidarian evolution must have been pre-Ediacaran and unknown on present fossil evidence.

The abundance of Ediacaran cnidarians at this apparently early stage of metazoan evolution might appear to accord with the general concensus regarding their primitive status. Notwithstanding unresolved questions of their role, if any, in the evolution of triploblastic metazoans, it should be stressed that the major evolutionary events that led to the origin of other lower invertebrates may have been largely restricted to small metazoans, especially if the paedomorphic process of progenesis (Gould 1977) is invoked as a significant mechanism. Fossilization of such minute animals would require exceptional conditions that are unavailable in typical Ediacaran deposits, even assuming that such hypothetical creatures occupied these habitats. Exceptionally preserved Precambrian microfloras are well known from fine-grained cherts and more unusually other rocks such as shales, but no cellular remains that could be unequivocally referred to metazoans appear to have been identified. S. M. Awramik (personal communication) has, however, identified trace fossils in late Precambrian cherts ($c.$ 700 Ma) from China which he regards as possibly representing the activities of a meiofauna.

Common experience suggests that soft tissues cannot survive for protracted periods in sedimentary environments in the face of decay and scavenging. Ediacaran fauna are almost entirely soft-bodied, yet are widespread and locally abundant; such a distribution would appear to require some special explanation. Reasons advanced revolve around an apparent absence of scavengers and predators, combining with limited bioturbation of the sediment by a restricted infauna (Sepkoski 1979). The reduced frequency of younger soft-bodied faunas contrasts strongly with Ediacaran occurrences and preservation of the former faunas evidently require a highly unusual sequence of events that served to exclude scavengers and arrest decay. In consequence, our knowledge of post-Ediacaran soft-bodied animals is largely dependent on a sequence of extraordinary fossil faunas, scattered through the stratigraphic column, which can only hint at the former diversity of fossil lower invertebrates.

The relative scarcity of post-Ediacaran soft-bodied faunas makes it difficult to assess later stages of cnidarian evolution, but even the Cambrian record of medusoids (Scrutton 1979; Pickerill 1982) suggests no striking continuity with Ediacaran fauna. The issue of post-Ediacaran cnidarian evolution is further confused by the probable misidentification of trace fossils as medusoids (Häntzschel 1970; Osgood 1970; Für-sich and Kennedy 1975) and sea-pens (Bayer 1955, see Häntzschel 1975).

Stanley and Stürmer (1983) have recorded a single example of a fossil ctenophore from the Early Devonian Hunsrückschiefer of West Germany, a deposit long celebrated for its exquisite preservation of fossils as thin pyritic films. This Devonian ctenophore, *Paleoctenophora brasseli*, shows little morphological change in comparison with living examples and appears to belong within the Cydippida. A curious organism from the mid-Cambrian Burgess Shale, *Fasciculus vesanus*, was interpreted by Simonetta and Cave (1978) as a possible cnidarian, but it may conceivably represent another fossil ctenophore, albeit with less similarity to *Paleoctenophora* and extant forms (Collins, Briggs, and Conway Morris 1983). A medley of other fossils that have been interpreted as cteno-phores with varying degrees of optimism do not stand up to critical examination (Caster and Brooks 1956; see also Häntzschel 1975; Smith 1982; Stanley and Stürmer 1983).

Platyhelminths

Parasitic platyhelminths appear to lack a fossil record, although indirect lines of evidence may provide some clues (Conway Morris 1982). Amongst the turbellarians Pierce (1960, 1964) identified a variety of supposed Miocene examples from California, but my re-examination of this material does not support such a determination. Pierce (1960) also illustrated material of the same age which he regarded as possibly representing turbellarian eggs. This identification is dubious, but Binford's (1982) description of sub-fossil turbellarian (Neorhabdocoela) eggs from Venezuelan lake sediments emphasizes the likelihood that more ancient lacustrine sequences will provide further examples. More significant in the context of this review is a claim that some late Precambrian (Ediacaran) worms (*Dickinsonia, Palaeoplatoda*) represent turbellarians (Palij et al. 1979; Fedonkin 1981). *Dickinsonia* has traditionally been regarded as a giant polychaete, closely allied to the aberrant genus *Spinther* (Wade 1972b). Such similarities that exist with *Spinther* are surely convergent (Conway Morris 1979), but Fedonkin's assignment of *Dickinsonia* and *Palaeoplatoda* to the turbellarians has not won the support of Runnegar (1982b) who presents various lines of evidence to reaffirm some sort of annelid relationship. A compromise position is that these Ediacaran worms are in some way intermediate between turbellarians and annelids (Palij *et al.* 1979). Other candidates for fossil turbellarians are at best dubious; *Platydendron ovale* from the mid-Cambrian Burgess Shale is interpreted by Simonetta and Cave (1978) as a polyclad, but I regard the only known specimen as indeterminate. A supposed flatworm from the late Precambrian or early Cambrian of Alaska (Allison 1975) appears misidentified and may represent a sponge spicule (Cloud, Wright, and Glover 1976).

Nemerteans

A number of examples have been proffered as fossil nemerteans (Conway Morris 1977a). However only the Carboniferous *Archisymplectes rhothon* from the Mazon Creek ironstone nodules of Illinois (Schram 1973) and possibly Bear Gulch Limestone in Montana (Schram 1979) has a claim to probable legitimacy, although identification is based on gross outline. A previously favoured candidate, *Amiskwia sagittiformis* from the Burgess Shale, was identified by Owre and Bayer (1962) as similar to bathypelagic forms such as *Nectonemertes*. This similarity, however, is almost certainly a result of convergence and *Amiskwia* cannot be regarded as a nemertean (Conway Morris 1977a). Barring the discovery of other soft-bodied fossil nemerteans, their geological history may be further elucidated from the remains of stylets. Nemertean stylets have a characteristic ultrastructure and, moreover, contain significant quantities of calcium phosphate (Stricker and Cloney 1981, 1982; Stricker 1983). Such a composition, perhaps augmented by post-mortem phosphatization when buried in sediment, suggests that isolated stylets could be recognized in the fossil record (S. Bengtson, personal communication). One horizon that may repay study is the Lower Cambrian Tommotian stage where a wide variety of phosphatic micro-fossils is known. They are of special significance because they represent some of the earliest metazoan hard parts (Matthews and Missarzhevsky 1975). The affinities of many of these microfossils are enigmatic, although none are directly comparable to known nemertean stylets.

Nematodes, Nematomorphs, Rotifers, and Priapulids

The fossil record of nematodes rests on moderately secure ground with a number of well attested examples, especially from Tertiary ambers (reviewed in Conway Morris 1982; see also Poinar 1981). The earliest record is generally given as Lower Carboniferous and is based on Størmer's (1963) description of nematodes (*Scorpiophagus*) in apparent saprobiotic association with a carcass of a scorpion. Some authorities on recent nematodes (e.g. Poinar 1983), however, have doubted this identification. Schram (1973) interpreted *Nemavermes mackeei* from the Upper Carboniferous Mazon Creek nodules as a large nematode, although this identification also remains equivocal. Further examples from the Bear Gulch Limestone (Carboniferous) were described by Schram (1979). Still more controversial are the remains of nematodes and their reproductive organs from Lower Carboniferous limestone of South Wales (Xian-tao Wu 1983). Minute borings in recent and fossil foraminifers were interpreted by Sliter (1971) as representing the activities of predatory nematodes. Moussa (1970) described convincing examples of sinusoidal nematode trails from Eocene lacustrine sediments in Utah, while Runnegar (1982b) expanded this concept by regarding other sinusoidal trace fossils, e.g. *Cochlichnus*, as representing nematode locomotion. *Cochlichnus* occurs as far back as the late Precambrian (Webby 1970; Palij *et al.* 1979) and if accepted as reflecting the presence of nematodes would have important implications on their evolution.

Nematomorph worms are known only in the fossil record from one example (*Gordius tenuifibrosus*) recorded in Eocene lignite (Voigt 1938). Commenting on this occurrence, Sciacchitano (1955) suggested that the specimen was in fact synonymous with an extant species (*G. albopunctatus*) and proposed further that as *Gordius* is regarded as a relatively advanced genus this pointed to an extended pre-Eocene history for the phylum.

The best documented occurrence of a fossil rotifer is ?*Keratella* sp. from Eocene sediments in South Australia (Southcott and Lange 1971). The only known specimen is represented by the lorica, which was presumably more resistant to decay, and one or more associated structures that appear to represent parthenogenetic eggs. Far less credible is Pierce's (1964) description of *Pararotifer enigmatica* from the Miocene of California; identification of this supposed rotifer must be regarded as highly suspect following my re-examination.

Our knowledge of fossil priapulids is largely dependent on a rich fauna (seven monospecific genera representing at least five families) from the mid-Cambrian Burgess Shale (Conway Morris 1977b). The most abundant species, *Ottoia prolifica*, is not dissimilar to the extant *Halicryptus spinulosus*, but the remaining forms are not directly comparable to living priapulids although their place within the phylum is established on the basis of homologies within the proboscis. *Selkirkia columbia* inhabited a finely annulated organic tube which is neither directly comparable to the lorica of larval priapulids nor the tube of *Maccabeus tentaculatus* (Por and Bromley 1974). The tube of *S. columbia* apparently possessed a moderate resistance to decay and has been recorded from other Middle Cambrian rocks (R. A. Robison, personal communication) but it is only within the Burgess Shale where exceptional preservation conditions prevailed that the associated soft parts are preserved. The largest Burgess Shale priapulid, *Louisella pedunculata*, sometimes reached 20cm in length. Direct evidence of its mode of life is not available, but the flattened body with two strands of elongate papillate gills running along much of the trunk recalls anatomically similar polychaetes that live infaunally and ventilate themselves by undulatory movements (Clark 1964). The proboscis of *L. pedunculata* bore lobate scalids that perhaps had a secretory function and helped to shore up the burrow walls. The great diversity amongst Burgess Shale priapulids is consistent with their being in the throes of a major adaptive radiation, with modern priapulids representing the evolutionary relic of a once far more diverse group. The fossil record of priapulids is silent between the Middle Cambrian and Carboniferous, but *Priapulites konecniorum* (Schram 1973; Conway Morris 1977b) from the Upper Carboniferous Mazon Creek nodules had a modern aspect and might even belong within the Priapulidae. To what extent, however, living priapulids may be regarded as living fossils effectively unchanged since the Carboniferous is uncertain. Further adaptive radiations may have occurred, and little is known about the origins of the extant genera *Maccabeus* (Por and Bromley 1974), *Tubiluchus* (Calloway 1975) and *Meiopriapulus* (Morse 1981).

Annelids

Amongst annelids, by far the best fossil representation is amongst the polychaetes. The fossil record of tubicolous polychaetes, especially those that secrete a calcareous tube, is relatively well documented. Questions, however, surround some early Phanerozoic worm tubes. Glaessner (1976) regarded the late Precambrian or early Cambrian worm tube *Cloudina* as representing an early polychaete, broadly comparable to serpulids. This worker also identified the enigmatic conical fossil *Volborthella* as belonging to the polychaetes, although Yochelson (1977) preferred to regard this Lower Cambrian genus, and the apparently related *Salterella*, as representatives of an extinct phylum named the Agmata. Another area of confusion arises from the misidentification of loosely coiled gastropods as polychaete tubes (e.g. Burchette and Riding 1977). Notwithstanding these difficulites most polychaete worm tubes are easily identifiable, and may in rare cases even contain the remains of the original inhabitant (Hayward 1977; for an example in a boring see Cameron 1967). Preservation of non-tubicolous soft-bodied polychaetes is scattered but examples are known from a number of horizons. Mention was made above of the late Precambrian (Ediacaran) worm *Dickinsonia*, supposedly allied to the aberrant polychaete *Spinther* (Wade 1972b). The coeval worm *Spriggina* has been compared with tomopterid polychaetes, but this resemblance is superficial (Conway Morris 1979) and Birket-Smith (1981) has presented a novel reconstruction which indicates that *Spriggina* has a closer affinity with the arthropods. Un-doubted polychaete worms, however, are known from the Cambrian. Glaessner (1979b) has described worm material that he regards as representing polychaetes. One form, *Myoscolex ateles*, is particularly interesting because the 'chaetae' are composed of calcium phosphate, a composition otherwise unknown amongst polychaetes. The Middle Cambrian Burgess Shale has a considerable polychaete diversity (Conway Morris 1979), and although one species (*Canadia spinosa*) has resemblances to the extant chrysopetalids the similarities may be due to convergence and the other Burgess Shale genera cannot be placed in recent families. Later faunas showing soft-part preservation also contain polychaetes, some of which have jaw apparatuses *in situ* (Schram 1979; Thompson 1979). It is, however, the isolated teeth, dispersed during post-mortem destruction of the polychaete, that are a neglected source of information. These teeth, referred to as scolecodonts, may occur in great abundance in some sediments and occasionally are recovered as associated apparatuses. The situation is complicated because although a considerable number of polychaete families possess teeth, in many cases their composition is not robust enough to withstand diagenesis within the sediment. Experiments with immersing polychaete jaws in potassium hydroxide (Szaniawski 1974) and sodium hypochlorite or bleach (Thompson 1979) have shown that teeth of certain polychaete families, e.g. nereids, are susceptible to destruction and their apparent absence from the fossil record might be due to a low preservation potential. A simple correlation, however, between resistance to alkaline fluids and long-term survival within chemically complex sediments cannot easily be drawn. In a number of cases, e.g. nephtyids, the geological record is sparse, in spite of the resistance of modern jaws. Isolated scolecodonts of nephtyids apparently remain unrecognized despite the occurrence of soft-bodied remains in the Carboniferous (Schram 1979; Thompson 1979). The relative neglect of scolecodont studies, despite substantial efforts by the Polish school, stems in part from a confused taxonomy and problems in reconstructing original apparatuses, combined with an apparently limited biostratigraphic potential. Kielan-Jaworowska (1966) presented evidence for an adaptive radiation within the lower Palaeozoic Eunicea which included the evolution of extinct major groups. Evidence on Mesozoic scolecodonts is more restricted, but this is probably as much a failure to look as any decline in polychaete diversity. However, with many families lacking jaws, the available record of scolecodonts can give only partial insights into former levels of polychaete diversity.

The poor fossil record of oligochaetes (reviewed by Conway Morris, Pickerill, and Harland 1982) and leeches (Kozur 1970) must be due in part to the relative paucity of terrestrial (s.l.) as against marine

sediments in the geological column. Examples of Carboniferous oligochaetes are moderately well documented, but whether their origins
may be traced to the earlier invasions of land by plants and animals in
the Silurian and Devonian is still speculative Traces of oligochaete
activity, including burrows and faecal pellets, have been recognized
in ancient Carboniferous and Triassic soils (palaeosols) and various
aspects of oligochaete history may be unravelled by further investigation of Palaeozoic palaeosols.

As yet, the fossil record is silent on the suggestion that the ancestral
annelid was oligochaete-like rather than being represented by a polychaete (Clark 1964). Attention, however, should be drawn to the
Lower Palaeozoic worms *Palaeoscolex* and *Protoscolex* (in part) (Conway
Morris 1977b; Glaessner 1979b; Conway Morris *et al.* 1982) which
appear to represent a distinct annelid group with metameric chaetigerous bands. Relationships to oligochaetes may be distant, but this need
not disbar them from a position relatively close to the ancestral annelid
(Conway Morris 1977b; Runnegar 1982b).

The ectocommensal and parasitic myzostomids, generally regarded
as a separate annelidan class, may produce galls and other disturbances
within the stereom tissue of their echinoderm hosts. Although similar
pits and galls are well known in the fossil record, especially amongst
crinoids, relatively few can be unequivocally ascribed to the activities
of myzostomids (Conway Morris 1982). The earliest well documented
examples appear to be Carboniferous (Welch 1976).

Echiurans

Amongst echiurans both body and trace fossils have been identified
with varying degrees of certainty. In the latter category echiuran activity has been identified in Silurian sandstones (Risk 1973) and within
the test of a Cretaceous echinoid (Joysey 1959), while Runnegar (1982b)
attributed the trace fossil *Plagiogmus*, common in many lower Cambrian rocks, to 'an animal of the echiurid grade of organization'. *Pro-
techiurus edmondsi* from the late Precambrian of Namibia was identified
by Glaessner (1979c) on the basis of a single specimen as an echiuran.
Abundant specimens of the Carboniferous *Coprinoscolex ellogimus* occur
in the ironstone nodules of the Mazon Creek area in Illinois. They
generally contain abundant faecal pellets, but apart from occasional
gut traces little is known of the internal anatomy (Jones and Thompson
1977). Neither occurence throws any light on the origin of echiurans
nor on the vexed question of whether their unsegmented coelom is
original or derived from a segmented ancestor.

Pogonophores

Fossil tubes, e.g. *Sabellidites* and *Saarina*, that have been widely inter-preted as belonging to pogonophores are abundant in some rocks of late Precambrian/early Cambrian age, especially from eastern Europe and parts of the U.S.S.R. (Sokolov 1972). Despite gross similarities between these tubes, generally referred to as sabelliditids, and those of modern pogonophores, doubt has been thrown on a relationship because of apparent differences in ultrastructure. In comparision with extant pogonophores the sabelliditid tube wall shows a more complex layering and also differences in the mode of construction of the outer wrinkles (Urbanek and Mierzejewska 1977). These Polish workers stress, however, that their comparisons are based on limited material and they further question the role of ultrastructural studies in phylogenetic studies. Other Lower Cambrian tubes that have been compared with pogonophores include *Hyolithellus* (Poulsen 1963) and *Coleoloides* (Brasier 1979), but the thick calcareous tube of the latter genus throws considerable doubt on this comparison.

If any of these late Precambrian/early Cambrian tubes are accepted as pogonophores, their abundance in various shallow-water sediments stands in apparent contrast to the general restriction of recent pogono-phores to deeper water (see also Poulsen 1963), although it should be noted that little is known of Cambrian deep-sea biotas. If this contrast in habitat is genuine it could indicate a subsequent migration of pogonophores into deeper water, perhaps under the pressure of in-creased competition and predation (Sokolov 1972). Such an explana-tion has been widely invoked for other deep-sea groups with Palaeozoic and Mesozoic shallow water ancestors but, as Lipps and Hickman (1982) stress, the colonization of the deep-sea is neither well understood nor a simple one-way process. Younger fossils that have been allied with the pogonophores include material from the Ordovician (Kozlow-ski 1967), Silurian (Sokolov 1972), Oligocene (Adegoke 1967) and Pliocene (de Heinzelin 1964), but this sparse record gives no firm guide on any possible dates of migration into deeper water.

Sipunculans

Soft-bodied remains of sipunculans appear to be unknown and those examples listed by Howell (1962) are not acceptable on present evidence. Some modern sipunculans live in borings, typically carbonate reef-rock, that appear to have been excavated by the animals themselves (e.g. Rice 1969; Bromley 1970). Fossil borings in hard substrata are well known, but few examples can be unequivocally attributed to sipunculan activity. Seilacher (1969) illustrated a boring (diameter 52mm) from the Pleistocene of Peru which he regarded as being produced by a giant sipunculan, while Pemberton, Kobluk, Yeo, and Risk (1980) present a reasonable case for specimens of *Trypanites* in the Devonian of Ontario representing sipunculan borings. Within soft sediments examples of the infaunal trace *Zoophycos* from the continental slope off north-west Africa were attributed to the feeding activities of sipunculans (Wetzel and Werner 1981). *Zoophycos* is known as a trace fossil from the Ordovician onwards, although it would be premature to ascribe all fossil examples to sipunculans. In discussing substrate specificity of the Devonian tabulate coral *Pleurodicytum* Brett and Cotterell (1982) presented indirect but convincing evidence that gastropod shells encrusted by this coral had been occupied by sipunculan worms rather than the original inhabitant. *Pleurodicytum* itself shows a frequent association with the spiral worm tube *Hicetes* which may in turn have been occupied by a sipunculan (Gerth 1952).

Acknowledgements

I thank R. G. Bromley (University of Copenhagen) for drawing my attention to evidence of fossil sipunculan borings, S. Bengtson (University of Uppsala) for critically reading the manuscript and B. West for typing several drafts of the manuscript. This is Cambridge Earth Sciences Contribution 548.

References

Adegoke, O. S. (1967). A probable pogonophoran from the early Oligocene of Oregon. *J. Paleont.* 41, 1090-4.

Allison, C. W. (1975). Primitive fossil flatworm from Alaska: new evidence bearing on ancestry of Metazoa. *Geology*, 3, 649-52.

Bayer, F. M. (1955). Remarkably preserved fossil sea-pens and their Recent counterparts. *J. Wash. Acad. Sci.* 45, 294-300.

Binford, M. W. (1982). Ecological history of Lake Valencia, Venezuela: Interpretation of animal microfossils and some chemical, physical, and geological features. *Ecol. Monogr.* 52, 307-37.

Birket-Smith, S. J. R. (1981). A reconstruction of the Pre-Cambrian *Spriggina*. *Zool. Jb. Anat.* 105, 237-58.

Brasier, M. D. (1979). The Cambrian radiation event. *Syst. Ass. Spec. Vol.* 12, 103-59.

Brett, C. E. and Cottrell, J. F. (1982). Substrate specificity in the Devonian tabulate coral *Pleurodictyum*. *Lethaia* 15, 247-62.

Bromley, R. G. (1970). Borings as trace fossils and *Entobia cretacea* Portlock, as an example. In *Trace fossils* (eds T. P. Crimes and J. C. Harper) pp. 49-90. Seel House Press, Liverpool.

Burchette, T. P. and Riding, R. (1977). Attached vermiform gastropods in Carboniferous marginal marine stromatolites and biostromes. *Lethaia* 10, 17-28.

Calloway, C. B. (1975). Morphology of the introvert and associated structures of the priapulid *Tubiluchus corallicola* from Bermuda. *Mar. Biol.* 41, 161-74.

Cameron, B. (1967). Fossilization of an ancient (Devonian) soft-bodied worm. *Science, N. Y.* 155, 1246-8.

Caster, K. E. and Brooks, H. K. (1956). New fossils from the Canadian-Chazyan (Ordovician) hiatus in Tennessee. *Bull. Am. Paleont.* 36, 157-99.

Clark, R. B. (1964). *Dynamics in metazoan evolution.* Clarendon Press, Oxford.

Cloud, P., Wright, J. and Glover, L. (1976). Traces of animal life from 620-million-year-old rocks in North Carolina. *Am. Scient.* 64, 396-406.

Collins, D. H., Briggs, D. E. G., and Conway Morris, S. (1983). New Burgess Shale fossil sites reveal Middle Cambrian faunal complex. *Science, N. Y.* 222, 163-7.

Conway Morris, S. (1977a). A redescription of the Middle Cambrian worm *Amiskwia sagittiformis* Walcott from the Burgess Shale of British Columbia. *Palaeont. Z.* 51, 271-87.

—— (1977b). Fossil priapulid worms. *Spec. Pap. Palaeontology.* 20, 1-95.

—— (1979). Middle Cambrian polychaetes from the Burgess Shale of British Columbia. *Phil. Trans. R. Soc.* B285, 227-74.

—— (1982). Parasites and the fossil record. *Parasitology* 82, 489-509.

——, Pickerill, R. K., and Harland, T. L. (1982). A possible annelid from the Trenton Limestone (Ordovician) of Quebec, with a review of fossil oligochaetes and other annulate worms. *Can. J. Earth Sci.* 19, 2150-7.

Fedonkin, M. A. (1981). White Sea biota of Vendian (Precambrian nonskeletal fauna of the Russian Platform North). *Akad. Nauk. S.S.S.R. Trudyi* 342, 1-100 (in Russian).

—— (1982). Precambrian soft-bodied fauna and the earliest radiation of invertebrates. *Third N. Am. Paleont. Conv. Proc.* 1, 165-7.

Fürsich, F. T. and Kennedy, W. J. (1975). *Kirklandia texana* Caster—Cretaceous hydrozoan medusoid or trace fossil chimaera? *Palaeontology* 18, 665-79.

Gerth, H. (1952). Die von Sipunculiden bewohnten lebenden und jungtertiaren Korallen und der wurmförmige Körper von *Pleurodictyum. Palaont. Z.* 25, 119-26.

Glaessner, M. F. (1971). The genus *Conomedusites* Glaessner & Wade and the diversification of the Cnidaria. *Palaeont. Z.* 45, 7-17.

—— (1976). Early Phanerozoic annelid worms and their geological and biological significance. *J. geol. Soc. Lond.* 132, 259-75.

—— (1979a). Precambrian. In *Treatise on invertebrate paleontology* (eds R. A. Robison and C. Teichert) Part A, pp. A79-A118. Geological Society of America and University of Kansas Press, Lawrence, Ka.

—— (1979b). Lower Cambrian Crustacea and annelid worms from Kangaroo Island, South Australia. *Alcheringa* 3, 21-31.

—— (1979c). An echiurid worm from the Late Precambrian. *Lethaia* 12, 121-4.
—— and Wade, M. (1966). The late Precambrian fossils from Ediacara, South Australia. *Palaeontology* 9, 599-628.
—— and Walter, M. R. (1975). New Precambrian fossils from the Arumbera Sandstone, Northern Territory, Australia. *Alcheringa* 1, 59-69.
Gould, S. J. (1977). *Ontogeny and phylogeny.* Belknap Press, Cambridge, Mass.
Häntzschel, W. (1970). Star-like trace fossils. In *Trace fossils* (eds T. P. Crimes J. C. Harper) pp. 201-14. Seel House Press, Liverpool.
—— (1975). Trace fossils and problematica. In *Treatise on invertebrate paleontology* (ed. C. Teichert) Part W, Supplement 1, pp. W1-W269. Geological Society of America and University of Kansas Press, Lawrence, Ka.
Hayward, B. W. (1977). Lower Miocene polychaetes from the Waitakere Ranges, North Auckland, New Zealand. *Jl. R. Soc. N.Z.* 7, 5-16.
Heinzelin, J. de (1964). Pogonophores fossiles? *Bull. Soc. belg. Geol. Paleont. Hydrol.* 73, 501-9.
Howell, B. F. (1962). Worms. In *Treatise on invertebrate paleontology* (ed. R. C. Moore) Part W, pp. W144-W177. Geological Society of America and University of Kansas Press, Lawrence, Ka.
Jenkins, R. J. F. and Gehling, J. G. (1978). A review of the frond-like fossils of the Ediacara assemblage. *Rec. S. Aust. Mus.* 17, 347-59.
Jenkins, R. J. F., Plummer, P. S. and Moriarty, K. C. (1981). Late Precambrian pseudofossils from the Flinders Ranges, South Australia. *Trans. R. Soc. S. Aust.* 105, 67-83.
Jones, D. and Thompson, I. (1977). Echiura from the Pennsylvanian Essex fauna of northern Illinois. *Lethaia* 10, 317-25.
Joysey, K. A. (1959). Probable cirripede, phoronoid, and echiuroid burrows within a Cretaceous echinoid test. *Palaeontology* 1, 397-400.
Kauffman, E. G. and Steidtmann, J. R. (1981). Are these the oldest metazoan trace fossils? *J. Paleont.* 55, 923-47.
Kielan-Jaworowska, Z. (1966). Polychaete jaw apparatuses from the Ordovician and Silurian of Poland and a comparison of modern forms. *Palaeont. Pol.* 16, 1-152.
Kozlowski, R. (1967). Sur certains fossiles Ordoviciens à test organique. *Acta palaeont. pol.* 12, 99-132.
Kozur, H. (1970). Fossile Hirudinea aus dem Oberjura von Bayern. *Lethaia* 3, 225-32.
Lipps, J. H. and Hickman, C. S. (1982) Origin, age, and evolution of Antarctic and deep-sea faunas. In *The environment of the deep sea* (eds W. G. Ernst and J. G. Morin) pp. 324-56. Prentice-Hall, Englewood Cliffs, NJ.
Matthews, S. C. and Missarzhevsky, V. V. (1975). Small shelly fossils of late Precambrian and early Cambrian age: a review of recent work. *Jl. geol. Soc. Lond.* 131, 289-304.
Morse, M. P. (1981). *Meiopriapulus fijiensis* n. gen., n. sp.: An interstitial priapulid from coarse sand in Fiji. *Trans. Am. microsc. Soc.* 100, 239-52.
Moussa, M. T. (1970). Nematode fossil tracks from the Green River Formation (Eocene) in the Uinta Basin, Utah. *J. Paleont.* 44, 304-7.
Osgood, R. G. (1970). Trace fossils of the Cincinnati area. *Palaeontogr. Am.* 6, 279-444.

Owre, H. B. and Bayer, F. M. (1962). The systematic position of the Middle Cambrian fossil *Amiskwia* Walcott. *J. Paleont.* 36, 1361-3.

Palij, V. M., Posti, E., and Fedonkin, M. A. (1979). Soft-bodied Metazoa and trace fossils of Vendian and Lower Cambrian. In *Upper Precambrian and Cambrian paleontology of East-European platform* (eds B. M. Keller and A. Yu. Rozanov) pp. 49-82. Akad Nauk S.S.S.R. (in Russian).

Pemberton, S. G., Kobluk, D. R., Yeo, R. K. and Risk, M. J. (1980). The boring *Trypanites* at the Silurian-Devonian disconformity in southern Ontario. *J. Paleont.* 54, 1258-66.

Pflug, H. D. (1972). Systematik der jung-präkambrischen Petalonama Pflug 1970. *Palaont. Z.* 46, 56-67.

Pickerill, R. K. (1982). Cambrian medusoids from the St. John Group, southern New Brunswick. *Curr. Res. Part B. Geol. Surv. Can. Pap.* 82-1B, pp. 71-6.

Pierce, W. D. (1960). Silicified Turbellaria from Calico Mountains nodules. *Bull. S. Calif. Acad. Sci.* 59, 138-43.

—— (1964). Three new types of invertebrates extracted from Miocene petroliferous nodules. *Bull. S. Calif. Acad. Sci.* 63, 81-5.

Poinar, G. O. (1981). Fossil dauer rhabditoid nematodes. *Nematologica* 27, 466-7.

——(1983) *The natural history of nematodes.* Prentice-Hall, Englewood Cliffs, NJ.

Por, F. D. and Bromley, H. J. (1974). Morphology and anatomy of *Maccabeus tentaculatus* (Priapulida: Seticoronaria). *J. Zool., Lond.* 173, 173-97.

Poulsen, V. (1963). Notes on *Hyolithellus* Billings, 1871, Class Pogonophora Johannson, 1937. *Biol. Meddr.* 23, 1-15.

Rice, M. E. (1969). Possible boring structures of sipunculids. *Am. Zool.* 9, 803-12.

Risk, M. J. (1973). Silurian echiuroids: possible feeding traces in the Thorold Sandstone. *Science, N.Y.* 180, 1285-7.

Runnegar, B. (1982a). A molecular-clock date for the origin of animal phyla., *Lethaia* 15, 199-205.

—— (1982b). Oxygen requirements, biology and phylogenetic significance of the late Precambrian worm *Dickinsonia*, and the evolution of the burrowing habit. *Alcheringa* 6, 223-39.

Schram, F. R. (1973). Pseudocoelomates and a nemertine from the Illinois Pennsylvania. *J. Paleont.* 47, 985-9.

—— (1979). Worms of the Mississippian Bear Gulch Limestone of central Montana, USA. *Trans. S. Diego Soc. nat. Hist.* 19, 107-20.

Sciacchitano, I. (1955). Su un Gordio fossile. *Monitore zool. ital.* 63, 57-61.

Scrutton, C. T. (1979). Early fossil cnidarians. In *The origin of major invertebrate groups* (ed. M. R. House), Systematics Association Special Publication, Vol. 12, pp. 161-207. Academic Press, London.

Seilacher, A. (1969). Paleoecology of boring barnacles. *Am. Zool.* 9, 705-19.

Sepkoski, J. J. (1979). A kinetic model of Phanerozoic taxonomic diversity II. Early Phanerozoic families and multiple equilibria. *Paleobiology* 5, 222-51.

Simonetta, A. and Cave, L. D. (1978). Notes on new and strange Burgess Shale fossils (Middle Cambrian of British Columbia). *Memorie Soc. tosc. Sci. nat.* A85, 45-9.

Sliter, W. V. (1971). Predation on benthic foraminifers. *J. Foram, Res.* 1, 20-9.

Smith, A. B. (1982). The affinities of the Middle Cambrian Haplozoa (Echinodermata). *Alcheringa* 6, 93-9.

Sokolov, B. S. (1972). Vendian and Early Cambrian Sabelliditida (Pogonophora) of the USSR. *Int. geol. Congr.* 23 (1968, Proc. Int. Paleont. Union), 79-86.

Southcott, R. V. and Lange, R. T. (1971). Acarine and other microfossils from the Maslin Eocene, South Australia. *Rec. S. Aust. Mus.* 16, 1-21.

Stanley, G. D. and Stürmer, W. (1983). The first fossil ctenophore from the Lower Devonian of West Germany. *Nature, Lond.* 303, 518-20.

Størmer, L. (1963). *Gigantoscorpio willsi*, a new scorpion from the Lower Carboniferous of Scotland and its associated preying microorganisms. *Skr. Norske Vidensk-Akad.* 8, 1-171.

Stricker, S. A. (1983). S.E.M. and polarization microscopy of nemertean stylets. *J. Morph.* 175, 153-69.

—— and Cloney, R. A. (1981). The stylet apparatus of the nemertean *Paranemertes peregrina*: its ultrastructure and role in prey capture. *Zoomorphologie* 97, 205-23.

—— —— (1982). Stylet formation in nemerteans. *Biol. Bull. mar. biol. Lab., Woods Hole* 162, 387-403.

Szaniawski, H. (1974). Some Mesozoic scolecodonts congeneric with recent forms. *Acta palaeont. pol.* 19, 179-99.

Thompson, I. (1979). Errant polychaetes (Annelida) from the Pennsylvanian Essex fauna of northern Illinois. *Palaeontographica* Abt A 163, 169-99.

Towe, K. M. (1981). Biochemical keys to the emergence of complex life. In *Life in the universe* (ed. J. Billingham) pp. 297-306. M.I.T. Press, Cambridge, Mass.

Urbanek, A. and Mierzejewska, G. (1977). The fine structure of zooidal tubes in Sabelliditida and Pogonophora with reference to their affinity. *Acta palaeont. pol.* 22, 223-40.

Voigt, E. (1983). Ein fossiler Saitenwurm (*Gordius tenuifibrosus* n. sp.) aus der eozänen Braunkohle des Geiseltales. *Nova Acta Leopoldina*, 5, 351-60.

Wade, M. (1969). Medusae from uppermost Precambrian or Cambrian sandstones, central Australia. *Palaeontology* 12, 351-65.

—— (1971). Bilaterial Precambrian chondrophores from the Ediacaran fauna, South Australia. *Proc. R. Soc. Vict.* 84, 183-8.

—— (1972a). Hydrozoa and Scyphozoa and other medusoids from the Precambrian Ediacara fauna, South Australia. *Palaeontology* 15, 197-225.

—— (1972b). *Dickinsonia*: Polychaete worms from the late Precambrian Ediacara fauna, South Australia. *Mem. Queensland Mus.* 16, 171-90.

Webby, B. D. (1970). Late Precambrian trace fossils from New South Wales. *Lethaia* 3, 79-109.

Welch, J. R. (1976). *Phosphannulus* on Paleozoic crinoid stems. *J. Paleont.* 50, 218-25.

Wetzel, A. and Werner, F. (1981). Morphology and ecological significance of *Zoophycos* in deep-sea sediments off northwest Africa. *Palaeogeog. Palaeoclimat. Palaeoecol.* 32, 185-212.

Xian-tao Wu (1983). Origin and significance of constant-size fenestrae associated with calcispheres from the Lower Carboniferous of the Gower Peninsular, South Wales. *Palaeogeog. Palaeoclimat. Palaeoecol.* 41, 139-51.

Yochelson, E. L. (1977). Agmata, a proposed extinct phylum of early Cambrian age. *J. Paleont.* 51, 437-54.

23. Current perspectives on the origins and relationships of lower invertebrates

ROBERT D. BARNES

Department of Biology, Gettysburg College,
Gettysburg, Pennsylvania, USA

Abstract

Over the past twenty years studies in functional morphology, comparative anatomy, and ultrastructure have contributed greatly to our current perspectives on the origins and relationships of lower invertebrates. Six phylogenetic assumptions which would now probably find general acceptance among zoologists are described along with the principal areas of controversy. New and persisting ideas about the relationships of phyla to each other are summarized.

Our conception of animal phylogeny has slowly evolved since the days of Haeckel, Hatschek, and Sedgwick over a hundred years ago. Some early ideas are still with us, at least in modified form; others have made a brief appearance and then disappeared. The changes reflect, of course, the steady increase in our understanding and appreciation of the great diversity of the Animal Kingdom. In these concluding pages I have attempted to construct a current perspective of early animal phylogeny, one based on the ideas and evidence that are presented in the papers and discussion in this volume.

To appreciate better where we now stand, it would be helpful to know from where we have come. In 1960 a symposium similar to this one was held at Pacific Grove, California, subsequently published under the editorship of Ellsworth Dougherty (1963). The principal concerns of that meeting are reflected by the speakers and topics of the papers from the general section on origins and affinities of animal phyla.

Homologies and the ciliate origin of the Eumetozoa by Earl Hanson elaborates the syncytial theory, which holds that the multicellular metazoan condition evolved by compartmentation of a multinucleate ciliate ancestor, the acoelomate flatworms being closest to the base of the metazoan phylogenetic tree.

The evolution of the Metazoa from colonial flagellates vs. plasmodial cilates by Adolf Remane argues against the syncytial theory in favour of a flagellate origin of metazoans, the multicellular condition arising by increasing interdependence of cells in some ancestral colonial species.

The early worm: a planula by Cadet Hand argues for a planula-like ancestral metazoan and the importance of the cnidarians in the early evolutionary history of the Animal Kingdom.

Origin and affinities of the lower Metazoa by Otto Steinböck defends the syncytial theory and an acoelomate flatworm at the evolutionary base of the Animal Kingdom.

The enterocoelic origin of the coelom by Adolf Remane proposes that the coelom evolved from the gastric pouches of some cnidarian ancestor.

A critique of the enterocoel theory by Willard Hartman reviews the enterocoel theory as well as alternative ideas, i.e. gonocoel, nephrocoel and schizocoel theories. He comes to the conclusion that none are really tenable.

The evolution of the coelom and metamerism by Robert Clark postulates that the segmented coelom of annelids evolved as a localized hydrostatic skeleton adapted for peristalic burrowing in soft marine sediments.

About the same time Libbie Hyman (1959), at the end of the fifth volume of her *Invertebrates* wrote a section called Retrospects. In that section she reviewed some of the ideas about phylogeny that had appeared in the literature since she had finished her earlier volumes. Never one to mince words she wrote:

The author hoped that the enterocoel theory was dead and buried, as it deserves, but it is being kept alive by a group of zoologists, notably Remane, Jägersten and Marcus. . . . This bright idea is said to have originated with Leuckart but was especially promulgated by Sedgwick (1884). It follows as necessary corollaries of the enterocoel theory that all Bilateria are coelomate, whether or not they have a coelom, and that all coeloms are of the enterocoel variety, whether or not they are seen to originate by this method.

The author regards the enterocoel theory as fantastic nonsense, for which there does not exist a single scrap of genuine evidence. The theory is a pure fabrication. It requires that the Anthozoa be regarded as the basic cnidarians. This is in itself implausible since they are anatomically the most complicated members of the phylum. . . . The supporters of the enterocoel theory find it very difficult to account for the Turbellaria. They can explain them only as degenerate annelids, and this view is totally unacceptable to the author. We are asked to believe that the Turbellaria have lost their coelom, lost their anus, lost their nephrostomes. This is asking too much of one's credulity. . . . The author regards such phylogenetic questions as the origin of the Metazoa from the Protozoa or the origin of the Bilateria from the Radiata as insoluble on present information. Also insoluble are such questions as to whether entoderm, mesoderm, and coelom have or have not some original mode of formation from which other modes are derived. Anything said on these questions lies in the realm of fantasy.

However, here we are over two decades later still asking those questions. And of course they will continue to be asked as long as there are curious, speculative zoologists. I wish that Libbie Hyman could have been with us, for despite her apparent impatience, I think she would have had much to contribute.

So where do we stand almost 25 years since that symposium at Pacific Grove. Do we know more about the evolutionary relationship of animals? I think we do because we have a better knowledge of the biological framework within which speculations on evolutionary origins and relationships must rest. And I think contemporary approaches to questions regarding phylogeny are more sophisticated. Thus our speculations, our guesses are better educated.

I believe contemporary approaches to be more sophisticated in three ways. There is much more concern for functional morphology and the adaptive significance of structures or conditions. This demands that phylogenetic speculations take into account preadaptive and postadaptive states and the possible selection factors that could have been operative in their evolution. This is not to say that there have not been some excellent studies in the past which have utilized functional morphology as a primary source of evidence. Clark's hypothesis regarding the origin of the annelid coelom and metamerism is a good case in point. But I think function is much more of a general concern in current phylogenetic speculations.

A second approach to phylogenetic problems which has provided more educated speculation is related to the first and that is careful comparative analysis of characters, especially the recognition of homologous and analogous structures. The homology theory as elaborated by Remane (1952) and the cladistic procedures of Hennig (1966) have been major contributions to a better understanding of phylogenetic relationships. Both involve a rigorous application of criteria for recognition of homology, such as (1) similar anatomical positions of structures, (2) the presence of transformation series between structures, and (3) an intelligible distribution of the character among known species (Rieger and Tyler 1979).

The third contribution to our current understanding of animal phylogeny is largely a new one and, I believe, the most significant.

This is the structural detail revealed by electron microscopy. The revelations provided by EM studies over the past twenty years have been enormous and have increased our knowledge of almost every group of animals. Some important contributions to comparative morphology include studies on ciliation, cuticles, duo-gland systems, body cavity linings, protonephridia, and sperm ultrastructure.

What has palaeontology contributed in the interval? Certainly our knowledge of fossil species is greater. Williams and Hurst (1977), in an analysis of brachiopod genera, record 323 genera for 1899 and 2400 genera for 1975, a 35-fold increase. We have also extended our knowledge of early animal life, for the Ediacaran assemblage of fossils gives us some hints of Precambrian metazoans (Cloud and Glaessner 1982). Also, palaeontology has taken great strides in using the fossil record to provide information about past climatic and ecological conditions. Moreover many palaeontologists are examining fossil species from a functional standpoint and have produced some very fine work on adaptive biology.

However, the discovery of transitional species which could provide information about the origin of groups is still lacking. At the generic and species level there are graded series of forms which when combined with the stratigraphic sequence are palaeontology's most important contribution to our understanding of animal evolution (Yochelson 1979; Olson 1981). However, at the level of family and above, especially classes and phyla, the fossil record provides little information about relationships and origins (House 1979; Conway Morris, this volume). *Knightoconus*, a septate monoplacophoran which may be on the line leading to cephalopods (Yochelson, Flower, and Webers 1973), is perhaps one of the few so called 'missing links' that the known fossil record has to offer about the evolution of higher categories (but see Runnegar 1983 for a radical alternative on early mollusc evolution as revealed by the fossil record).

That the fossil record has so little to contribute at present is not really surprising. Olson (1981) in his paper on missing links points out that the chances of finding useful transitional fossils:

(a) increase as the categorical level goes down. However, we are concerned here not with the lower taxa but with linkage between phyla and the origin of classes and phyla;

(b) increase the more recent the geological time of origin. But many of the evolutionary events that concern us are Precambrian;

(c) increase with the fossilization potential of the organism; but we are largely concerned with soft-bodied forms;

(d) increase with the morphological complexity of the organism, but we are dealing in many cases with less complex forms.

The significance of these factors is illustrated by the Cambrian record, where most metazoan groups make their appearance. Olson (1981) points out that without the well preserved fossils from the mid-Cambrian Shale of British Columbia our knowledge of early metazoan diversity would be greatly reduced (Conway Morris 1979; Collins, Briggs, and Conway Morris 1983).

So it is unlikely that the fossil record is ever going to provide many answers to the puzzles of invertebrate origins, and the bulk of evidence will continue to come from comparative morphology, development and biochemistry of living forms. Such has been the nature of the evidence presented here.

An appreciation of our current perspective regarding the evolution of lower animals can be gained in part by determining those ideas for which there is now some concensus. I would suggest six phylogenetic assumptions with which most zoologists here and elsewhere would agree.

1. The first is that primitively metazoans are monociliated, i.e., there is only one cilium (or flagellum) per cell. This condition is true of *Trichoplax*, sponges, cnidarians, gnathostomulids, some gastrotrichs and the deuterostome line in general. Recognition of the distribution of the monociliated condition has been one of the contributions of EM over the past ten years (Rieger 1976), and the implications of such a primitive condition are far reaching.

2. This leads to a second assumption, that metazoans, being monociliated, must have evolved from some colonial flagellated protozoan ancestor. Choanoflagellates appear to be the best candidates for that ancestral position, for their flagellatory and mitochondrial ultrastructure is similar to that of metazoans. Moreover, many metazoans, such as cnidarians and gastrotrichs, have a low ring of microvilli around the cilium, which is comparable to the collar microvilli of choanoflagellates. Note that the debate between the colonial flagellate theory and the syncytial theory of metazoan origin seems resolved. Flatworms have multiciliated cells, as indeed they would be expected to have if they evolved from a ciliated protozoan ancestor as proposed by the syncytial theory.

3. Metazoans, including sponges, appear to be monophyletic. There is no real evidence for a separate evolution of either sponges or cnidarians, as has been proposed by some (Sleigh 1979; and see reviews by Bergquist 1978 and this volume). However, they certainly must have diverged early, especially sponges, in metazoan evolution.

4. Early metazoans were radially symmetrical. Such a primitive symmetry is reflected by the larvae of sponges and by cnidarians. The biradiality of anthozoans is probably a secondary specialization, but there may be more dissenters on this assumption than some others. Grell (1981) would disagree, for he considers the benthic creeping

Trichoplax adhaerens (Placozoa), which has the upper and lower surfaces differentiated, to be near the ancestral metazoan form.

5. Bilaterality and cephalization evolved as adaptations for mobility over some sort of substratum. Association with the substratum most easily accounts for differentiation of dorsal and ventral surfaces.

6. The protostome and deuterostome assemblages continue to be recognized as major lines of animal evolution.

These six assumptions represent about the extent of our current consensus. But I think they are significant and although still limited, would have found less general support twenty years ago.

The principal area of current debate about animal phylogeny is the origin of the coelom and acoelomate bodyplan. If the acoelomate condition is primitive, then flatworms retain their central position at the base of the Bilateria. If the acoelomate condition is derived from a coelomate condition, then flatworms must be removed and we can hardly escape the assumption that all or most of the living Bilateria are coelomate or at least derived from coelomates. This is the very position that so agitated Hyman over twenty years ago. The two sides of the current debate are reflected in the papers of Ax and Rieger in this volume.

We still have unanswered the questions as to how the coelom evolved and if it evolved once or several times. Several papers here have implied an archicoelomate at the base of the Bilateria. So the enterocoel theory is still not 'dead and buried' as Libbie Hyman wished.

The papers of this volume have provided some excellent reviews of various metazoan phyla. They have also provided some new ideas or confirmed old ones about the phylogenetic relationships of these groups. Such ideas should be included as part of our current perspective and might be summarized in the following way.

(i)   The hexactinellid, or glass, sponges are perhaps not true poriferans.

(ii) Ctenophores may not be closely related to cnidarians.

(iii) The aschelminths encompass two lines of evolution, one embracing the rotifers and acanthocephalans and the other the gastrotrichs and nematodes. The aschelminth complex also contains a new phylum, the Loricifera (Kristensen 1983).

(iv) The gnathostomulids, in spite of a number of specialized features, appear to lie somewhere near the base of the Bilateria.

(v)   The phylogenetic position of the priapulids continues to be puzzling and at the present time there is no good evidence for allying them with any other group.

(vi) All available information continues to place the sipunculans somewhere near the stem of the annelids and molluscs, and the pogonophorans near the annelids.

(vii) The concept of the Archiannelida has been finally laid to rest, the last 'archiannelid' worms now being placed with the polychaetes.

What does the future hold for our understanding of the origins and relationships of animal phyla? I am optimistic. I believe that the steady accumulation of knowledge of comparative morphology, biochemistry and development will indicate the limits of possible pathways of evolution. For example, I believe that comparative ultrastructural research will resolve in the not too distant future the question as to whether the flatworm acoelomate condition is primitive or secondarily derived. We will probably never be able to make more than educated guesses about the ancestral metazoans, but we should eventually have a fair understanding as to which evolutionary pathways were or were not probable.

References

Bergquist, P. R. (1978). *Sponges*. Hutchinson, London.

Cloud, P. and Glaessner, M. F. (1982). The Ediacarian Period and System: Metazoa inherit the earth. *Science, N.Y.* 217, 783-92.

Collins, D., Briggs, D., and Conway Morris, S. (1983). New Burgess Shale fossil sites reveal Middle Cambrian faunal complex. *Science, N.Y.* 222, 163-7.

Conway Morris, S. (1979). The Burgess Shale (Middle Cambrian) fauna. *A. Rev. Ecol. Syst.* 10, 327-49.

Dougherty, E. C. (ed.) (1963). *The lower Metazoa*. University of California Press, Berkeley, CA.

Grell, K. G. (1981). *Trichoplax adhaerens* and the origin of the Metazoa. In *Origine dei grandi phyla dei Metazoi*. pp. 107-21. Atti Accad. naz. Lincei Convegni Rome.

Hennig, W. (1966). *Phylogenetic systematics*. University of Illinois Press, Urbana, IL.

House, M. R. (ed.) (1979). *The origin of major invertebrate groups*. Systematics Association Special Vol. 12, Academic Press, London.

Hyman, L. H. (1959). *The invertebrates*. Vol. 5. *Smaller coelomate groups*. McGraw-Hill, New York.

Kristensen, R. M. (1983). Loricifera, a new phylum with Aschelminthes characters from the meiobenthos. *Z. zool. Syst. Evolutionsforsch.* 21, 163-180.

Olson, E. C. (1981). The problem of missing links: to-day and yesterday. *Q. Rev. Biol.* 56, 405-42.

Remane, A. (1952). *Die Grundlagen des naturlichen Systems, der vergleichenden Anatomie und der Phylogenetik*. Geest und Portig K.-G., Leipzig (Reprinted 1971 by Otto Loeltz, Konigstein).

Rieger, R. M. (1976). Monociliated epidermal cells in Gastrotricha: Significance for concepts of early metazoan evolution. *Z. zool. Syst. Evolutionsforsch.* 14, 198-226.

—— and Tyler, S. (1979). The homology theorem in ultrastructural research. *Am. Zool.* 19, 655-64.

Runnegar, B. (1983). Molluscan phylogeny revisited. *Mem. Ass. Australas. Palaeont.* 1, 121-44.

Sleigh, M. A. (1979). The radiation of eukaryote Protista. In *The origin of major invertebrate groups* (ed. M. R. House) Systematics Association Special Vol. 12, pp. 23-54, Academic Press, London.

Williams, A. and Hurst, J. M. (1977). Brachiopod evolution. In *Patterns of evolution* (ed. A. Hallam) pp. 79-121. Elsevier, Amsterdam.

Yochelson, E. L. (1969). Early radiation of Mollusca and mollusc-like groups. In *The origin of major invertebrate groups* (ed. M. R. House) Systematics Association Special Vol. 12, pp. 323-58. Academic Press, London.

——, Flower, R. H. and Webers, C. F. (1973). The bearing of the new Late Cambrian monplacophoran genus *Knightoconus* upon the origin of cephalopods. *Lethaia* 6, 275-309.

Author Index

Subject Index

Index of Genera and Species

Text references are shown in ordinary type, figure references in **bold** type, and table references in *italics*.

Systematics Association Publications

1. Bibliography of key works for the identification of the British
    fauna and flora *3rd edition* (1967)
    Edited by G. J. Kerrich, R. D. Meikle and N. Tebble
    Out of print
2. Function and taxonomic importance (1959)
    Edited by A. J. Cain
3. The species concept in palaeontology (1956)
    Edited by P. C. Sylvester-Bradley
4. Taxonomy and geography (1962)
    Edited by D. Nichols
5. Speciation in the sea (1963)
    Edited by J. P. Harding and N. Tebble
6. Phenetic and phylogenetic classification (1964)
    Edited by V. H. Heywood and J. McNeil
    Out of print
7. Aspects of Tethyan biogeography (1967)
    Edited by C. G. Adams and D. V. Ager
8. The soil ecosystem (1969)
    Edited by J. G. Sheals
9. Organisms and continents through time (1973)[†]
    Edited by N. F. Hughes

Published by the Association

Systematics Association Special Volumes

1. The new systematics (1940)
    Edited by Julian Huxley (Reprinted 1971)
2. Chemotaxonomy and serotaxonomy (1968)*
    Edited by J. G. Hawkes
3. Data processing in biology and geology (1971)*
    Edited by J. L. Cutbill
4. Scanning electron microscopy (1971)*
    Edited by V. H. Heywood

*Published by Academic Press for the Systematics Association
[†]Published by the Palaeontological Association in conjunction with
the Systematics Association
[‡]Published by Oxford University Press
for the Systematics Association

F